THE ASHGATE RESEARCH COMPANION
TO PLANNING AND CULTURE

ASHGATE
RESEARCH
COMPANION

The *Ashgate Research Companions* are designed to offer scholars and graduate students a comprehensive and authoritative state-of-the-art review of current research in a particular area. The companions' editors bring together a team of respected and experienced experts to write chapters on the key issues in their speciality, providing a comprehensive reference to the field.

The Ashgate Research Companion to Planning and Culture

Edited by

GREG YOUNG

Principal Editor

University of Sydney, Australia

and

DEBORAH STEVENSON

UWS, Australia

Routledge
Taylor & Francis Group

LONDON AND NEW YORK

First published in paperback 2024

First published 2013 by Ashgate Publishing

Published 2016 by Routledge
4 Park Square, Milton Park, Abingdon, Oxon OX14 4RN

and by Routledge
605 Third Avenue, New York, NY 10158

Routledge is an imprint of the Taylor & Francis Group, an informa business

British Library Cataloguing in Publication Data
The Ashgate research companion to planning and culture.
  1. City planning--Social aspects. 2. Regional planning--
  Social aspects.
  I. Planning and culture II. Young, Greg, 1947-
  III. Stevenson, Deborah, 1958-
  307.1'2-dc23

Library of Congress Cataloging-in-Publication Data
The Ashgate research companion to planning and culture / [edited] by Greg Young and Deborah Stevenson.
     pages cm
  Includes bibliographical references and index.
  ISBN 978-1-4094-2224-2 (hardback)
1. City planning--Social aspects.
2. Regional planning--Social aspects. 3. Urban policy. 4. Cultural
policy. I. Young, Greg, 1947- II. Stevenson, Deborah, 1958-
  HT166.A822 2013

ISBN 13: 978-1-40-942224-2 (hbk)
ISBN 13: 978-1-03-291943-0 (pbk)
ISBN 13: 978-1-31-561339-0 (ebk)

DOI: 10.4324/9781315613390

Contents

List of Figures

List of Tables

Acknowledgements

Grateful thanks go to Valerie Rose, Publisher, Ashgate for her keen understanding and vision for the book and for graciously facilitating it at all stages. Associate Professor Glen Searle, Director of Planning, University of Queensland, Australia provided valuable advice on potential authors and other substantive issues.

We also extend our thanks to Vibha Bhattarai Upadhyay for diligent and engaged research assistance and the University of Western Sydney and Macquarie University for providing resources at critical stages of the project.

Permission from the late Dr Jeffrey Smart, AO to use his painting *The Dome* as the book's cover image is gratefully acknowledged by the editors and thanks go to his archivist Stephen Rogers for facilitating details of the arrangement.

Last, but by no means least, the gratitude of the editors goes to all contributors who committed themselves to a lengthy production process and other complexities inherent in an ambitious, global volume of this kind.

Greg Young
Deborah Stevenson

Notes on Contributors

Gregory Ashworth was educated in Geography at the Universities of Cambridge, Reading and London (PhD 1974). He has taught at the Universities of Wales, Portsmouth and since 1979 Groningen, the Netherlands. Since 1994, he is Professor of Heritage Management and Urban Tourism in the Department of Planning, Faculty of Spatial Sciences, University of Groningen. His main research interests focus on the interrelations between tourism, heritage and place marketing, largely in an urban context. He is the author of around 15 books, 100 book chapters, and 200 articles. He received honorary life membership of the Hungarian Geographical Society in 1995, an honorary doctorate from the University of Brighton in 2010 and was knighted for services to Dutch Science in 2011.

Tüzin Baycan is Professor of Urban and Regional Planning at Istanbul Technical University. She is a Fellow of the Academia Europaea and a Panel Member of ERC Advanced Grants for Social Science and Humanities: Environment and Society. She is an Editorial/Advisory Board Member of *International Journal of Sustainable Society, Studies in Regional Science, Romanian Journal of Regional Science, Journal of Independent Studies, Research-Management: Social Sciences and Economics* and *A/Z ITU Journal of Faculty of Architecture*, and co-editor of *Sustainable City and Creativity: Promoting Creative Urban Initiatives* (2011) and *Classics in Planning: Urban Planning* (2008). Tüzin is the author of many scientific papers and book chapters. Her main research interests cover urban and regional development and planning; urban systems; environment; sustainable development; creativity, innovation and entrepreneurship; diversity and multiculturalism.

Franco Bianchini is Professor of Cultural Policy and Planning at Leeds Metropolitan University. From 1992 to 2007 he was Reader in Cultural Planning and Policy and Course Leader for the MA in European Cultural Planning at De Montfort University, Leicester. His books include *Urban Mindscapes of Europe* (co-edited by Godela Weiss-Sussex with Franco Bianchini, 2006), *Planning for the Intercultural City* (with Jude Bloomfield, 2004), *Culture and Neighbourhoods: A Comparative Report* (with L. Ghilardi, 1997), *The Creative City* (with C. Landry, 1995), *Cultural Policy and Urban Regeneration: The West European Experience* (co-editor, with M. Parkinson, 1993) and *City Centres, City Cultures* (with M. Fisher et al., 1988).

Kim Dovey is Professor of Architecture and Urban Design at the University of Melbourne where he has also served as Associate Dean and Head of Architecture and Urban Design. He has a PhD from the University of California, Berkeley and has long researched social issues in architecture and urban design mostly focused on understandings of 'place' at a range of scales and types. Books include *Framing Places* (2nd edition 2008), *Fluid City* (2005) and *Becoming Places* (2010). He leads research projects on urban intensification, creative clusters and informal settlements. Theoretical interests include understanding urban places as complex, resilient and self-adaptive assemblages.

Nancy Duxbury is a Senior Researcher and Co-coordinator of the Cities, Cultures and Architecture Research Group of the Centre for Social Studies, University of Coimbra, Portugal. Her research focuses on culture in sustainable development, and the integration of cultural considerations within sustainability planning initiatives internationally. She is Chair of the Policies Working Group of the European COST Action on 'Investigating Cultural Sustainability' and an Adjunct Professor of the School of Communication, Simon Fraser University, Canada. She is internationally published, and has guest edited issues of *Culture and Local Governance, Society and Leisure* and the *Canadian Journal of Communication*. She was a co-founder of the Creative City Network of Canada.

Graeme Evans is Chair of Design, Brunel University, London and Special Chair of Culture and Urban Development at Maastricht University where he has been contributing to the European Capital of Culture 2018 programme. He was the founding director of the Cities Institute at London Metropolitan University (2003–12). He has undertaken numerous cultural plans and strategies for arts councils, local and regional government and developed a national Cultural Planning and Mapping toolkit for the UK Department for Culture, Media and Sport and regional cultural agencies. Key publications include *Cultural Planning: An Urban Renaissance?* (2001) and *Designing Sustainable Cities* (2009). He advises the Council of Europe, OECD and UNESCO on cultural creative city policy and programme evaluation.

Luigi Fusco Girard is Professor of Economics and Environmental Evaluations and Professor of Urban Economics at the Faculty of Architecture, University of Naples Federico II, Italy. He is the joint editor of numerous volumes including *Città attrattori di Speranza* (2006), *Energia, Bellezza, Partecipazione* (2004), *The Human Sustainable City: Challenges and Perspectives from the Habitat Agenda* (2003), *Cultural Tourism and Sustainable Local Development* (2009) and *Sustainable City and Creativity* (2011). He is Chair of the International Scientific Committee on Economics of Conservation, ICOMOS International Council of Monuments and Sites, Paris, France.

Clara Greed is Emerita Professor of Inclusive Urban Planning at the University of the West of England, Bristol. She holds an MBE for her services to urban design, is a chartered town planner and a Fellow of the Chartered Institute of Building. Her main interest is in the 'social aspects of planning' and she has written more than 12 books, and many refereed articles on gender and planning, accessibility, equality issues, and professional education and practice. Her teaching has included planning history, theory, and practice, and urban sociology. Her research has included ESRC-funded research on women in the construction professions, gender mainstreaming research for the RTPI and Nuffield-funded research projects on accessible school buildings and the differences between user and provider perspectives on public toilet provision.

Jean Hillier is Professor and Associate Dean of Sustainability and Urban Planning at RMIT University in Melbourne, Australia, where she teaches planning theory. Her research interests lie in developing Deleuzean-inspired planning theory and methodology for strategic spatial planning practice in conditions of uncertainty. She is also interested in narrative complexity theories and draws inspiration from Deleuze, Guattari and Foucault in analysing planning decisions and processes. Recent publications include the *Ashgate Research Companion to Planning Theory: Conceptual Challenges for Spatial Planning* (2010, co-edited with Patsy Healey) and the three volumes, also edited with Healey, *Critical Essays in Planning Theory* (2008).

M. Sharon Jeannotte is Senior Fellow, Centre on Governance, University of Ottawa. From 2005 to 2007 she was Senior Advisor to the Canadian Cultural Observatory, and from 1999 to 2005 she was the Manager of International Comparative Research in the Department of Canadian Heritage. She has published research on provincial cultural policies, cultural policy and social cohesion, cultural citizenship, and the role of culture in building sustainable communities. In 2005, she co-edited with Caroline Andrew, Monica Gattinger and Will Straw a volume entitled *Accounting for Culture: Thinking Through Cultural Citizenship*. In 2011, she co-edited with Nancy Duxbury a special issue of the journal *Culture and Local Governance* on culture and sustainable communities.

Setha Low is Professor of Environmental Psychology, Geography, Anthropology, and Women's Studies, and Director of the Public Space Research Group at The Graduate Centre, City University of New York. She has been awarded a Getty Fellowship, a NEH fellowship, a Fulbright Senior Fellowship and a Guggenheim for her ethnographic research on public space in Latin America and the United States. Her most recent books include *Politics of Public Space* (2006, with Neil Smith), *Rethinking Urban Parks: Public Space and Cultural Diversity* (2005, with S. Scheld and D. Taplin) and *Behind the Gates: Life, Security and the Pursuit of Happiness in Fortress America* (2004). Her current research is on the impact of private governance on cooperative and condominium housing regimes.

Toby Miller is Professor and Chair, Department of Media and Cultural Studies, University of California, Riverside, USA. He is the author and editor of over thirty books and has written hundreds of journal articles and book chapters. His work has been translated into Portuguese, German, Chinese, Spanish and Swedish. His latest books are *Greening the Media* (2012) and *Blow up the Humanities* (2012).

John Montgomery is an urban economist and town planner who has specialized over many years in the creative industries, economic development, urban regeneration and urban design. Much of his experience is drawn from the United Kingdom and Ireland, dating from the mid-1980s, and with his London-based firm Urban Cultures Ltd since 1991. He emigrated to Australia in 2002 and has worked on projects in most of the major cities. He is an Adjunct Professor with the Queensland University of Technology. His research interests lie in the field of evolutionary economics and the growth of cities over long waves of economic development. He is a Fellow of the Royal Society in the United Kingdom. Dr Montgomery's most recent books are *The New Wealth of Cities* (2007) and *Upwave* (2011).

Peter Newman is Professor of Comparative Urban Planning at the University of Westminster. He has published numerous books and articles on urban governance and planning. Recent research has focused on the strategic planning challenges facing world cities in Europe, North America and Asia and on the particular challenges of planning and delivering major infrastructure.

Torill Nyseth holds an MA and PhD in planning and is a full professor at the Department of Sociology, Political Science and Community Planning at the University of Tromsø. Her research includes place management and place making, local democracy, governance and urban development and design. She has also been a college lecturer and worked as a municipal planner. She has published numerous articles on political science, entrepreneurship, and urban planning and governance in journals such as *Local Government Studies, Planning Theory and Practice, Planning Theory, Town Planning Review* and *European Urban and Regional Studies*. She is joint editor of *Place Reinvention: Northern Perspectives* (2007, with Arvid Viken).

Justin O'Connor is Professor in the Creative Industries Faculty, Queensland University of Technology, Brisbane, Australia and visiting Chair, Department of Humanities, Shanghai Jiaotong University. He was director of Manchester Institute of Popular Culture from 1997 until September 2006. He was then appointed Professor of Cultural Industries at the School of Performance and Cultural Industries, University of Leeds, where he led an MA in Culture, Creativity and Entrepreneurship. His main areas of interest are contemporary urban cultures, cultural and creative industries, cultural policy, urban regeneration and aesthetics, especially in the area of music. He has recently completed a lengthy report on Arts and Creative Industries for the Australia Council.

Ronan Paddison is Emeritus Professor of Geography at the University of Glasgow. He has researched and published widely in the politics of urban and regional change. His recent research has centred on the role of public participation in the installation of public art and, in a separate project, the problems encountered in engineering public space to foster inter-cultural dialogue. Recent edited volumes include *The Handbook of Urban Studies* (2001) and *Culture-Led Urban Regeneration* (2009). He is Managing Editor of two journals, *Urban Studies* and *Space and Polity*.

Edgar Pieterse holds the National Research Foundation Research Chair in Urban Policy, directs the African Centre for Cities and is Professor in the School of Architecture, Planning and Geomatics at the University of Cape Town. His most recent book is *City Futures: Confronting the Crisis of Urban Development* (2008). He has also edited or co-edited the following recent works: *African Cities Reader II: Mobilities and Fixtures* (2011), *Counter-Currents: Experiments in Sustainability in the Cape Town Region* (2010); *African Cities Reader: Pan-African Practices* (2010); *Consolidating Developmental Local Government* (2008) and a notable earlier book: *Voices of the Transition: The Politics, Poetics and Practices of Development in South Africa* (2004). Edgar's research stems from the borderzone between geography, planning and cultural studies with a strong orientation towards political philosophy.

Masayuki Sasaki is Professor of the Graduate School for Creative Cities and Director of the Urban Research Plaza at Osaka City University, Japan. He is a global pioneer in the theory and practices of the Creative City, an adviser of UNESCO Creative Cities in Japan and Editor in Chief of the journal *City, Culture and Society*. He has published many books and articles including *The Prospects to the Creative City* (2007), *Challenge for the Creative City* (2001) and *Economics of the Creative City* (1997). His most recent main article was 'Urban Regeneration through Cultural Creativity and Social Inclusion: Rethinking Creative City Theory through Japanese Case Studies', *Cities* (2010). His research interests are creative cities, cultural economics and urban economics.

Glen Searle is Associate Professor in Planning in the School of Geography, Planning and Environmental Management at the University of Queensland, Queensland, Australia. His research has mainly focused on an institutionally-based understanding of planning outcomes in Australian cities, especially Sydney. A second focus has been planning and the contemporary economy of Australian cities, with the monograph *Sydney as a Global City* (1996). In the first half of his career he held several urban policy-related positions with the New South Wales and UK governments. He is Chief Editor of the journal *Urban Policy and Research*.

Deborah Stevenson is Professor of Sociology and Urban Cultural Research in the Institute for Culture and Society at the University of Western Sydney. Her many publications include the books: *The City* (2012); *Tourist Cultures: Identity, Place and the Traveller* (2010); *Cities and Urban Cultures* (2003); *Art and Organisation: Making Australian Cultural Policy* (2000); *Agendas in Place: Urban and Cultural Planning for Cities and Regions* (1998); and *Cities of Culture: A Global Perspective* (forthcoming). In addition, she is co-editing *Culture and the City: Creativity, Tourism, Leisure* and sits on the editorial boards of several journals including the *International Journal of Cultural Policy*.

Andy Thornley is Emeritus Professor of Urban Planning at the London School of Economics and Political Science (LSE). He has written extensively on the relationship between politics and urban planning, especially from a comparative perspective. He has a particular interest in World Cities, London governance and mega-projects. His most recent book is *Planning World Cities: Globalization and Urban Politics* (2011) written with Peter Newman. He is co-author of *Struggling Giants: City-Regional Governance in London, New York, Paris and Tokyo*, to be published in 2012. Other books include *Urban Planning under Thatcherism: The Challenge of the Market* (1993) and *Urban Planning in Europe* (1996), with Peter Newman.

James A. Throgmorton is Emeritus Professor in the School of Urban and Regional Planning at the University of Iowa. His scholarly work has focused on the roles of rhetoric and persuasive storytelling in planning practice, especially in relation to the construction of sustainable places. His published works include *Planning as Persuasive Storytelling: Constructing Chicago's Electric Future* (1996) and (co-edited with Barbara Eckstein) *Story and Sustainability: Planning, Practice, and Possibility for American Cities* (2003). Much of his theoretical work is based on his experiences as an elected member of the Iowa City's City Council in the mid-1990s. In 2011 he was elected for a new term to the Council.

Stephen V. Ward is Professor of Planning History at Oxford Brookes University and is a well-known international authority on the history of planning. He is the author or editor of several books, including *Planning the Twentieth Century City: The Advanced Capitalist World* (2002) and *Planning and Urban Change* (2nd edition, 2004), along with many articles and book chapters. He is a former President of the International Planning History Society and former Editor of *Planning Perspectives*. His current research and most of his recent publications focus on the processes by which planning ideas and practices have spread between different countries. He is also writing a book on British garden cities and new towns.

Sophie Watson is Professor of Sociology, Open University, UK and Co- Director of ESRC Centre for Research into Socio-Cultural Change. She is the co-editor of *The New Blackwell Companion to the City* (2011) and *The Blackwell City Reader* (2nd edition, 2010) and joint editor of *Security: Sociology and Social Worlds* (2008). She is the author of *City Publics: The (Dis) Enchantments of Urban Encounters* (2006) and *Markets as Sites of Social Interaction: Spaces of Diversity* (2007). She has been a Consultant to Regeneris on the LDA report on London's street markets, a Special Adviser to the Department of Communities and Local Government Inquiry into Traditional Retail Markets, a Consultant to NVAO Commission Brussels and Consultant to the BBC *Thinking Allowed* programme.

Vanessa Watson is Professor in the School of Architecture, Planning and Geomatics at the University of Cape Town, South Africa, Deputy Dean of the faculty and founder member of the African Centre for Cities. Her research focus is on planning in a global context and developing planning ideas from a global South perspective. She is an editor for *Planning Theory* and on the editorial boards of *Planning Practice and Research*, *Progress in Planning* and the *Journal of Planning Education and Research*. She is co-ordinator of the Association of African Planning Schools and represents them on the Global Planning Education Association Network, which she has also chaired and co-chaired.

Greg Young is Adjunct Associate Professor, Urban and Regional Planning, Faculty of Architecture, Design and Planning, University of Sydney, Australia. He is the author of numerous academic publications including the recent *Reshaping Planning with Culture* (2008) and many landmark planning publications and strategies for Australian Governments encompassing heritage conservation, the planning of Sydney Harbour, cultural mapping and cultural tourism, and contributions to Australia's first National Cultural Policy *Creative Nation* (1994). He has consulted with governments, the private sector and NGOs internationally and is a former Chair, Sydney Harbour National Park Advisory Committee, the inaugural NSW Premier's Scholar to Venice, Italy and a US Getty Scholar (2013). His research interests include cultural theory and culturized planning and governance internationally.

Introduction: Culture and Planning in a Grain of Sand

Greg Young

I ... am the gazer and the land I stare on.
Judith Wright, *For New England*, 1994

In an era commonly viewed through the lenses of culture as much as those of the environment, ideology or politics a publication tackling the twin themes of culture and planning in conceptual, theoretical and practical terms would seem to be an essential requirement for academics, graduate students, public policy makers and planners. This volume is designed to address the challenge through its relevance to multiple disciplines, roles and sectors. In this introductory framing, however, while I present a brief outline of the book's contents and refer to the theoretical and conceptual perspectives of the contributors' chapters, my principal goal is to indicate an overall point of view on culture and planning. This not only serves to illustrate the perspective that shaped the volume from its inception, but also accommodates the considerable diversity, as well as inevitable similarities in points of view, that characterize an edited collection.

In proposing to present an overall point of view on culture and planning, I also note that the shaping vision for the volume is most rooted in the theory, concepts and practices of the global North and in the experience of Anglophone cultures. Yet at the same time, it is also clear that these elements themselves include postcolonial understanding as well as three other key aspects closely related to universality. The first of these is the fact that in planning 'the issues of power, transparency, flexibility, objective standards, accountability and commitment to negotiating competing interests transcend national boundaries' (Shmueli 2005: 512). The second is the contradiction that exists between the practical and the utopian and abstract in planning and in geography that according to Phelps and Tewdwr-Jones is a function 'of the impossibility of separating an understanding of space and time in the social sciences' (Phelps and Tewdwr-Jones 2008: 573). The third aspect is what Glacken described in 1967 as the 'expression of ever-recurring form of the quest for meaning in man, in nature, and in the relationship between the two' (Phelps and Tewdwr-Jones 2008: 572). Subsequently, this perspective is added to by material in the Prefaces to each of the book's six Parts and in an Afterword reflecting on the themes and issues raised in the contributors' chapters.

Culture in a Grain of Sand

Sketching a brief profile of culture requires perhaps as a point of departure, a resonant metaphor, such as that of the epigraph from the poem by Judith Wright – infused with an Australian Aboriginal perspective on place and 'country' – or a familiar story or observation known to most readers. One such observation, is Raymond Williams's description of culture in *Keywords*, as 'one of the two or three most complicated words in the English language' (1983: 87). Like other keywords, Williams viewed the word 'culture' as charting the course of significant social and intellectual changes in history, a role that has not been surrendered since first publication of the book in 1976, although it has been added to in ever more problematical terms. In a more recent opinion, for example, in a prominent dictionary of human geography, culture is described as 'one of the most influential, yet elusive, concepts in the humanities and social sciences' (Barnett 2009: 135). Under these circumstances, little is it to be wondered how and why culture has so often eluded planning conceptualization, or planning inclusion, whether in terms of culture's tangible and intangible dimensions or across the full spectrum of its social, environmental and historical axes. In addition, while there is a key body of writing on planning and culture there are few texts that take a comprehensive perspective on their entanglement, although arguably such holism may be the most likely opportunity for the creative development of planning as a practice.

Yet since the late nineteenth century, a number of key conceptualizations of culture have crystallized and perhaps more importantly have assumed a practical social dominance at different times and places, including in planning (Young 2005) and these are variously characterized throughout the volume. (See Miller, Greed, O'Connor, Bianchini and Young, Chapters 3, 5, 10, 22 and 23.) For the most part, however, the picture is frequently one of a number of concepts jockeying for priority at any one point in time, such that culture may be seen as dialectically related – as the relationship between culture and international development is nowadays widely perceived (Radcliffe 2006) – or alternatively, as a form or indeed chain of hegemonic and counter-hegemonic interplay. Added to this, over the twentieth century, broad trends and perspectives on culture have ranged far and wide, including those associated with the cultural norms of architectural and planning modernism, the dogmas of communist and fascist ideologies, the Frankfurt School's critique of the cultural industries, the approach of British Marxism and the influential Birmingham School of Cultural Studies, post-structural and postmodern approaches to culture and the impact on scholarship of the so-called 'cultural turn' identified by sociologists in the late twentieth century.

Typically, however, in contemporary terms cultural knowledge and understanding is applied across diverse sectors, disciplines and practices and a cultural perspective is increasingly embedded in considerations of economic development, institutional and organizational capacity-building and in the characterization of social well-being. In these contexts, culture can be variously seen as a foundational research resource and planning tool, as the hinge of power and identity and as a key index of social development. While the relationships between these sectoral paradigms may not be mutually supportive – as for example in the case of the historical relationship 'between classical economics and utilitarian philosophy' (Archer 2005: 27) – similar preoccupations and the potential for theoretical and methodological exchange, joint learning, and ultimately points of convergence are possibilities in each case.

The Dynamics of Culture

In order to set the conceptual scene for this discussion I begin with a brief sketch of the key trends I see operating in global culture followed by an outline of the main concepts of culture that can be distinguished historically and in contemporary usage. The goal here is to indicate why culture has come to be viewed by many as 'humanity's most important intellectual resource' (Chaney 1994). To begin with, culture has long been recognized as an expanding phenomenon (Williams 1966) and this expansion is accelerating as a result of the convergence of the economic and cultural spheres (Scott 2000, Soja 1996, 1993). Within the context of late capitalist civilization, this includes the shift from liberal to monopoly capitalism identified by Horkheimer and Adorno (1997: ix). As a defining aspect of the cultural economy, Scott describes a double process in which culture becomes more of a commodity at the same time as commodities themselves acquire greater cultural and symbolic content (Scott 2000). As a result 'economic and organizational life is increasingly "culturalized"' (du Gay and Pryke 2000: 6) and in urban terms great metropolitan cities become 'the flagships of a new global capitalist cultural economy' (Scott 2000: 3; see also O'Connor, Chapter 10, this volume). In spite of this, a range of the more positive aspect of culture's dialectic can be found in the chapters to follow, including for example in respect of global cultural governance policy (see Duxbury and Jeannotte, Chapter 21, this volume), in the assertion of the social and ethical dimensions of culture (see O'Connor, Chapter 10), in more inclusive and holistic levels of place reinvention (see Nyseth, Chapter 19) and through the concept of 'culturization' and culturized planning based on the ethical, critical and reflective appreciation and utilization of culture in discourse, governance and planning (Young, 2008a, 2008b, see also Young, Chapter 23).

The antinomies of economic culturalization, as a form of late capitalist commodification, and culturization as an approach to humane governance and planning, both however reflect what Jameson described as a pattern in which everything in our social life 'from economic value and state power … to the very structure of the psyche itself – can be said to have become "cultural" in some original and yet untheorized sense' (1984: 87). In addition to Jameson's over-arching framing, sociologists in the late twentieth century identified the presence of a 'cultural turn' (Chaney 1994) referencing a new focus on everyday experience and the practicalities of lived social life, the impact of which continues to be felt on scholarship in the social sciences and humanities.

In the case of planning, however, I single out six factors in terms of the dynamics of culture as they together reflect the themes that run through the book's chapters as well as current and potential impacts on planning viewed both as opportunities and threats.

Six cultural factors and planning
The six factors consist of 1) cultural diversity, 2) local place and global flows, 3) the cultural and creative industries, 4) public space and citizenship, 5) cultural planning and sustainability perspectives, and 6) social and cultural theories and concepts of culture, history and heritage.

First, cultural diversity is recognized in the UNESCO *Universal Declaration on Cultural Diversity* (2001) in terms that encompass respect for differences based on ethnicity, race, religion, gender, age and sexual preference, and varied combinations of these. At the same time the Declaration recognizes that 'cultural diversity is necessary for humankind as biodiversity is for nature' (UNESCO 2001: 11). This presents the growing multicultural cities of the world with many opportunities and challenges, including the opportunity for planning to contribute to living with difference and to the reduction of fear and bias. At the same time, cultural diversity is continuously constituted and re-constituted in a kaleidoscope of new and hybrid phenomena around the globe. Not only is this the case, but also the 'Internet Galaxy' (Castells 2001) and other media give direct experience of both contemporary manifestations of diversity as well as its historical expressions in many civilizations including Arabic, Chinese, Indian and Western.

Second, local place and global flows are co-configured in an era described by Castells (1991: 350) as a global 'space of flows' generated by the Information Economy. Under this pattern, local societies are obliged to 'preserve their identities, and build upon their historical roots regardless of their economic and functional dependence on the space of flows' (Castells 1991: 350). In the process, however, local places may reimage themselves to induce the flow of globalization's favours. (See also Nyseth, Chapter 19, this volume.) At the same time, commodities and the products of lightly remunerated labour flow northwards from the global South, while industrial wastes including e-wastes are re-exported south. (See Miller, Chapter 3, this volume.)

Third, the cultural and creative industries have become self-perpetuating. According to Sassoon, for example, the cultural industry 'feeds on itself and is limitless' (2006: xvi). Similarly, creativity is seen as a key driver of the modern economy (Florida 2004, see also Sasaki, and Montgomery, Chapters 12 and 20, this volume) and is widely recognized by writers on urbanism and cultural planning such as Landry (2006). All the while, the information economy expands through online culture, with Internet research and creative web practices fostering the creation of new knowledge. The potential to place 'everything we call knowledge' (McNeely and Wolverton 2009: xi) on the World Wide Web encourages creativity, with planning benefiting from online research and through tools such as citizens' data banks that are able 'to enhance citizen participation' (Castells 1991: 353).

Fourth, public space and citizenship are theorized in terms tantamount to each being viewed as the proxy of the other, with the dynamic between them the site of competing theory, practice and discourse. Public space is seen to have negative and positive potentialities for citizenship and identity formation according to the degree to which it is shaped in democratic and equitable terms. (See Stevenson, Dovey, Low, Paddison, Chapters 9, 15, 17 and 18, this volume.)

Fifth, cultural planning and sustainability perspectives have significant potential as well as possibilities for alignment. Cultural planning and cultural policy have the opportunity to assist in the co-production of broader public policy and to 'play the public policy game' recognized by Bianchini (see Chapter 22, this volume). At the same time, sustainability policy, for example, is increasingly perceived in cultural terms, to the extent that culture has become the 'core question' (UNESCO 2009: 6) for sustainability and its framing. The *Agenda 21 for Culture* places culture at the centre of sustainability and argues that 'the role of culture in sustainable development is mainly about including a cultural perspective in all public policy' (UNESCO 2009: 6). Most recently in 2010, the global United Cities and Local Governments organization (UCLG 2010) adopted culture as 'The Fourth Pillar of Sustainability', adding it as a term to the earlier social, economic and environmental dimensions of the sustainability equation.

Sixth, social and cultural theories and concepts of culture, history and heritage are the contexts in which planning evolves. These include theories about the nature and role of culture as well as of planning and varied concepts of culture, history and heritage such as in Ashworth's chapter in this volume (see Chapter 11) where a concept of heritage is defined as 'the contemporary uses of the past'. Social theories including structural, post-structural and postmodern theories have all had an impact on planning understanding and new theory in the social sciences and humanities develops as if in chains of ongoing hegemonic and counter-hegemonic iteration. This is reflected in Harvey's argument, in the need to develop new concepts where old are found wanting, and to the priority maintained by Marx, to use concepts and categories, rather than be used by them (Harvey 1988: 298). In these senses perhaps theoretical and conceptual developments are a form of soft power (Nye 2004), heightened by the elevation of the cognitive dimension in contemporary social and economic life. I would include here concepts of recent origin described in this volume such as those of planning cultures (Newman and Thornley, Chapter 4), place reinvention (Nyseth, Chapter 19) and culturized planning and governance (Young, Chapter 23).

The six factors I have cited above are a narrative to facilitate interpreting, comprehending and characterizing planning's operating environment as much as that of other key social technologies.

Concepts of Culture

I believe the main conceptualizations of culture may be seen in terms of three dominant approaches relating to first, the arts and high culture, second, an anthropological conception of culture and third, a concept of culture based on ontological holism.

The first concept of culture, as comprising the arts, views culture in aesthetic terms. This is nowadays frequently allied with the view that 'artistic output emerges from creative people' (Miller and Yudice 2002: 1), an idea taken further in the concept of the cultural economy and the creative city. Similarly, the nineteenth-century concept of high culture based on the arts was most famously defined by Matthew Arnold as 'contact with the best that has been thought and said' (1979: 6) in his 1882 book *Culture and Anarchy*.

The second concept is that of an anthropological conception of culture, first developed in the nineteenth century by Edward Tylor, who defined culture in his influential *Primitive Culture* published in 1871 to mean in its broad ethnographic sense 'that complex whole which includes knowledge, belief, art, morals, law, custom, and any other capabilities and habits acquired by man as a member of society' (Tylor 1903: 1). Tylor concluded his book by famously asserting that 'in aiding progress and in removing hindrance, the science of culture is essentially a reformer's science' (Tylor 1924: 453). Tylor's metaphor reverberated in the twentieth century where it was utilized by the most prominent British cultural theorist Raymond Williams and later adopted as the name of a well-known book *Culture: A Reformer's Science* (Bennet 1998). I am of the opinion, however, that such a striking and suggestive metaphor may also be applied to a strand in the history and theory of planning itself. On one level at least, planning also originated as a reformer's science late in the nineteenth century, designed to address the impacts of industrialization, rapid urbanization and the emergence of modern mass society. A number of writers such as Dear have also called for planning theory to be better connected with its 'progressive utopian roots' (2000: 135). These utopian roots, however, sit alongside planning's darker role. This was manifest in European colonial rule and the conduct of a 'civilizing mission' that lead to the usurpation of indigenous cultures in the process of implanting foreign governance and urbanism and an extractive developmental infrastructure. (See also Ward, Chapter 2, this volume.)

In the twentieth century Raymond Williams built on Tylor's pioneering perspective and furnished key disciplines such as cultural studies with the modern definitional tools to approach culture. Williams's description of culture as 'a whole way of life, material, intellectual and spiritual' (1966: 16) became the shorthand definition of choice woven through the disciplines of the humanities and the social sciences. In viewing culture as a 'reformer's science' Williams and other writers from the Birmingham School of Cultural Studies proposed that through new intellectual and conceptual tools, culture could be a means to promote social justice and development. In a sense this approach recapitulated Tylor's goal to use humanity's history and prehistory as a basis for the reform of British society (Lewis 1998: 727).

In this vein, Williams (1966) also recognized the gaps between what he described as 'dominant', 'residual' and 'emergent' forms of culture as opportunities for the reformer's social project. Interpreted in today's terms examples of these contrasts and inequalities include those that exist between middle-class culture and mainstream lifestyles as against the multicultural complexities of less powerful groups, between the male gender and heterosexuality and women and sexual minorities and in post-colonial settings between 'settler' culture and indigenous cultures and values. I also note that Williams's concept of 'emergent culture' is close to the idea and practices the planning theorist Leonie Sandercock (1998) identifies as 'insurgent culture'.

The third approach to culture is based on an analysis of its ontological structure. The French philosopher Henri Lefèbvre (1992) identified a 'trialectics of being' based on a cultural triad of 'spatiality', 'historicality' and 'sociality' that equates to the environmental, historical and social categories of culture and the key related disciplines of geography, history and sociology. This holistic approach to culture has the advantage of accommodating all aspects of culture, including its tangible and intangible dimensions and historical and contemporary axes. (See also Young, Chapter 23, this volume.) As an approach it is consistent with the idea of culture as 'the very medium of lived experience' (Jacobs and Hanrahan 2005: 1) although it is weaker in conveying the idea that Barnett (2009) argues, that culture is 'best thought of as a process, not a thing'. In spite of this, it is also the case that the embodied and material aspects of culture are important in each of the three concepts of culture I describe and play a key role in planning. In respect of the first two approaches to culture, viewed as the arts and in anthropological terms, Miller and Yudice recognize that these are brought together in cultural policy through the 'institutional supports that channel both aesthetic creativity and collective ways of life' (2002: 1).

The Global, National and Local State

The role of the international state in relation to culture in the period following the Second World War needs to be singled out for its historical relevance and growing contemporary importance. In terms of global governance, culture is espoused in multiple contexts and considered in both basic ethical terms and as a resource. Culture as a human right received its categorical expression in the 1947 *Universal Declaration of Human Rights* (United Nations 1948). In the 1990s the 'Agenda 21' of the UNCED 'Rio Summit' of 1992 nominated local cultural awareness as the foundation for sustainable strategies for cultures and their environments (United Nations 1993), and in 1995 the World Commission on Culture (WCC) in its report *Our Creative Diversity* argued that 'the relationship between culture and development should be clarified and deepened, in practical and constructive ways' (WCC 1996: 8). In 2010, as a further example of local and supranational policy coordination, the United Cities and Local Governments (UCLG 2010) – representing cities with populations totalling several billion inhabitants – adopted culture as 'The Fourth Pillar of Sustainability'.

In terms of national cultural policy and planning I single out Australia as an example and illustrate it for each of the three tiers of Australia's federal system of government. In 1994 the Australian Government released the country's first national cultural policy *Creative Nation: Commonwealth Cultural Policy* (DOCA 1994) and in 1995 published an influential model for cultural mapping under the title *Mapping Culture: A Guide for Cultural and Economic Development in Communities* (Young et al. 1995). Both documents featured a holistic approach to culture, referencing among other definitions that of Raymond Williams. The cultural mapping model spelt out in detail cultural and ethical criteria for mapping that included intellectual property and confidentiality guidelines, and provided proposals for recommendations for cultural integration across governance and business functions and activities. In general, it operated in a spirit that celebrated cultural diversity and the power and relevance of local place. At the Australian state level in the 2000s, the development of cultural plans was mandated by the New South Wales Government, as a precondition for state funding for local government cultural facilities (Young 2008b: 22), and at the local government level cultural diversity and the recognition of cultural holism were embedded in the policy of many local councils. For example, the City of Fairfield in Sydney described its culture as consisting of 'personal histories and stories, relationships with people and places, diverse communities, interesting shopping centres, waterways and wetlands, heritage, stories of migration, places of worship, Aboriginal heritage, arts, sport, rural lands, celebrations, hotted up cars and distinctive architecture' (Fairfield City Council 2005: 6)

In respect of these approaches Yudice points out that 'The notion of culture as a resource entails its management, a view that was not characteristic of either high culture or everyday culture in the anthropological sense and to further complicate matters, culture as a resource circulates globally, with ever increasing velocity' (2003: 4). The governance of the international, national and local states all play their parts in this pattern of circulation, but in the process may also contribute to the safeguarding of cultural standards in land use, conservation and community empowerment.

Creativity and the Arts

The arts play many social roles. They help to reduce and soften our instrumental sense of the world, they are a source of social distinction and preferment, based on what Pierre Bourdieu (2001) termed 'cultural capital', and they are the basis of the growth and development of the cultural industries, often promoted as the key to regional economic and social development. The cultural industries, however, are ultimately based upon the creativity of individual artists whose artistic creativity spills or is diffused outwards (Throsby 2008) into dependent industries, where it has the potential to promote employment and wealth creation through the generation and exploitation of intellectual property (Johnson 2006). Thus the arts have profound urban and regional implications, not only in terms of city economies, but also in respect of urban lifestyles and in the competitive marketing of place. Recognition of the role of the arts and of creativity in the urban sector has at times resulted in quixotic attempts to replicate their urban preconditions. Beyond this, as Hall notes (1998) in the process of generating a cultural economy, cultivated cities and nations sell their cultural products as much as their own virtue and aesthetics to the rest of the world.

Planning in a Grain of Sand

Planning has had a complex history since its modern inception in the late nineteenth century. In a heady span of history, congested with radical social, economic, political and intellectual changes, the role of the planner and the nature of planning have varied widely. Key influences on planning have ranged from modernism before the Second World War, cybernetics systems thinking in the 1960s, Marxist-inspired political economy and equity and advocacy planning movements in the 1970s, the impacts of communicative theory and postmodern philosophies in the 1980s and 1990s, post-structural approaches in the 1990s and the current and ongoing impacts of neoliberal economics and complexity theory. As against this, for much of the twentieth century the culture of planning was rooted in the Western Enlightenment of the eighteenth century, with Healey arguing in the 1990s that planning had been '"trapped" inside a modernist instrumental rationalism for many years' (1997: 7). In addition, Sandercock recognized that modernist planning 'whether socialist or capitalist came from the same epistemological roots' (Sandercock 1998: 21). Contemporary planning instrumentalism, however, is probably best represented in the marketing and promotion of global cities and their cultures and economies through the development of architectural icons and pliant planning regimes as urban mega-regions compete internationally and assume a powerful role in national affairs. (See also Newman and Thornley, and Searle, Chapters 4 and 8 this volume.)

In political administration a stronger instrumentalism was represented by the rescaling of governance that occurred in many countries in the 1980s and 1990s, such as the United Kingdom, the United States and Australia, as neo-liberal doctrines came increasingly into fashion. Under the rescaling process, the shift from government to governance shaped a greater role for government as a developer, and this was implemented through regional strategic planning and partnerships with private corporations and non-government organizations (NGOs). Ironically, this rescaling of governance led frequently not only to a reduced role for planning, but also repressed community knowledge and subtler possibilities for culture. For example, in the case of the role of tangible and intangible heritage in large-scale redevelopments by the state, such as at Darling Harbour, Sydney, Australia (see Searle, Chapter 8, this volume) and the Isle of Dogs, London, UK, heritage items, historical images and nomenclature, and representations of intangible social history, such as community memories, images and attachments to place, were transformed into developmental commodities for the purposes of design, branding and promotional 'collateral'. Opportunities for culturally responsive planning were downplayed.

At the same time, alongside such developments more traditional urban planning co-existed taking less instrumental and more domesticated forms. As an example of these traditional horizons, the American Planning Association (APA) describes planning as bringing together 'data, citizens' ideas and opinions, civic leaders' goals, and good planning practice into a deliberative process of community decision making' (APA 2012). Contemporary planning has come to dwell, on the whole, on rather modest and circumspect beliefs 'fuelled not only by the neo-liberal disdain for planning, but also by postmodernist scepticism, both of which tend to view progress as something which, if it happens, cannot be planned' (Albrechts 2010: 216). On a larger and more inspirational note, however, Hillier has defined the broad mission of spatial planning, while recognizing the potential and limitation of such a mission, as being 'stewardship of the future well-being of the planet – comprising humans, non-humans and their natural and constructed environments' (Hillier 2010: 2). This holistic perspective sits alongside a consensus on planning monopolized by the concept of sustainable development. Not for the first time perhaps, a specific planning doctrine has come to be viewed as the optimal goal for planning. Gunder and Hillier describe sustainable development as the now dominant spatial planning narrative, although they argue that it is 'perhaps implicitly trumped in achievement at the city-region level by the desire to be a "globally competitive city"' (Gunder and Hillier 2009: 20).

In spite of this narrowing, planning stubbornly retains 'dimensions of both art and science' (Gunder and Hillier 2009: 2). On this basis, viewed as an expression of technical, political, artistic and scientific contributions and processes, Sandercock (2004) proposes a different sensibility for planning from the regulatory planning that dominated twentieth-century practice. Planning as Sandercock (2004: 134) describes it would be based on a sensibility as much alert to the emotional economy of the city as to its political economy, as aware of the sensory landscape of the city as of the city census, as much alert to the desires of its citizens as to the need for hard infrastructure, able to see the ludic as inseparable from the city's productive spaces, capable of being critical of capitalist excesses and able collectively to forge new hybrid cultures and places.

In short, this is a planning alert to all of the registers and dimensions of culture, with the potential to challenge and modify the impacts of neo-liberal governmentalities that limit the possibilities for cultural integrity and planning coherence. A perspective such as this would play a part in beneficially modulating the dialectic between culturized planning and economic culturalization.

Cultural Planning, Culturized Planning and Planning with Culture

Over the last several decades cultural planning by name, has been a significant trend in planning notably in Australia, Canada, Spain, the United Kingdom and the United States, although it has a proto-history in the early twentieth century most notably in the work of Patrick Geddes. Emerging from developments in cultural theory and cultural studies in the 1970s, and from pressures for participatory democracy cultural planning was conceived as a means of improving the public realm, often hand in hand with initiatives to empower local communities (Bianchini and Schwengel 1991, see also Bianchini, Chapter 22, this volume). In Australia, for example, cultural plans have been developed as strategic planning exercises by local authorities, and are the vehicles for assessing or mapping local culture in qualitative and quantitative terms through a facilitated process of structured community engagement. Cultural plans typically set out administrative responsibilities, resource needs and timelines for implementation and are intended for integration with other planning undertaken by a local authority. Planning for the culture of cities and regions is emphasized, for example, by the parliaments of Scotland (Scottish Executive 2002) and Wales, and numerous Australian and Canadian local governments.

More recently, however, cultural planning and planning for the arts have grown to reflect 'the growing attention paid to the cultural economy and the commodification of the arts as urban cultural assets' (Evans 2001: 16) and are bolstered by the agenda of the 'creative city' and the 'creative class'. Cultural planning has been criticized in an Australian context, for example, for the fact that its 'central assumptions are not about using the arts or cultural activity to achieve social justice, but are concerned with social control, place management, and the achievement of conservative forms of citizenship and community' (Stevenson 2004: 125).

In contrast to the all too frequently arts-biased interventions of cultural planning is the concept of culturized planning, in which the goal is to 'mainstream' the use and consideration of culture throughout all spatial and non-spatial strategic planning using planning approaches that are specifically creative, critical, ethical and reflective in their techniques (Young 2008b). In this perspective, culture is theorized as a contested arena and set of practices, but with a focus on the cultural enrichment and sensitization of planning while recognizing the risks of its uptake in processes of state legitimation, in the economy of spectacle and in the differentiation of urban products for the market place. In common with 'planning culturally' the objective of culturized planning is to put culture on the table first in all planning projects regardless of type. Culturized planning reflects the safeguards inbuilt in the Australian model for community cultural mapping (Young et al. 1995) that acknowledge the right of each community and sub-group to define its own culture, and to protect it in the mapping process through an ethical system based on confidentiality, the protection of intellectual property and copyright and related measures. Regardless of the source of its inspiration perhaps, a revitalized and re-thought cultural planning has numerous strands, models and practices to draw on for further development.

Planning Theory

It has been claimed that planning has no endogenous body of theory 'unlike other areas of the social sciences or other professions including medicine' (Allmendinger 2002a: 6). Yet planning requires and utilizes theoretical systems and insights in order to develop and refine its own versions of theory and practice. Campbell and Fainstein have described the role of planning theory as intended to include the updating of planning thinking in order to accommodate new 'urban phenomena ... or social theories from other fields' (2003: 12). The examples of new urban phenomena the authors cite include globalization and cyberspace, while their examples of relevant social theories include postmodernism and critical theory. As Sandercock has succinctly observed, 'the need for different kinds of theories shifts as societies change' (Sandercock 1998: 104) and this is no less so for planning than for other sectors.

In characterizing and accessing planning theory Allmendinger has helpfully divided it into neo-modern and postmodern wings. In addition (Allmendinger 2002b: 7) he has neatly condensed a general characterization of the postmodern wing in social theory around five principles: the breakdown of transcendental meaning; the existence of a discursively created subject; the role of cultural influences in ordering society; considerations of fragmentation and dispersal; and Foucault's and Baudrillard's notions of power. In this diverse context, operating with an 'eclectic "pick and mix" basis to theory development and planning practice' (Allmendinger 2002a: 84) would seem to be appropriate, with the goal to locate the right theory according to the research, planning and community contexts and purposes in mind. As an example, Flyvbjerg and Richardson in searching for the dark side of planning theory contrast the use of a Foucauldian perspective as against that of 'Habermasian normativism' (2002: 61). They argue that Foucault's emphasis on conflict and power is more relevant to an understanding of planning that is 'practical, committed and ready for conflict' and provides a superior paradigm to 'planning theory that is discursive, detached and consensus-dependent' (2002: 62).

Depending on context and contextualization, these and other cultural and social theories are capable of infusing the planning imagination with fresh inspiration and new opportunities. At the same time, theory is challenged to consider cultural issues and cultural framing and to develop creative approaches for a 'transformative practice' (Albrechts 2010: 216). Albrechts argues that such practice is required, as the challenges that places face cannot be adequately addressed 'either with the neo-liberal perspective or with the intellectual technical-legal apparatus and mind-set of traditional land use planning' (2010: 216).

I myself perceive limitations in most current planning approaches, but consider that these could be ameliorated through the greater integration and assimilation of culture in critical, ethical and reflective terms. This may be best indicated by using four key planning theories and planning practices as examples, consisting of 'planning physicalism', collaborative planning, postmodern theories and approaches, and the creative city approach. First, 'planning physicalism', or a dominant physical approach to planning, identifies the provision of hard infrastructure as the overriding economic and social priority to provide planning with a panacea. This approach can be de-cultured and socially unresponsive. In spite of this, it is possible to re-conceptualize hard infrastructure in more balanced terms and to re-frame its delivery and urban and

11

social interface in more socially sensitive and responsive terms. Second, collaborative planning often focuses on local issues and micro-perspectives, at the expense of broader issues, and involves unrealistic levels of social engagement, yet it has the potential for enlargement to address wicked problems and the 'big picture' in cultural terms. Third, postmodern social and cultural theories are often constrained by high levels of abstraction, elite accessibility and the need for specialist interpretation. In spite of this, the interpretive 'unpacking' of such theories and insights into accessible themes and practical linkages to planning reasoning and planning techniques could result in a re-worked theory for planning action. Fourth, the 'creative city' approach focuses on the arts and technology and exhibits a bias to 'cool' aesthetics, social winners and private and privatized heritage, but could learn to accommodate broader considerations in the mix, such as intercultural and non-mainstream forms of cultural capital and issues of cultural equity.

To achieve a repositioning, in each of the four examples I cite, would be to develop a more sensitive and nuanced approach to culture, regardless of the spatial scale, planning mode or type. It would also be to repudiate what Gunder and Hillier (2009: 194) term 'copy-paste planning templates' which are all too often based on the 'mechanistic pre-shaping of planning issues – through application of universal planning master signifiers (or those of economic development) and their dominant supporting knowledge sets', the effects of which are to frame discussion in a predetermined way (2009: 194).

Outline of the Book

The book consists of six Parts with this Introduction and an Afterword. Each Part is divided into four chapters and concludes with a Case Study Window. The six Parts are 'Global Contexts', 'Planning and Its Dimensions', 'Culture and Its Dimensions', 'Planning Practices', 'Cultural Practices' and 'Cultural and Planning Dynamics'. Each Case Study Window is a concrete case study with illustrations.

Themes of the Chapters and Book

In this section I relate the key overall themes of the book to its individual chapters. While most chapters inevitably reference a number of themes in most cases a single theme dominates. As set out previously, the six themes consist of 1) cultural diversity, 2) local place and global flows, 3) the cultural and creative industries, 4) public space and citizenship, 5) cultural planning and sustainability perspectives, and 6) social and cultural theories and concepts of culture, history and heritage.

First, cultural diversity is considered as reflected in urban and especially metropolitan populations and in planning policy and practices. In this way, cultural diversity is encompassed in chapters on:

* living with difference in multicultural cities including the expression of street markets and religious sites as discussed by Sophie Watson in Chapter 1;

- the various national strands contributing to the history of Western planning practice documented by Ward in Chapter 2;
- the diverse, specific national planning cultures of global cities outlined by Newman and Thornley in Chapter 4;
- the fundamental dimension of gender diversity and historically gendered planning approaches to citizenship and public and private space discussed by Greed in Chapter 5.

Second, local places and global flows can be seen as presenting an interpenetrated dynamic with consequences reflected in:

- the dynamics of the theoretical and practical bridges that exist between global cultural governance and cultural planning practices at the local level traced by Duxbury and Jeannotte in Chapter 21;
- the varied cultural needs and uses for public parks including the cultural rights of migrant groups identified in New York by Low in Chapter 17;
- the complexities of place reinvention in Nordic and other cultures described by Nyseth in Chapter 19;
- the limited relevance of the transfer of Western planning theory to the global South argued by Vanessa Watson in Chapter 7.

Third, the cultural and creative industries are self-perpetuating and exist as highly mobile frontiers. Their impacts cycle through many dimensions discussed in terms of:

- toxic e-wastes shipped to the global South, comprising a dark side to the cultural economy, targeted by Miller in Chapter 3;
- favouring the new economy in inner urban redevelopment in planning discourse in Sydney, and the use of cultural theory in understanding the process, as discussed by Searle in Chapter 8;
- the sometimes regressive signifier of the cultural industries argued by O'Connor in Chapter 10;
- a Japanese perspective on the creative city and its cultural economy illustrated by Sasaki in Chapter 12;
- the planned role of cultural quarters in urban redevelopment models presented by Montgomery in Chapter 20.

Fourth, public space and citizenship are intertwined in normative terms and find reflection in each other as illustrated through:

- evocations of citizenship as they occur in cultural planning and their specific privileging explored by Stevenson in Chapter 9;
- contradictions and ambiguities contained in the concept of place and in its social mobilization discussed by Dovey in Chapter 15;
- the reaestheticization of the city, as a threat and opportunity to the material and symbolic meaning public space gives to urban life, discussed by Paddison in Chapter 18.

Fifth, the further development of cultural planning and sustainability perspectives has great potential including cross-fertilization illustrated through:

- the possibilities for cultural planning to contribute to sustainable development in communities through a whole population approach and mainstream embedding in planning, outlined by Evans in Chapter 13;
- recognition of the fundamental need in the global South to address the crisis of urban life through a radically democratic planning recast with cultural understanding, discussed by Pieterse in Chapter 14;
- the growth of the slow city movement as a cultural movement in civil society, with the objectives of local sustainable development and the promotion of a global manifesto, outlined by Baycan and Girard in Chapter 16;
- expansion of the concept and role of cultural planning across public policy in broad terms and in political and global terms through an international perspective critically considered by Bianchini in Chapter 22.

Sixth, social and cultural theories and concepts of culture, history and heritage in relation to planning encompass a range of ideas and perspectives including:

- the social and planning role of heritage as a construct observed by Ashworth in Chapter 11;
- the limitations of the model of communicative planning rationality in the context of the contemporary United States as perceived by Throgmorton in Chapter 6;
- the value of a new cultural paradigm for governance in widespread terms and specifically for culturized governance and planning developed by Young in Chapter 23;
- a Deluezean inspired re-theorization and reconceptualization of planning activities and practices outlined by Hillier in Chapter 24.

Conclusion

The extensive series of essays in this volume illustrate the fact that the place of culture is highlighted in contemporary planning practice and theory as never before, whether in the spatial expressions and planning implications of cultural practices and cultural diversity, the role of cultural equity in just urban development, the place of creativity and the cultural industries in urban development, and through culturally sensitive and responsive planning concepts of place reinvention and culturized planning. In all of these modes the role of culture and of imagination is to deepen and sensitize planning theory and planning practice to assist planning in its development and transformation.

In an age characterized by cultural diversity, global flows, environmental stress and planning scepticism, there is a heightened need to recognize that local and regional patterns of ways-of-life, histories, and cultural landscapes are dynamic cultural constellations created by people 'suspended in webs of significance spun by themselves' (Geertz 1973: 5). Planning sustainable cultures rests, therefore, on discovering, interpreting and building the mosaics of these lives, landscapes and legacies with the soft power of culture. In this task, our very ideas about culture are themselves constitutive of the planning and governance fields, as through a positive governmentality reconstituted from the progressive elements of the 'reformer's science'. For culture is the home of being and the foundation of the planner's house.

References

Albrechts, L. 2010. Enhancing creativity and action orientation in planning, in *The Ashgate Research Companion to Planning Theory: Conceptual Challenges for Spatial Planning*, edited by J. Hillier and P. Healey. Farnham: Ashgate, 215–33.

Allmendinger, P. 2002a. Towards a post-positivist typology of planning theory. *Planning Theory*, 1(1), 77–99.

——. 2002b. The post-positivist landscape of planning theory, in *Planning Futures: New Directions for Planning Theory*, edited by P. Allmendinger and M. Tewdwr-Jones. London: Routledge, 3–17.

American Planning Association. 2012. *What is Planning?* [Online]. Available at: http://www.planning.org/essay/resources/pdf/planit.pdf [accessed 3 August 2012].

Archer, M. 2005. Structure, culture and agency, in *The Blackwell Companion to the Sociology of Culture*, edited by M. Jacobs and N. Hanrahan. Malden, MA: Blackwell, 17–34.

Arnold, M. 1979. *Culture and Anarchy*. Cambridge: Cambridge University Press.

Barnett, C. 2009. Culture, in *The Dictionary of Human Geography*, edited by D. Gregory, R. Johnston, G. Pratt, M. Watts and S. Whatmore. Chichester: Wiley-Blackwell, 135–138.

Bennett, T. 1998. *Culture: A Reformer's Science*. Sydney: George Allen and Unwin.

Bianchini, F. and Schwengel, H. 1991. Re-imagining the city, in *Enterprise and Heritage: Crosscurrents of National Culture*, edited by J. Corner and S. Harvey. London: Routledge, 212–34.

Bourdieu, P. 2001. The forms of capital, in *The Sociology of Economic Life*, edited by M. Granovetter and R. Swedberg. Boulder, CO: Westview Press, 96–111.

Campbell, S. and Fainstein, S. 2003. Introduction: The structure and debates of planning theory, in *Readings in Planning Theory*, edited by S. Campbell and S. Fainstein. Malden, MA: Blackwell, 19–20.

Castells, M. 1991. *The Informational City: Information Technology, Economic Restructuring, and the Urban-Regional Process*. Oxford: Blackwell.

——. 2001. *The Internet Galaxy: Reflections on the Internet, Business and Society*. New York: Oxford University Press.

Chaney, D. 1994. *The Cultural Turn*. London: Routledge.

Dear, M. 2000. *The Postmodern Urban Condition*. Malden, MA: Blackwell.

Department of Communications and the Arts, DOCA. 1994. *Creative Nation: Commonwealth Cultural Policy*. Canberra: Commonwealth of Australia.

du Gay, P. and Pryke, M. 2000. Cultural Economy: Cultural Analysis and Commercial Life. London: Sage.

Evans, G. 2001. *Cultural Planning: An Urban Renaissance?* London: Routledge.

Fairfield City Council. 2005. *Making the Most of Our Culture: Fairfield City's Cultural Plan, 2005–2009*. Fairfield, NSW: Fairfield City Council.

Florida, R. 2004. *Cities and the Creative Class*. New York: Routledge.

Flyvbjerg, B. and Richardson, T. 2002. Planning and Foucault: in search of the dark side of planning theory, in *Planning Futures: New Directions for Planning Theory*, edited by P. Allmendinger and M. Tewdwyr-Jones. London: Routledge, 44–62.

Geertz, C. 1973. *The Interpretation of Cultures*. New York: Basic Books.

Glacken, C. 1967. *Traces on the Rhodian Shore: Nature and Culture in Western Thought from Ancient Times to the End of the Eighteenth Century*. Berkeley, CA: University of California Press.

Gunder, M. and Hillier, J. 2009. *Planning in Ten Words or Less: A Lacanian Entanglement with Spatial Planning*. Farnham: Ashgate.

Hall, P. 1998. *Cities in Civilisation: Culture, Innovation and Urban Order*. London: Phoenix.

Harvey, D. 1988. *Social Justice and the City*. Oxford: Blackwell.

Healey, P. 1997. *Collaborative Planning: Shaping Places in Fragmented Societies*. Houndmills: Palgrave.

Hillier, J. 2010. Introduction, in *The Ashgate Research Companion to Planning Theory*, edited by J. Hillier and P. Healey. Farnham: Ashgate, 1–34.

Horkheimer, M. and Adorno, T. 1997. *The Dialectic of Enlightenment*. New York: Continuum.

Jacobs, M. and Hanrahan, N. 2005. Introduction, in *The Blackwell Companion to the Sociology of Culture*. Malden, MA: Blackwell, 1–14.

Jameson, F. 1984. Postmodernism, or the cultural logic of late capitalism. *New Left Review*, 146, 53–92.

Johnson, L. 2006. Valuing the arts: theorising and realising cultural capital in an Australian city. *Geographical Research*, 44(3), 296–309.

Landry, C. 2006. *The Art of City-Making*. London: Earthscan.

Lefèbvre, H. 1992. *The Production of Space*. Oxford: Blackwell.

Lewis, H. 1998. The misrepresentation of anthropology and its consequences. *American Anthropologist*, 100(3), 716–31.

Miller, T. and Yudice, G. 2002. *Cultural Policy*. London: Sage.

McNeely, I. and Wolverton, L. 2009. *Reinventing Knowledge: From Alexandria to the Internet*. New York: Norton.

Nye, J. 2004. *Soft Power: The Means to Success in Politics*. New York: Public Affairs.

Phelps, N. and Tewdwr-Jones, M. 2008. If geography is anything, maybe it's planning's alter ego? Reflections on policy relevance in two disciplines concerned with place and space. *Transactions of the Institute of British Geographers*, 33(4), 566–84.

Radcliffe, S. (ed.) 2006. *Culture and Development in a Globalising World: Geographies, Actors and Paradigms*. Abingdon: Routledge.

Sandercock, L. 1998. *Towards Cosmopolis*. Chichester: John Wiley.

—— . 2004. Towards a planning imagination for the 21st century. *Journal of the American Planning Association*, 70(2), 133–41.

Sassoon, D. 2006. *The Culture of the Europeans: From 1800 to the Present*. London: HarperCollins.

Scott, A. 2000. *The Cultural Economy of Cities: Essays on the Geography of Image-Producing Industries*. London: Sage.

Scottish Executive. 2002. Implementation of the National Cultural Strategy: *Draft Guidance for Scottish Local Authorities* [Online]. Available at: www.scotland.gov.uk [accessed 7 April 2013].

Shmueli, D. 2005. Is Israel ready for participatory planning? Expectations and obstacles. *Planning Theory and Practice*, 6(4), 485–514.

Soja, E. 1993. *Postmodern Geographies: The Reassertion of Space in Critical Social Theory*. London: Verso.

——. 1996. *Thirdspace: Journeys to Los Angeles and Other Real-and-Imagined Places*. Cambridge: Blackwell.

Stevenson, D. 2004. Civic gold rush: cultural planning and the politics of the third way. *International Journal of Cultural Policy*, 10(1), 119–31.

Throsby, D. 2008. The concentric circles model of the cultural industries. *Cultural Trends*, 17(3), 147–64.

Tylor, E. 1903. *Primitive Culture*. 4th Edition. Volume 1. London: John Murray.

——. 1924. *Primitive Culture*. 7th Edition. Volume 2. New York: Brentano's.

UCLG. 2010. *Culture: The Fourth Pillar of Sustainable Development*. UCLG. [Online]. Available at: www.uclg.org/en/resources/policy-statementsonline [accessed 7 April 2013].

UNESCO. 2001. *Universal Declaration on Cultural Diversity*. Paris: UNESCO.

——. 2009. *Culture and Sustainable Development: Examples of Institutional Innovation and Proposal of a new Cultural Policy Profile*. Culture 21. UNESCO. [Online]. Available at: www.agenda21culture.net [accessed 7 April 2013].

United Nations. 1948. *Universal Declaration of Human Rights*. [Online]. Available at: www.un.org/en/udhr [accessed 7 April 2013].

——. 1993. *United Nations Conference on Environment and Development, Agenda 21*. New York: UN Department of Public Information.

Williams, R. 1966. *Culture and Society, 1780–1950*. Harmondsworth: Penguin.

——. 1983. *Keywords*. London: Flamingo.

World Commission on Culture. 1996. *Our Creative Diversity*. 2nd Edition. Paris: UNESCO.

Wright, J. 1994. *Collected Poems*. Sydney: Angus and Robertson.

Young, G. 2005. Concepts of culture in society and planning in 20th to 21st century Australia and Britain. *Planning History*, 27(1–2), 15–19.

——. 2008a. The culturization of planning. *Planning Theory*, 7(1), 71–91.

——. 2008b. *Reshaping Planning with Culture*. Aldershot: Ashgate.

Young, G., Clark, I. and Sutherland, J. 1995. *Mapping Culture: A Guide for Cultural and Economic Development in Communities*. Canberra: AGPS.

Yudice, G. 2003. *The Expediency of Culture: Uses of Culture in the Global Era*. Durham, NC: Duke University Press.

PART 1
Global Contexts

Preface to Part 1

Deborah Stevenson

The chapters that comprise this Part of the book present penetrating insights into the global contexts within which city building and urban planning occur. Cities, their cultures, planning practices and economies increasingly are enmeshed in an array of global processes that include transnational flows and multifaceted circuits of capital, people, ideas and services. With increased globalization has come a fascination not only with its contours and contradictions, but also with the situation of different cities within and outside influential networks. It was in this context, for instance, that the notions of the 'global' and 'world' city gained currency. Globalization and associated categorizations of cities are founded on the existence of a core and a marginalized periphery with cities increasingly being defined in terms of their locations *vis-à-vis* the global. A consequence of the fixation on global networks, status, contexts and processes is countervailing interest in the 'other' of the global – the local. And so it is with the chapters here, each of which in some way seeks to understand local conditions and processes not only in the context of, but frequently in opposition to, the global.

The task of interrogating the global contexts which shape and are shaped by the local, and very directly mould the lives of urban dwellers, begins with Sophie Watson's chapter examining the challenges planning faces as a result of increased urban diversity that is an outcome, in part, of global migration and mobility. Watson is keen to probe the role that planners can play in shaping spaces, and public spaces in particular, which will not only accommodate this diversity but do so in ways that support a multiplicity of cultures. This is a significant undertaking but Watson is sanguine, suggesting that urban complexity and uncertainty may actually provide the conditions for the development of a 'radical and creative imaginary' required for a rethinking of planning practice. The starting point is to reconceptualize many of the entrenched processes and cultures of planning and to see planning not solely as a technical exercise but as a set of engaged knowledges and practices. It also means scrutinizing the often limited assumptions that planners make about people, their cultures and the ways in which they relate to and use the city.

Stephen Ward shares some of Watson's concerns regarding the culture of contemporary planning and its ability to contribute to the development of socially progressive landscapes in the context of global influences. But where Watson sees cause for optimism Ward is more circumspect, leaving open the question of whether a 'cultural renewal' of planning is possible. He suggests that in order to understand both the limits and potential of contemporary urban planning, it is necessary first to examine its cultural and historical dimensions and, to this end, he traces the development of planning from its origins as a highly localized, progressive activity to one that has become largely a mediator of competing interests. Ward suggests that although urban planning is a 'cultural form in its own right, with distinct national, regional and local characteristics' it is shaped by values and protocols that are international and universalizing and it is these that have undermined its engagement with the political, social and cultural dimensions of the local and diminished its reformist capacity as a result.

A growing awareness of the cultural dimensions, consequences and intentions of urban planning has led many to suggest that greater links between urban and cultural policy and planning are necessary. Indeed, some contend that urban planning is already an aspect of cultural planning because of the all-embracing way in which culture has come to be defined. Under such conditions, cultural policy increasingly focuses more on the economic potential of the culture and its strategic use including in the revitalization of cities and less on aesthetics, meaning and memory. It is these trends and their consequences that are of concern to Toby Miller whose chapter maps the historical, conceptual and ideological contexts of cultural policy studies and the emergence of the creative industries agenda that has come to dominate both it and cultural planning. Miller argues that in privileging industrial and economic priorities rather than the cultural, cultural policy has lost its progressive agenda and this, in turn, has significant consequences for the possibility of developing reformist urban policies and creating socially equitable and environmentally sustainable urban landscapes. He suggests that the shipping of toxic e-wastes from the global North to the global South represents a troubling outcome of the cultural economy.

The final chapter in this Part is by Peter Newman and Andy Thornley who take as their starting point the proposition that globalization is leading to the homogenization of cities and urban cultures. They acknowledge, for instance, that city form and ideas about architecture and planning are significantly shaped by globalized ideas and trends and in response to economic globalization. Evidence of this include striking similarities between city skylines and architectural forms as well as the replication of planning approaches, blueprints and fashions, with skyscrapers, mega-malls and themed waterfront developments being obvious examples. With reference to four (global/world) cities, however, Newman and Thornley assert that the extent of global standardization and seriality is overstated and that each city (and its responses to the challenges of globalization) is a product of its specific histories, cultures, traditions and forms of governance while the interplay of these factors are 'balanced' through the 'political culture' of the national state.

In addressing the theme of 'Global Contexts' the chapters in this Part of the book in different ways highlight not only significant macro trends and influences but also the contingent, fractured and, indeed, local dimensions of the global. Processes of globalization have indeed transformed cities and their study and every city is now in some way locked into (or out of) significant global circuits of information, capital, people and ideas. But each city is also a lived space, shaped by its histories and cultures including those of planning, politics and everyday life.

Global Futures: Reflections on Culture, Diversity and Planning for the Twenty-first Century

Sophie Watson

All planning involves imagination and speculation about the future, since there is an inevitable time lag between the conception of the plan and its execution and an even longer period during which the plan has effects, particularly where the plan involves infrastructures and the built environment. Planning is simpler then in periods of social, economic, political and cultural stability, where greater levels of certainty as to the future trajectories of the city and its people are in place. Instead, a plethora of events during the first decade of the twenty-first century indicate that we have entered a period of increasing uncertainty, complexity and cultural change. Global warming at unprecedented levels shows signs of increasing further bringing potentially devastating effects; reliance on oil for virtually every activity from industry to transport looks increasingly precarious and irrational; population shifts from political unrest, environmental disaster, or the search for employment or shelter look set to continue apace; financial collapse at institutional and sometimes national levels with its devastating consequences for the less well-off is a growing feature of capitalist economies; and socio-cultural changes in the form of new family formations, lifestyle choices or religious affiliations are impossible to predict. This is to mention only some of the contours of a future, particularly an urban future, marked by unpredictability and often rapid change with very uneven effects at a global level.

Here my intention is to focus on questions that arise as a consequence of growing cultural diversity in many cities which should be of central concern to planners in the coming years. (See also Stevenson, Low, and Duxbury and Jeannotte, Chapters 9, 17, and 21, this volume.) I do this through looking at a variety of sites and spaces where diverse cultural practices are enacted, and consider the planning implications, constraints and possibilities which follow. Before moving to this more pragmatic set of concerns, let me indulge in a moment of polemic. It seems to me that we have entered dangerous times, times where radical visions and radical imaginaries of a future of greater social equality and redistribution from the rich to the poor, care for the environment, democratic politics, decent social welfare provision, are held by an ever decreasing minority. Yet notions of this kind precisely have underpinned the rationale for planning since its inception, and with their demise, the role of planning looks ever more uncertain. In this context, my point is that there is a far greater imperative to embrace a radical and creative imaginary, one that lends itself to openness, collectivity, plurality and optimism rather than closure, singularity, negativity and individualism. Castoriadis is helpful here:

> I think that we are at a crossing in the roads of history, history in the grand sense. One road already appears clearly laid out, at least in its general orientation. That's the road of the loss of meaning, of the repetition of empty forms, of conformism, apathy, irresponsibility, and cynicism at the same time as it is that of the tightening grip of the capitalist imaginary of unlimited expansion of 'rational mastery', pseudorational pseudomastery, of an unlimited expansion of consumption for the sake of consumption, that is to say, for nothing, and of a technoscience that has become autonomized along its path and that is evidently involved in the domination of this capitalist imaginary. The other road should be opened: it is not at all laid out. It can be opened only through a social and political awakening, a resurgence of the project of individual and collective autonomy, that is to say, of the will to freedom. This would require an awakening of the imagination and of the creative imaginary. (Castoriadis 2007: 146)

Though the material world undoubtedly constrains, makes possible and performs the social worlds we inhabit (Latour 2005), there is still room for resistance, manoeuvre, change, perhaps (idealistically) even revolution. So also imaginaries and discourses frame and perform realities. As planners in these precarious times, new imaginaries that address complexity and diversity and mobilize a more democratic politics are urgently needed.

Public Space

Core to this new world of complexity and cultural diversity is public space. (See also Stevenson, Low, and Paddison, Chapters 9, 17, and 18, this volume.) The notion of public space is conceived in a multiplicity of ways from Habermas' (1989) formal public sphere of rational debate and communication conducted in the eighteenth century coffee house to Arendt's (1958) idealized notion of the agora as a space where people come together freely as non-economic slaves and the life of labour which has the capacity to oppress them or Young's (1990) idea of the city as a space of 'unassimilated otherness'. In my own work (S. Watson 2006a) I have emphasized the significance of more marginal and often hidden spaces as key sites of 'rubbing along' and encounters with 'different others'. My concern, like Connolly (1995) and also Deutsche (1999) has been how to conceive of democratic public space which is not predicated on the exclusions of others who are different from ourselves. Planners have a role to play here, in resisting those developments which keep people out, those gated communities and residential enclaves based on the fear of mixing with others who are different which embody Bauman's (2003) notion of 'mixophobia'. All theories of public space concern, or should concern, the mediation, accommodation and resolution of difference in public to a greater or lesser extent, yet such concern has not been a core preoccupation of planning theory until fairly recently. Sandercock (2003) has repeatedly argued that the question of the integration of migrants and living together as strangers without conflict and fear is central to the concerns of urban planners, designers and engineers, as well as governments more broadly. As Bollens (2004: 212) puts it: 'differing planning systems are a defining characteristic of ethnically polarized cities and also appear to be an increasing attribute of planning and resource allocation debates in North America and Western Europe.'

Nevertheless, the growing cultural diversity in cities has had considerable impact on city spaces, some of which are contested and contentious hence becoming a matter of concern to the day to day life of city planners. Other multicultural spaces occur in a hotchpotch kind of way, operating as key sites in the mediation of differences. My argument in this chapter, endorsing a similar plea by others over the last two decades (Sandercock 2003) is that planning has an ever more important role to play in this complex terrain of multicultural differences in the city. Planners need to be involved in creating new and innovative spaces where different cultural practices can be enacted and performed, supporting those that exist already, and at the same time intervening to find solutions where conflicts and tensions arise. Nevertheless, as Vanessa Watson (2003) argues, we are seeing increasingly irreconcilable gaps between communities who regard each other from different rationalities, and between planners and those that are planned for, where the possibility of reaching a consensus looks virtually improbable and where views as to what constitutes progress or development differ dramatically. There may not always be a clear resolution. With these considerations in mind, this chapter considers two kinds of spaces – street markets and religious sites – which I suggest are crucial spaces for living with difference that planners should take seriously. There are many others.

Street Markets

In recent years, street markets have increasingly been recognized as places which matter for social, economic, environmental and health reasons, reversing the view of their inevitable decline faced with competition from supermarkets and other cheap retail outlets. In *Markets as Sites of Social Interaction:Spaces of Diversity* (S. Watson 2006b), based on research in eight London street markets, I argued that markets fulfilled a number of key policy agendas: they provided sites for social encounters, for the mediation of social and cultural differences, for social inclusion, for urban regeneration, for the revitalization of city centres, for economic and social innovation – for example for start-up businesses, for healthy eating through the provision of fresh and cheap food, for environmental agendas – through reduction in packaging and local access by foot or public transport (in contrast to out of town shopping centres and supermarkets), and for the promotion of city liveability. (See also Ashworth and Montgomery, Chapters 11, and 20, this volume.) Arguments such as these have increasingly been taken up in national policy agendas. In 2008–9 in the United Kingdom, the House of Commons

Department of Communities and Local Government Committee established an inquiry into markets including traditional street markets, farmers' and specialist markets (CLGC 2009). The Government response published in October 2009 agreed with the committee's conclusions and in particular the role of markets in contributing to the vitality of streets and regeneration of the wider area, seeing the key source of their support as the responsibility of the local authority. However, the response to the Department of Communities and Local Government (DCLG) recommendations has been very uneven across the United Kingdom.

Of the different agendas that markets connect to, my argument here is that markets have particular significance in assisting migrants to establish themselves in the city and that for this reason alone, they should be of concern to planners and policy makers. Markets can provide opportunities for survival and also for different communities to encounter one another in public spaces, mediating racial/ethnic cultural differences and mitigating inter-ethnic and racial conflict in the global city. However, their success as spaces of conviviality, encounter and sociality is by no means assured, indeed it is contingent on serendipitous factors, such as the market's location, history, design, as much as it is on local authority strategy, planning and investment. Markets can operate as spaces of social homogeneity, closure and exclusion, just as much as they can operate as spaces of mixing, encounter across differences, and social inclusion, and planners can play a part in enhancing one outcome rather than another. Three different market sites will be explored briefly here to illustrate my argument.

At one end of the spectrum is Four Tigers market in Budapest. This is a market which is home to one of the largest Chinese communities in Europe. In brief, Asian immigrants began to arrive in the city before the fall of the Communist regime, when visa requirements were abolished between the two countries in 1988. Within three years the number of registered Chinese rose from zero to 30,000, with the importation of garments in suitcases via the Trans-Siberian railway in the initial phase later to be replaced by containers. Within a few years, Hungary became the hub of Chinese imports to Central and Eastern Europe. The majority of the Chinese community in Budapest live around the Four Tigers market, which is squeezed between the railway tracks and a large trunk road out of town, and is a sprawling complex of shacks and warehouses located in Budapest's eighth district where hundreds of family businesses are occupied with the daily wholesale and retail grind of providing cheap commodities to a diversity of mainly Chinese consumers. The market is a teeming city comprised of shipping containers and tin shacks. At the entrance to the market a billboard illustration of a gun, dogs and a camera, each dissected by a large red cross graphically reveals the contours of the market as a space of illicit and illegal activity, under the protection of private security guards employed by the traders. All the elements of a Chinese market in mainland China are in place: the noodle bars, the men playing mahjong, the beautifully laid out mass produced designer clothes on mannequins and the cheap alcohol stalls. According to the figures from the Immigration and Naturalisation Office, there are currently 11,000 legal Chinese residents in Hungary, but the real figure is estimated at 20,000 or even 30,000, most of whom live in Budapest. The Four Tigers represents a key site of connection and survival for many of this group.

Despite its economic and social benefits to the Chinese community, by all accounts there is very little assimilation of this group into the city and antagonisms are easily fuelled in the media by the inspectors' reports of irregularities and illicit trading revealed in their surprise visits. High-tech machines and electronics worth some $7.5 billion account for 80 per cent of the trade between Hungary and China of which the informal and unregulated sector represents an unknown proportion. Processes of inclusion and exclusion are hard to unravel without extensive research, as is the extent to which the Chinese choose, or are forced to live and work in an area cut off from the mainstream of Budapest social and economic life. What is clear though is that this market, like many markets across the globe, represents a strategy for survival for a large group of migrants whose opportunities for integration or economic success in a new place are somewhat limited.

What are the implications then for planners? In the specific case discussed here, it could be argued that it is a poorly planned space, which has limited visibility or accessibility from the outside, and which thus spatially as well as seemingly intentionally from the insiders' perspective, serves to keep people out and exclude outsiders. As such it does not operate as a space of mixing across differences and inclusion. But as already suggested, it is a significant space in the lives of the Chinese residents of Budapest. In this sense then, there are strong arguments for vacant sites being made available to migrant groups and for allowing a high degree of autonomy for the traders and consumers of the market, within reasonable parameters of adherence to statutory health and other regulations. What we see here is a market that appears to have been left to its own devices by the local city authority (at least as far as I could ascertain). As a consequence it is a place which offers protection, employment, consumption opportunities and sociability, to one specific community. But it is not a space which encourages intermingling across cultural differences – many Budapest residents never have visited the site – nor mutual understanding or recognition. In contrast, outsiders were viewed with suspicion.

Queen's Crescent market offers a different model. This is a market which runs through a very racially and ethnically mixed locality of Gospel Oak in the London borough of Camden. The local population is 10,900 of which Black minority ethnic groups constitute 41 per cent; many of these households are refugees or migrants with few prospects of stable employment. Gospel Oak has three small areas that fall into the 20 per cent most deprived nationally. The market has a total of 77 pitches but in 2011 many of these sites stood empty during the Thursday and Saturday market. The market has a bedraggled air, the stalls selling cheap clothes and domestic commodities, with one fruit and vegetable stall and a flower stall as reminders of what the market once was in the days when Emecheta described it in 1974 as a lively and friendly place full of the sounds of voices and laughter and the smells of ripe tomatoes and flowers. Since that time the market has seen almost steady decline as supermarkets moved into the locality draining money out of the local economy, while the local authority turned a blind eye. Rather than reinvesting the rents collected from stalls, the council saw the market as a 'cash cow' and for the better part of 50 years had limited strategic vision and no clear management structures for the market. As a result, what once was a vibrant space which the long term stall holders remember as full of life and activity, with a mix of stalls selling a wide range of products, by the late 1990s had become a site of inter-racial disharmony (including open racism from the long term white stall holders and shop owners towards the new arrivals), neglect and a public life characterized by crime and violence rather than sociality and mutual support (Watson and Wells 2005, Wells and Watson 2005). Testimony to the impact of local authority neglect and indifference,

27

and the potential for reversing such a radical decline in a market, was a brief period in mid-2005–6, when under the Single Regeneration Budget highly successful attempts to regenerate the locality, and the market in particular, had visible effects. Under the strategic leadership of an energetic market manager, traders selling a diversity of products were encouraged back into the market selling jewellery, Italian cheeses, breads, fish and South Asian ethnic foods, and for a brief period the market was a place of sociality, conviviality and inclusion across ethnic and racial differences. With the manager's departure, its return to decline by the end of the decade has led to the complete separation of the market from council control.

What are the planning implications of this brief narrative? Queen's Crescent represents a market whose success as a local commercial and social space, and a space of potential intermingling across different cultures was sadly relinquished by the local authority through lack of investment, planning and management. During the research (Watson and Wells 2005) many of the traders referred to this lack of investment, poor design, the lack of suitable parking for shoppers, the lack of amenities such as seating places for the elderly as responsible for the market's decline. At the same time they saw the decision by the local planning department to allow the construction of two large supermarkets in the locality as responsible for the market's failure to attract customers.

The third market for my argument is Ridley Road market in the London Borough of Hackney. This is one of the most ethnically diverse areas in the country with a non-white population of 40 per cent including, in particular, a large number of Afro-Caribbean, African and Asian households. The local authority is notable for its proactive engagement in diversity policies supporting its multicultural population in a myriad of ways. Ridley Road, close to Dalston Junction, has a long history from at least the mid-nineteenth century of being a focal point for migrants to the area, and until recent years the majority of the stall holders came from the Jewish community who were numerous in the borough. As this population moved north and East into the outer suburbs of London, the market became increasingly ethnically diverse, with a minority of Jewish traders remaining. Unlike Queen's Crescent this is a market where inter cultural mixing is celebrated and lived, and Ridley Road is well known across the country for being a place to meet up with long lost friends and relatives from countries of origin, where virtually anything can be found, from unusual spices to tropical fish, snakes and goat (S. Watson 2006b). Though for many years the market effectively ran itself, albeit successfully, it is now recognized and supported by the council as a key public space and site of inter-ethnic mixing, as well as a successful commercial centre and focus for tourism. Plans have been put in place to provide seating areas, local parking, transport facilities and amenities.

Across the world there are a plethora of markets such as these representing different models, cultural practices, represented discursively by different interest groups in a myriad of ways. Yet despite support for markets from many quarters, and in the UK, from such bodies as the DCLG as we have seen, the future of markets in many places remain uncertain. In the United Kingdom for example, the National Market Traders Association recently reported that markets are battling for survival as shoppers turn to online shopping and supermarkets (Baron 2011: 12). From the traders' viewpoint a large part of the problem is the raising of rents – by 400 per cent in Mexborough, Yorkshire for example. While in Friends of Leeds Kirkgate Market there is fear that the council is looking to gentrify facilities and change its character. My point here is that for markets to survive, local government support and intervention can be crucial, which includes planners, both at the level of strategy, and at the micro level of planning regulations concerning parking, design and the provision of facilities. Given the importance of markets for economic survival and for intercultural mixing and social encounters in the global multicultural city, planning has a crucial role to play.

Religious Spaces

Diverse religious practices of different migrant groups in cities represent one of the most visible forms of cultural difference enacted in city spaces (see also Greed, Chapter 5, this volume). Yet arguably religion represents a key site of belonging and connection for new migrants to the city, especially in the context of cutbacks in expenditure on welfare and community services. Religious difference contains elements of the symbolic, the cultural and the material and these have different implications in the city, yet any generalization of how religious difference has been accommodated in cities inevitably would be both inaccurate and simplistic. However, in the post-2001 political climate across the Western world we have witnessed the exacerbation of a trend – which was already in evidence (Gale 2005, Hill 2011, Naylor and Ryan 2002) – of using planning discourse and policy as a means of opposition to the construction of mosques. Many mosques are located in sites that have undergone change of use – domestic residences, warehouses, factories – and are often difficult to recognize as mosques. Others have taken over religious buildings which have been vacated by other faiths whose congregations have declined, this is a pattern particularly common in the East End of London in the case of synagogues. Where opposition has tended to arise has been in the case of new constructions which are seen as visibly different from the local surroundings or vernacular architecture, either on account of their size or style – minarets, for example. As well as opposition to architectural style, common objections to the construction of mosques take the form of planning issues such as lack of parking or congestion in the locality. Many Muslim people would argue that objections mobilized in planning or architectural discourses simply mask a resistance to the visible presence of a religious faith in the city which has come to be associated with terrorism at worst, or the view that an immigrant group is alien. Such a reaction has its most extreme representation in the British National Party whose website has vituperative and provocative advice from Nick Griffin, their national figurehead, on how to campaign against a mosque planning application.

The picture on the ground is a mixed one, with some localities embracing a multicultural politics, and seeing the construction of a mosque in the area as a welcome addition, while other local councils have used planning as a tool to resist unpopular applications from Muslim groups. A recent application for extension of the Al-Jamaat-ul-Muslimin of Bangladesh mosque into a neighbouring building in Northampton, United Kingdom, for example, met with a positive response from the local council. As Councillor Clarke put it: 'There's still a need for a central mosque to keep up with the growing number of practising Muslims in the town. But the plans to extend the mosque in St George's Street are very welcome' (Bontoft 2011). According to an article in the *Northampton Chronicle and Echo* 'Documents submitted to Northampton Borough

Council said: 'The proposal will be a community resource in an accessible location with benefits to the whole community. It would provide a venue for social activities and strengthen links between different groups within the wider community''(Bontoft 2011). In contrast, the requested planning by the Bengali Welfare Association Surrey for a mosque in Camberley met with the planning decision that the benefits that it would bring to the town's Islamic congregation did not outweigh the potential loss of the 140-year-old Victorian structure (Engage 2011).

Not surprisingly, the most widespread resistance to a proposed mosque in Britain in recent years was focused on the proposal by Tablighi Jamaat, a strict Islamic association, to build a very large mosque next to the London Olympic site. Quite apart from the size of the development its symbolic presence close to London's impending flagship national event appeared to have challenged even liberal tolerance of difference. More than 48,000 people petitioned the Government to prevent the development, dubbed the 'mega-mosque'. Planning played a crucial role here as in early 2010 Newham Council (where the Olympic Park is located) were considering compulsory purchase of the site as the sect had failed to lodge a master plan, and had been operating illegally in a temporary mosque without planning permission thus facing eviction. From the Muslim Council of Britain's point of view, however, planning was not the issue, instead 'the group had fallen victim to "unfounded hostility and hysteria"'. Campaigners welcomed the outcome, saying that the proposed mosque – which would have held four times as many worshippers as Britain's largest Anglican cathedral – was inappropriate. Alan Craig, a councillor representing the Christian Peoples' Alliance, said: 'It would have given a huge national platform, right by the Olympics, for them to promote their ideology' (Hamilton and Gledhill 2010). Such disputes have been prevalent across Europe with a recent proposal by Turkish migrants living in a small Swiss village in the Alps objected to by local residents due to the minarets and potential noise disruption: 'It's the noise, and all the cars. You should see it on a Friday night', complained Roland Kissling, a perfume buyer for a local cosmetics company. 'I've got nothing against mosques, or even against minarets. But in the city. Not in this village. It's just not right. There's going to be trouble' (Traynor 2007). My own study of Marrickville in Sydney (Watson 2009) revealed a borough seemingly able and willing to embrace multicultural religious difference over many decades, with striking religious buildings offering various migrant communities a space to worship scattered through the locality.

It is not only visible material architectural structures that have exercised planners across the world, symbolic sites which challenge normative definitions of space represent another form of religious cultural intervention in the city which have become points of contention. The Jewish eruv represents one such intriguing example. First, what exactly is an eruv? For traditional Jews, Sabbath is the day which is set aside for rest and calm away from the fast pace of weekday life, which involves a cessation of labour of various kinds. Various restrictions are laid down in Jewish law that impose prohibitions on the Sabbath which include the carrying of objects from private domains to public domains and vice versa. These public domains include streets, thoroughfares, open areas, highways and so on. Private domains are homes and flats in residential areas which are enclosed and surrounded by a wall and thus closed off from the public areas. In these private areas, carrying of objects on the Sabbath is permitted.

The purpose of the eruv, which in Hebrew means mixing or joining together, is to integrate a number of private and public properties into one larger private domain or, to put this another way, to redefine the activities permitted in semi-public (or *karmelite*) space for the purposes of the Sabbath in order that activities normally allowed only in the private domain can be performed (Cooper 1996, Valins 2000: 579). This is a process of temporal spatial reordering. Once an eruv is constructed, individuals within the designated area are permitted to carry and move objects across what was hitherto a private/public boundary. This may include anything from the carrying of house keys, handbags (as long as no money is contained), a walking stick, to the pushing of a stroller or wheelchair. The construction of an eruv is thus of particular relevance to women with small children and people who are frail or disabled – those effectively excluded by age, gender or infirmity from public space on the Sabbath.

The practice of demarcating an eruv has been used by Orthodox Jews for 2000 years and is based on principles derived from the Torah, developed in the Talmud, and codified in Jewish Law. According to Talmudic law there is a very precise definition of an eruv. For an area to be reconstituted as a private domain it must cover a minimum of 12 square feet and be demarcated from its surroundings by a wall or boundary of some sort or by virtue of its topography. Already existing boundaries such as fences, rivers, or railways or even rows of houses can serve as the basis for the eruv, but where the boundary is not continuous – broken, for instance, by a highway – a boundary line must be constructed in order to maintain the enclosed space. The concept here is that, where a door separates two rooms in a house, the remaining structure on either side is still a wall, even if there are many openings. The eruv in the modern city is thus the limit case where the notional wall contains many openings with very little solid wall remaining. To construct the enclosure, there are clear vertical and horizontal elements which make up its parameters. To make acceptable the door/lintel combination, an eruv can use existing poles in the street – such as telephone, electric, cable poles – or new poles can be constructed. These are joined either by existing wires (usually the lowest in place) or by a new wire, in the case of the Barnet eruv in London this was a nylon fishing line, or plastic cable. For the pseudo-door to be acceptable the lintel (wire) must rest above the door posts, which can be made by attaching a thin vertical rod to the existing pole. In Hebrew these are 'lechi'. In the construction of eruvim in the United Kingdom and the United States it is these almost invisible objects which have become the site of contestation even though their visibility or intrusion on the street landscape is minimal.

There are eruvim in many urban areas across the globe including Canada (the eruv in Toronto has existed there for over 60 years), Australia (in Sydney, where the boundary is created from cliff faces, a golf course, and fences along the Bondi beach, and in Melbourne), Belgium (Antwerp), France (Strasbourg), Italy (Venice), South Africa (Johannesburg) and many in the United States. The Barnet proposal to construct an eruv (S. Watson 2005) represented the largest of its kind in the United Kingdom. They vary in size from a small front yard of a single household to a large building such as a hospital (allowing Orthodox Jewish medical staff to work on the Sabbath), to ones that match the boundaries of whole cities as is often the case in Israel. Even the White House is included in the boundaries of the Washington eruv (Vincent and Warf 2002: 35–6). Usually eruvim are distinct spatial entities, although where they overlap, because of the lack of recognition of each other's eruv by different communities, coloured ribbons are attached to the wires to avoid confusion for the members of synagogue communities. Ironically, though they themselves draw on an ancient concept, some eruvim are highly modern in their use of the Internet for keeping their members informed of the state of their local eruv through websites. Typically, eruvim are patrolled the day before the Sabbath to ensure that the enclosure is intact and wires are not broken, as they cease to function once a gap has emerged – often indicated on websites by a kind of traffic-light system of green and red lights. This eruv boundary is unlike other boundaries in that when it ruptures nowhere inside is safe or unaffected.

For an eruv to become operational, in the first instance a civil figure with jurisdiction over the prescribed area has to give permission for the eruv. In the American case a nominal fee is paid. In Britain an application is made through the planning system. At one level this requirement necessarily constitutes the group that is requesting the eruv as dependent and powerless in relation to a state, or other, official. For example, one Orthodox woman described how an Oxford quad was defined as an eruv for one evening for the purposes of a party on the Sabbath, for which permission was requested from the college rector. In the United States they are generally established by means of a ceremonial proclamation that the area is 'rented'. This is issued by municipal authorities.

According to two rabbis interviewed in Barnet, many eruvim have been established with minimal local objection, often barely entering the consciousness of local residents. However, in various instances the construction of an eruv has been hotly contested, sometimes over many years. Tenafly in New Jersey, United States, and Barnet are two such cases where opposition was intense but was articulated through different arenas of the state. In my own research on these two sites what was revealed were several things: first, the symbolic claim on the use of space by a particular religious group, though having minimal visible impact on a local space, nevertheless was highly contested by local groups, and confounded local planning authorities and other juridical concerns. Second, in Britain planning and related arguments represent the key site for opposition and resolution – unlike in the United States where constitutional law is deployed. What brings the eruv into being then is a series of rituals, a performativity where new identities, spaces, social practices and notions of the private are constituted. It is not simply a question of the construction of an eruv, rather it is the routinized and repetitive recognition of the boundary by its users and the vigilant maintenance required to keep it intact that maintain it and keep it alive.

My argument here is that the power of symbols in this conflict over space is precisely the point. The fact that legal or planning approval has to be granted for the eruv is, in some sense, to submit to it. The eruv thus raises the questions of whether competing meanings of space can, or should, be resolved in the policy or planning arena, and can policy initiatives confront the symbolic? There is no easy answer here, but, if the symbolic power of space is to be taken seriously, particularly in the context of competing cross-cultural claims, planning and policy responses will need to confront some rather thorny questions. One such issue is that minority groups requesting permission from the state and from planners in particular – in whatever form this takes – are differently (and less powerfully) positioned from those imbued in the dominant culture, of which the state may be imagined to be an integral part.

Conclusion

In conclusion, the context of increasing uncertainty can well be seen as a potential opportunity as much as it might be articulated as a problem, since in settled and prosperous times, many people are suspicious of change, whereas during periods of upheaval it becomes possible to do things differently, and to mobilize concern in productive ways. In this sense, planners perhaps will find that new ideas and ways of working may be far more possible than previously thought. In particular, as argued here, finding ways of empowering less powerful groups, including migrants and other minorities, to define the everyday uses of, and practices performed in, urban public spaces, has to be a matter of serious concern to those involved in planning and designing the contemporary city and new approaches will need to be found.

As Vanessa Watson (2006: 31) argues, there is an increasing incongruence between the realities that face planning and the philosophical roots of traditional planning thought, which is based on universalistic and rational assumptions about homogenous societies which resemble less and less the urban societies within which they operate. (See also, Watson, Chapter 7, this volume.) Differences are always embedded in relations of power, not open to reconciliation, and contestation and conflict inevitably will occur. But where possible deliberative democratic processes need to be fiercely defended and upheld, and planners should be held to account to intervene as much as is possible to confront deep divisions and inequalities in contemporary society, to challenge the neo-liberal agendas which perpetuate these, and to find creative solutions in mediating differences, particularly where these are embedded in relations of racism, sexism and other forms of power. As this chapter has shown, it is by no means a homogenous picture and at the local level across the globe, there are countless examples of local and metropolitan councils supporting and enhancing innovative and different economic, social and cultural practices. That said, Vanessa Watson's (2006: 50) argument that in the context of deepening differences, and the increasing difficulties in reaching consensus, philosophies based on universalizing liberal tendencies have to be rethought, and new moral philosophical sources found to confront the thorny issues which confront planners on a day to day basis in the multicultural city.

References

Arendt, H. 1958. *The Human Condition*. Chicago, IL: University of Chicago Press.

Baron, J. 2011. Markets under threat as shoppers go online for fruit and veg. *Guardian*, 3 October. [Online]. Available at: http://www.guardian.co.uk/uk/2011/oct/02/markets-under-threat-online-shopping [accessed 4 April 2013].

Bauman, Z. 2003. *Liquid Love: On the Frailty of Human Bonds*. Cambridge: Polity.

Bollens, S. 2004. Urban planning and inter-group conflict: confronting a fractured public interest, in *Dialogues in Urban and Regional Planning*, edited by B. Stiftel and V. Watson. London: Routledge, 209–46.

Bontoft, W. 2011. Expansion plan for Northampton town centre mosque. *Northampton Chronicle*, 15 July. [Online]. Available at: http://www.northamptonchron.co.uk/

community/local-information/planning-applications/expansion_plan_for_ northampton_town_centre_mosque_1_2867524 [accessed 4 April 2013].

Castoriadis, C. 2007. *Figures of the Thinkable* (translated by Helen Arnold). Stanford, CA: Stanford University Press.

Communities and Local Government Committee (CLGC). 2009. *Market Failure?: Can the Traditional Market Survive?*. London: House of Commons, CLGC.

Connolly, W. 1995. *The Ethos of Pluralisation*. Minneapolis, MN: the University of Minnesota Press.

Cooper, D. 1996. Talmudic territory? Space, law, and modernist discourse. *Journal of Law and Society*, 23(4), 529–48.

Deutsche, R. 1999. Reasonable urbanism, in *Giving Ground: The Politics of Propinquity*, edited by J. Copjec and M. Sorkin. London: Verso, 175–206.

Emecheta, B. 1974. *Second Class Citizen*. London: Allison and Busby.

Engage. 2011. Camberley mosque planning appeal rejected, 22 June. [Online]. Available at: http://www.iengage.org.uk/component/content/article/1-news/1411-camberley-mosque-planning-appeal-rejected [accessed 4 April 2013].

Gale, R. 2005. Representing the city: mosques and the planning process in Birmingham. *Journal of Ethnic and Migration Studies*, 31(6), 1161–79.

Habermas, J. 1989. *The Structural Transformation of the Public Sphere: An Inquiry into a Category of Bourgeois Society*. Cambridge, MA: MIT Press.

Hamilton, F. and Gledhill, R. 2010. Islamic sect's plan to build mega mosque next to Olympic site collapses. *Sunday Times*, 18 January.

Hill, A. 2011. The city, the psyche and the visibility of religious spaces, in *The New Blackwell Companion to the City*, edited by G. Bridge and S. Watson. Oxford: Wiley Blackwell.

Latour, B. 2005. *Reassembling the Social: An Introduction to Actor-Network Theory*. Oxford and New York: Oxford University Press.

Naylor, S. and Ryan, J.R. 2002. The mosque in the suburbs: negotiating religion and ethnicity in South London. *Social and Cultural Geography*, 3(1), 39–59.

Sandercock, L. 2003. *Cosmopolis 11: Mongrel Cities of the 21st Century*. London and New York: Continuum Publishing.

Traynor, I. 2007. The rise of mosques becomes catalyst for conflict across Europe. *Guardian*, 11 October. [Online]. Available at: http://www.guardian.co.uk/world/2007/oct/11/thefarright.religion [accessed 4 April 2013].

Valins, O. 2000. Institutionalised religion: sacred texts and Jewish spatial practice. *Geoforum*, 31(4), 575–86.

Vincent, P. and Warf, B. 2002. Eruvim: Talmudic places in a postmodern world. *Transactions of the Institute of British Geographers, New Series*, 27(1), 30–51.

Watson, S. 2005 Symbolic spaces of difference: contesting the eruv in Barnet, London and Tenafly, New Jersey. *Environment and Planning D: Society and Space* 23(4), 597-613

—— . 2006a. *City Publics: The (Dis) Enchantments of Urban Encounters*. London: Routledge.

—— . 2006b *Markets as sites of Social Interaction: Spaces of Diversity*. Joseph Rowntree Foundation and Policy Press

—— . 2009. Performing religion: migrants, the church and belonging in Marrickville, Sydney. *Culture and Religion*, 10(3), 317–38.

Watson, S. and Wells, K. 2005. Spaces of nostalgia: the hollowing out of a London market. *Journal of Social and Cultural Geography*, 6(1), 17–30.

Watson, V. 2003. Conflicting rationalities: implications for planning theory and ethics. *Planning Theory and Practice*, 4(4), 395–407.

—— . 2006. Deep difference: diversity, planning and ethics. *Planning Theory*, 5(1), 31–50.

Wells, K. and Watson, S. 2005. A politics of resentment: shopkeepers in a London neighbourhood. *Ethnic and Racial Studies*, 28(2), 261–77.

Young, I. 1990. *Justice and the Politics of Difference*. Princeton, NJ: Princeton University Press.

A Cultural History of Modern Urban Planning

Stephen V. Ward

Introduction

Few planning historians have explicitly labelled their work as *cultural* history. Yet most have in recent years tried to reveal the attitudes, values and goals which have moulded planning (Ward, Freestone and Silver 2011). They have explored the social and cultural milieu within which planning ideas were conceived and elaborated and the wider political and business networks that empowered them. Planning has also been seen as a cultural form in its own right, with distinct national, regional and local characteristics. Yet it has also reflected larger international patterns of dominance and deference in the cultural and other spheres.

This chapter uses existing planning historical writing to emphasize this cultural strand. It distils work already presented in other writings by the present author (Ward 2012, 2010a, 2010b, 2005, 2002 and 1992) where the reader will find full bibliographic information. (Only research not hitherto cited appears here.) This account is organized on broadly chronological lines, examining first the emergence of what is generally termed 'modern urban planning' during the later nineteenth century. Highlighted are the various national cultural roots of this new form of planning. All grew from broadly liberal reform movements though differently within each country. However, the chapter will show how these national movements soon also became part of an international movement, sharing technical knowledge and, to a large extent, attitudes, values and goals.

Urban planning then evolved as an international movement but with different national expressions. (See also Greed, Chapter 5, this volume.) Later conceptual and practical changes are noted and the increased role of planning during the long post-1945 boom. An important cultural theme was how knowledge and practice spread across the world. I emphasize particularly the role of imperialism and later post-colonial international agencies. In doing this the chapter will consider the tensions which arose when essentially Western practices shifted into sharply different cultural contexts. Here the extent of local political consent was typically much lower (especially under colonialism) than in the countries where planning had originated. Relatedly, I will also show how the professionalization and technicalization of planning partially detached it from its liberal reformist roots. This allowed it to be deployed in more authoritarian contexts, for example in totalitarian states.

The chapter also shows how, in spreading, what were originally Western practices have become more truly global. Originally the main innovatory countries were Germany, Britain, France and the United States. Since then, however, innovation has occurred across a much wider range of Western countries, to some extent the Communist world and, more recently, in other continents.

The Emergence of a Modern Urban Planning Culture

Many past civilizations had ordered the physical arrangement of their towns and cities, especially when they formed new settlements. Much pre-modern planning was the work of kings, princes, prelates, aristocrats or oligarchies, each powerful enough to define the urban order. From the late eighteenth century, however, this began to change quite rapidly. In Europe and North America, the combined effects of rapid economic growth and a dramatic increase in the number and size of cities created the broad pre-conditions for a 'modern' form of urban planning.

Modern urban planning embraced the city as a dynamic and capitalistic centre of production, distribution, consumption and reproduction. The new approach embraced the emergent technologies of the industrial era so that cities might function effectively as economic and social entities. It also sought variously to harness, tame, marginalize or even supplant the increasingly dynamic land markets of the capitalist city. Crucially important to these growing aspirations for the city was the growing political ascendancy of liberalism.

Modern planning marked a distinct shift from the laying out of fortifications and grand urban spaces or avenues typical of pre-industrial planning. The need for an urban symbolism that represented the modern city remained, often remembering urban forms of earlier eras. But more distinctively modern were newer functional priorities, concerned with formalizing land use, infrastructure, efficient circulation and, increasingly, promoting social welfare. To respond to these aspects, new ideas, techniques, policies and instruments were developed.

The shift took place in piecemeal fashion during the later nineteenth and early twentieth centuries, especially within France, Germany, Britain and the United States. In each (and soon also other countries) the bases of distinct national planning cultures appeared, albeit with many common features. Literature specifically about planning began to get published and circulated. Formative examples of the new planning were widely reported and visited. Discursive events including lectures, exhibitions, conferences and, in some countries, planning competitions, helped widen and deepen knowledge.

Such changes were nourished by the appearance of national and local organizations which became settings for knowledge exchange and debate. Everywhere, these organizations mirrored the dual nature of urban planning as both a movement for reform *and* an activity requiring expert knowledge and professional skill. This shift from a reformist to a technical and professional focus was critically important. It allowed planning to become a relatively 'neutral' practice increasingly detached from its reformist origins. In turn this meant that it could be reattached to other, less progressive values, for example in the service of more authoritarian regimes. Even within liberal democratic countries, the 'mystification' of planning's essentially political roots within a culture of professional expertise sometimes allowed it to become an insensitive, even brutal instrument of an avowedly progressive state.

Another important cultural consequence of professionalization was that it cast the 'planner' much more into a 'masculine' mould (Meller 2007, see also Greed, Chapter 5, this volume). In its original more reformist incarnation, much activism was voluntarist, charitable and philanthropic in nature. In some countries at least, these arenas provided rich opportunities for feminine endeavour. Figures such as Octavia Hill or Henrietta Barnett in Britain or Jane Addams, Florence Kelley and Mary Simkhovitch in the United States, played central roles in the formative years. However, professionalization, especially based around the specific 'parent professions' of planning, that is, architecture, engineering, surveying, marginalized these kinds of roles. Later, of course, initially from the 1930s, it became easier for women to pursue professional, official careers. In such roles some individual women came to wield great power and influence, albeit conforming to a 'masculine' template of the planner which had by then become established.

Collectively, however, reformist and professional organizations gave a framework within which emergent planning networks could grow. Planning historians have frequently stressed the individual 'inventors' of the new ideas of planning. Without a wider movement, however, any individual's ideas were likely to be ignored. The remarkable Barcelona engineer, Ildefons Cerdà was a telling case. In 1859 he had authored an astonishingly advanced plan for Barcelona and in 1867 added the first theoretical exposition of modern urban planning principles. Yet he was a lone figure, unpopular in his home city and without network organizations to circulate his ideas. As such they were simply forgotten and played almost no part in the formation of modern urban planning in later decades.

Emergent National Cultures of Planning

Most commentators see Germany as the country with the first recognizable planning movement. Key professional and reformist organizations appeared in the 1870s. Most prominent were the Verband deutscher Architekten- und Ingenieurvereine (Confederation of German Associations of Architects and Engineers) established in 1871 and the Deutsche Verein für offentliche Gesundheitspflege (German Association for the Promotion of Public Health) founded two years later. The former brought together the two key professions that fostered the emergent expertise of urban planning. The second combined a political and social reformist membership with municipal officials.

It was mainly in such settings that new principles were rehearsed, defined and reproduced, establishing a reformist and professional discourse of urban planning. The Verband published important studies on planning matters which began to shape theory and practice. Its members were actively involved in the growing number of competitions to plan urban extensions which further refined practice and informed theory. Also encouraged were education and training to develop and reproduce the new skills, starting with the world's first urban planning course at Aachen in 1880. All these developments were encouraged by an expanding German language literature on planning, increasingly labelled from around 1890 as *Städtebau* (town building). Distinctively German contributions to practice were land use and density zoning and land readjustment, the re-parcelling of land ownerships to facilitate orderly planning and provision of public land for streets, open spaces and so on.

The same broad tendencies were evident elsewhere, though manifest in different ways. In France, the Musée Social (literally Social Museum), established in 1894, was the principal reformist organization where new ideas and possibilities were rehearsed. It shared with the German movement a predominantly liberal reformist character, with progressive industrialists and professionals in prominent roles. In a predominantly Catholic country, many were also Protestants originating from those (at that time lost) parts of France, Alsace and Lorraine, that had most affinity with Germany.

Quite distinct from German *Städtebau*, however, was the strong French conviction that the new professional activity, around 1910 labelled as *urbanisme* (urbanism), was pre-eminently a branch of architecture, an exercise in grand design. The early work of another Alsatian protestant, Baron Haussmann, had left a powerful legacy. As prefect of Paris between 1853 and 1870, he had transformed the city under the Second Empire. Despite Haussmann's concerns with functional matters of drains and circulation it was a firmly architectural vision of the new Paris that he had created which dominated French thinking about how the modern city should be planned. Thus when a specific professional body for urban planning was created, in 1911, it was the Société Française des Architectes-Urbanistes. Although the architectural prefix was soon dropped, the underlying assumption remained.

Different again was Britain. The social and economic backgrounds of those reformers behind the early planning movement were similar to their equivalents in Germany and France. In contrast to France's protestant reformers, however, pioneers of British planning came disproportionately from non-conformist faiths, particularly Quakers, Congregationalists, Unitarians or others whose beliefs was not mediated through the established or Catholic churches. More so than in either Germany or France, British urban reformers emphasized social welfare, especially social housing and the ideal of community within an urbanized society. These and other concerns came together in the main British contribution to planning thought and action, the garden city, unveiled by Ebenezer Howard in 1898. It quickly formed the basis for a vigorous pressure group, the Garden City Association. Another strand, albeit with longer term rather than immediate impact, was the conception of planning as an expression of regional culture, pioneered by the inspirational figure of Patrick Geddes.

The professional side also differed from both France and Germany. The late nineteenth century strength of British local government institutions had given major importance to engineers and surveyors in the burgeoning municipal bureaucracies. As yet, the architectural profession had no comparable institutional position. Notwithstanding key contributions of individuals such as Raymond Unwin, architects were less dominant in forming modern planning culture in Britain than in France or Germany. Britain was also unusual in that the foundations of a separate profession of 'town planning' (from 1905 the preferred English label) appeared at a very early stage, in 1914, when the Town Planning Institute was founded.

The last of the innovative national cultures was that of the United States. In contrast to Europe, a boosterist ethic pervaded emergent American debates about planning. 'City planning' (as Americans termed the new activity from around 1909) was more actively shaped by mainstream business values than were European planning movements. Initially such interests supported the 'City Beautiful' movement which proposed grand parks, avenues and civic centres with major, classically inspired public buildings and other adornments. These mainly favoured the urban elites but also became a way of attracting business investment.

More consciously liberal reformers, representing the more socially progressive edge of bourgeois opinion, were also important. A group of largely New York social reformers, including prominent members of that and other cities' liberal Jewish elites founded the Committee on Congestion of Population in 1907. This articulated a social welfare agenda for the embryonic American city planning movement. Yet, until a revival of social progressivist sentiments in the 1920s and 1930s, this remained a lesser theme. As American business elites recognized the limitations of the City Beautiful they adopted, not a social welfare model, but one that emphasized the efficient and functional city. It was concerned with circulation systems, comprehensive zoning to allow the expansion of business and more economical and orderly residential growth. From a very early stage it also began to adopt the quantitative methods of increasingly scientific business management to make plans. The new thinking was heralded by the great 1909 plan for Chicago, directly sponsored by the city's business community. Though having many City Beautiful elements, it was also a highly functional plan of a city being shaped directly to service business interests.

In professional terms, the United States had more in common with Britain than with either France or Germany. Although the City Beautiful movement leaned heavily on the French *Beaux Arts* tradition of grand design, this was moderated by strong engineering and landscaping inputs. All these professional elements (and others including public administration and community development) were embraced in an integrative professional body, the American City Planning Institute, founded in 1917.

By 1920 therefore, each of the major countries most associated with the new field had established distinctive planning cultures. The networks, institutions, values and knowledge that framed these cultures had given each a rudimentary template on which planning in those countries could be (and was) subsequently reproduced. Even today, many distinctive national elements of planning directly reflect aspects that appeared at this time.

Internationalizing Planning Culture

Yet, only in the very earliest stages could the formation of planning cultures be understood solely in autonomous national terms. (See also Watson, Chapter 7, this volume.) The later nineteenth and early twentieth centuries also saw widening possibilities for international travel. Major technological changes in transport and communications hugely eased movement of people and knowledge. Before 1914 travel formalities were also minimal over many parts of the world.

The various national movements for modern urban planning benefited greatly from this. The international mobility of planning knowledge broadened and deepened the repertoire of ideas and practices available to individual countries. New thinking could be learnt, copied, adapted, hybridized or synthesized with distinctively national approaches. Key texts were translated into other languages. A small but increasing number of planners worked in countries other than their own. Discursive planning events such as conferences, exhibitions, competitions and so on increasingly transcended national boundaries. These and more deliberate fact-finding visits and study tours became staple activities of this increasingly internationalized planning culture.

The result was that the leaders of the new national movements could interact with their equivalents elsewhere. A wider group of political and other opinion formers could read about and see what was happening in other countries and make comparisons with their own cities. As early as 1867, the Exposition Universelle in Paris attracted many international visitors (the new Thomas Cook business arranged the travel of 10,000 from Britain alone) who saw Haussmann's transforming city.

By the early twentieth century several planning-related organizations appeared that were avowedly international. The Permanent International Association of Road Congresses was formed in 1909. The International Union of Local Authorities and, most important, the International Garden Cities and Town Planning Association followed in 1913. These organizations, the latter renamed as the International Federation of Housing and Town Planning (IFHTP), were increasingly significant after the First World War. Their annual conferences, held in various locations, were key events where national approaches and innovations could be inspected and discussed. Initially mainly European, their membership and range broadened. For example the 1938 IFHTP conference was held in Mexico City.

A new international architectural organization very important for planning appeared in 1928. This was the CIAM (Congrès International d'Architecture Moderne) founded in Switzerland. Within a few years it was attracting 'modernists' from German, French, Dutch, Spanish, Czech language groups. During the 1930s it widened to encompass British and later American members, along with the Nordic countries and some other parts of the Slavonic world. Its mission was to develop self-consciously 'modern' forms of architecture and planning based firmly on rational, functional principles. Though representing *avant garde* approaches in its early years, its ultimate impact on the mainstream of post-1945 architecture and planning in many countries was huge.

This trend to internationalization greatly strengthened the young national planning cultures. Thus British industrialists, reformers and professionals looked admiringly on German innovations in the urban sphere, while their German equivalents reciprocated. The Deutsche Gartenstadtgesellschaft in 1902 became the first foreign association to adopt Howard's garden city vision. Britain's first town planning legislation, in 1909, was effectively a synthesis of garden city principles with German town extension and zoning. Despite French enmity towards Germany after 1870, they learnt of Germanic efforts through displaced reformers from Alsace and Lorraine. Switzerland and Belgium also became intermediate areas where some indirect interchange occurred. The Anglo-French Entente Cordiale spawned a planning equivalent as French reformers created a Gallic movement for the *cité-jardin* in 1903.

Meanwhile many American reformers and professionals crossed the Atlantic to learn the latest European thinking, including British innovations in social housing or the garden city, German zoning techniques or French grand design. In turn the 1909 Chicago Plan impressed Europeans with its scale, vision and technique. There was also a more general European wonderment at the vast and highly complex buildings that were appearing in the biggest American cities.

Internationalization and New National Planning Cultures

These international interchanges had an even bigger impact elsewhere. Across Europe and increasingly beyond grew an eager trade in planning knowledge that intensified after 1918. Exogenous ideas and practices, often from multiple sources, were combined and merged with domestic approaches. Direct copyism sometimes occurred but more typically there was conscious or unconscious adaptation, involving hybridization or synthesis. In some cases, this mobility amounted to a process of reinvention with received knowledge being deployed in ways that masked exotic origins.

This largely voluntary proliferation increasingly spanned the globe, including Latin America and self-governing territories of European colonialism such as Australia or Canada. Space, however, allows consideration of only two, somewhat different, examples of how distinct new national planning cultures emerged. The Netherlands was a small country that, virtually alongside the 'big four', established a highly distinctive national planning culture, while remaining open to external influence. Linguistic affinities perhaps made it inevitable that Germanic innovations were initially dominant, not least in the Dutch neologism for urban planning, *stedebouw*, which appeared around 1900. Yet British ideas about the garden city and social housing were also much in evidence.

No external influences were, however, sufficient to shape a planning culture which also drew on strong national traditions. The longstanding Dutch imperative of managing water within the national territory was central. Nowhere was this more evident than in the ambitious Zuider Zee project begun after 1918 to create new polder areas and a large freshwater lake in what was formerly sea. By the interwar years, there was a strong Dutch movement for modern planning widely acknowledged for its innovation. The 1935 Amsterdam Plan expressed a scientific and modernistic approach to planning that anticipated wider post-1945 trends.

Perhaps the most interesting non-European planning culture based substantially on international contact was Japan. Following the Meiji restoration of 1868, the country underwent radical modernization. Western practices and innovations were carefully examined and, to varying extents, adopted and adapted. No one exemplar country was exclusively favoured with the result that the Japanese officials and professionals encountered all the main Western planning ideas at an early stage in their development.

German practices were most favoured in the formation of *toshi keikaku* (as planning was termed from 1913). The Prussian legal code had been the template for the new Japanese legal system and this precedent also shaped the planning system. This comprised various measures specific to Tokyo (especially following the 1923 earthquake) and a national planning act of 1919. This was not direct copying, however, and Japanese practice was in some respects more radical than German templates, for example in land readjustment. Partly because of misunderstanding, Japanese legislators adopted a draft measure that had proved too radical in Germany. Nevertheless the fit proved appropriate, arguably because it echoed certain traditional Japanese practices. Land readjustment has subsequently become a highly distinctive feature of Japan's planning culture. (See also Newman and Thornley, Chapter 4, this volume.)

Another Western idea with an early and enduring appeal for the Japanese was the garden city. At a time when the country was rapidly industrializing and urbanizing with very cramped accommodation, the image of a city of gardens, *den en toshi*, was a powerful one. Initially it shaped the creation of affluent commuter settlements by the railway companies.

Planning as a Practice of Empire

The Japanese leadership had embraced modernization to insulate the country from Western colonialism. Yet the latter became another powerful vehicle accelerating the global spread of Western modern planning, albeit in ways less controllable by receiving countries than the examples so far examined. Imperial experiments were also one of the principal test-beds for modern Western-style urban planning. The advantage was that planners were less compromised by citizen rights, vested interests or other governmental concerns than within the 'mother countries'.

Imperialist motives for turning to planning varied. In part, certainly, there was a simple need for spaces for governance and spacious residential districts for colonial elites. But imperial planning was also about more systematic exploitation of colonial resources through improved physical infrastructure. This same concern drew colonial administrations into health and welfare, to combat both economic disruptions caused by disease and, especially after 1945, the proliferation of informal housing on the edges of cities. At a symbolic level, planning also helped promote the notion of imperialism as a benevolent force. The widely-accepted progressive self-image of urban planning could help portray imperialism as an enlightened project of modernization.

A striking example was in the French world. The new imperial protectorate of Morocco, conceived very much as a model for a new, more progressive form of empire, became the first French territory to adopt planning as a state function from 1912. Six years before equivalent action in France itself, the Moroccan planning law of 1913 launched a national programme of *urbanisme*. Over the following years this template was widely applied, with variations, throughout the French Empire.

The British also used their own town planning from a very early stage in various colonial territories. Most notable were their efforts in India which included planning the new imperial capital at New Delhi from 1911 and measures to facilitate the renewal of Bombay (1898) and Calcutta (1912). By 1939 a growing number of British imperial possessions, including Malaya, Northern Rhodesia, Nigeria, Transjordan and Trinidad, had British-style planning powers. Broadly similar tendencies were evident in the Dutch Empire in the present Indonesia and the Japanese Empire in Taiwan, Korea and Manchuria. Germany and Italy also used planning in their own short-lived colonial empires. Even the (supposedly anti-colonial) United States deployed it in the Philippines.

Despite sometimes impressive physical results, colonial planning showed clearly how a progressive notion could be deployed as part of a process of external dominance. Colonial territories lacked mature civil societies with developed and articulate institutions which could impede the insensitive colonial imposition of urban planning. Indigenous land rights and customary usages were extinguished in the pursuit of a Western notion of order. Alien building types and functional land uses could be imposed, removing the diversity and richness of traditional cities or marginalizing them compared to grand new colonial districts. In extreme circumstances, colonized peoples might violently contest planning's subversion of their own valued places. More usually, they gradually changed the intended meanings of imperially planned spaces by the way they used them.

The actual technical process was usually led by professionals from the colonial power, and in earlier periods completely dominated by them. A handful of indigenous planners were trained before the end of colonial control in most empires. Usually this occurred either within the imperial 'mother country' or, more rarely, within the colonial territories themselves but following a template that was set by the colonial power. What was conspicuously lacking was the indigenous, more culturally appropriate approach to planning equivalent to the selectivity and synthesis which occurred within Western countries. A few individual colonial planners, most notably Patrick Geddes in India and Thomas Karsten in the Dutch East Indies, sought more sensitive approaches. These did not, however, prevent the emergent indigenous planning cultures of late colonialism showing excessive deference to external knowledge and expertise.

Planning in Authoritarian and Totalitarian Contexts

The wider diffusion of planning also took it to other settings with governments much less liberal or democratic than in its European and American heartlands. Even Germany, with a modern planning tradition which had largely emerged as part of a socially progressive trend of liberalization, moved sharply to the right in the 1930s. As in colonial contexts, the German planning movement began to serve ends quite different to those dominant before the First World War or in the years immediately afterwards. Many Jewish planners or those with left-wing views (often seen as synonymous with avowedly modernist approaches) found themselves unwelcome or facing persecution. Many emigrated, helping diversify the planning cultures of other countries including Britain, the United States, Mexico, Turkey, the future Israel and elsewhere.

The easy interchanges that a more liberally minded German planning movement had enjoyed with other mainstream planning traditions became more uncomfortable. Yet they did not cease and Hitler's *autobahn* programme attracted widespread admiration. However, the autonomous organizations and institutions which had comprised the planning movement were increasingly brought under state control or supplanted. International agencies were viewed with suspicion unless they could be Nazified. In 1938 control of the International Federation of Housing and Town Planning began to shift to Germany with the assumption of office by a Nazi President. Early German wartime successes completed this process so that it became a propaganda vehicle to promote Hitler's 'New Order' for Europe. A new democratic organization had to be reconstructed with American backing in the later war years.

Other European countries experienced similar fates, swelling the numbers of exiled drifters. Some with left-wing sympathies looked, at least for a time, to the Soviet Union. Although its collapse has now made it easy to overlook, it was the most important new planning culture to emerge after 1914. It exerted widespread influence amongst many Western planners from the 1930s to the 1950s and had a more direct quasi-imperial impact on its post-war satellites and more distant communist nations. Yet it too relied for a time on foreign examples. Even before the Revolution, Russian reformers had been attracted to transformative urban planning models, above all the garden city. Howard's book was translated into Russian in 1911. Some early Soviet planners continued to admire it as resolving the essential unevenness of capitalist spatial development, between urban and rural society.

Other proposed solutions were more radical though it was not until the first Five Year Plan (to achieve industrialization) in 1928 that settled Soviet urban planning policies appeared. These accepted that mass urbanization was necessary to facilitate large scale industrialization. However, Soviet planners sought to limit concentrated growth and mitigate its effects by decentralized urban forms. In these early years, there was considerable reliance on Western technology and planning expertise. Major new industrial plants used American technology, hiring its architects and engineers to plan entire workers' towns. European planners were also recruited, most famously in 1930 the largely German group led by former Frankfurt planner Ernst May.

Some Western urban planners admired Soviet commitment to planning in all areas of life. The acute problems of uneven development in all Western countries during the 1930s fostered strong interest in national and regional planning. Yet the Soviet Union appeared far bolder than elsewhere in how it tackled these matters. This boldness was also apparent at urban level, reflecting the complete absence of private land ownership and an apparently single-minded Soviet commitment to address the major needs of a rapidly urbanizing society. These included large scale housing construction within planned *microrayons* (neighbourhoods), major investments in transportation and other infrastructure and so on. The 1935 Moscow Plan was a showpiece, rated by authoritative external observers as the most advanced and comprehensive big city plan anywhere in the world.

The defining feature of Soviet urban planning culture was not primarily its substantive content, but that it was framed entirely by the state and Communist Party. It was infused, far more than was yet true of Western planning cultures, with a belief in 'scientific' knowledge and expertise. Though the Communist system was not completely monolithic, civil society was far less developed than in democratic societies. There were no equivalents of independent reformist bodies, pressure groups or professions that had shaped Western planning movements. This meant that disputes or debates about urban planning actions were kept within prescribed limits. Opportunities for real political contestation over planning matters such as might occur in democratic societies were absent.

Initially at least, many prominent Western observers regarded these aspects as less significant than the apparent material benefits and technical qualities of Soviet achievements. They admired the extensive state propaganda used to promote popular commitment to planning, yearning for something similar in their own countries. Approved of also were the Soviet commitment to intensive training of planners and especially the popular and political respect accorded to them. In the 1930s and 1940s many planners in Western countries genuinely felt that their own countries would be improved if they emulated Soviet planning.

Planning, Reconstruction and Modernization after 1945

In many respects, planning in all Western countries, especially within Europe, did move in this direction in the 1940s and 1950s. Though it nowhere matched the totality of the Soviet approach, planning everywhere acquired important new powers and responsibilities. In part this was a response to the pressing needs for reconstruction of cities. Yet it was also a delayed response to perceived shortcomings during the interwar years – the lack of real direction or control over urban growth; the degradation of rural areas; the persistence of large slum areas, the failure to provide for new transportation modes; and the decay of regions in the 1930s depression. The relative importance of these issues varied in different countries. The experience of an already heavily urbanized country such as Britain differed significantly from those with bigger proportions of their populations still dependent on agriculture such as France or Italy.

Everywhere in the West, the expertise of urban planning widened significantly from its original physical design focus, acquiring dimensions grounded within social science, economics and geography. There had been many earlier hints of this in the approaches advocated by Patrick Geddes (and his slightly later French equivalent, Marcel Poëte). They were expressed more tangibly in plans such as that for Amsterdam in 1935 or in President Roosevelt's 'New Deal' in the United States, especially the great regional development project for the Tennessee Valley.

Now, however, such approaches became more typical, with planners involved in questions of employment and social welfare. A stark example of this shift came in France where in the late 1940s a new term for planning, *l'aménagement du territoire*, took precedence. Compared to *urbanisme*, it signified a broader geographical approach, more social science-based and more closely linked with the wider economic planning and large scale public investment programmes pursued by successive post-war French governments. The shift was less dramatic elsewhere, evident more in a widening of planning's disciplinary basis.

The physical design core expertise of planning also shifted throughout the West and many other parts of the non-communist world, embracing ever more confidently the ideals of modernism that had been rehearsed by CIAM. Cities were to be designed much more on functionalist lines, with less emphasis on traditional artistic approaches or architectural adornment. New highway systems were to be inserted into cities, facilitating freer vehicle movements. These primary networks became also ways of demarcating cities into different functional zones or neighbourhoods, presuming rather fixed and predictable ways of using cities.

Increasingly from the later 1950s and especially in the 1960s and 1970s, this general trend was reinforced by serious attempts to change urban and regional planning into an increasingly quantitative science. Beginning with land use transportation modelling in the United States, other quantitative or quasi-quantitative methodologies were deployed. Numerical values were, sometimes dubiously, assigned to landscapes and places and inserted into planning equations to give a supposedly rational basis for reaching one answer or several options for political choice. Meanwhile the process of making choices about planning was itself beginning to be opened up to more popular participation. By the later 1960s there were widespread pressures across the West for governance to become more accountable and citizens to be given a more active voice. Again, though, planners viewed this as a further stage within the scientific planning process. It was seen as a way of creating a direct interface between rational planning methodology and public choice, uninterrupted by the 'irrational' unpredictability and vested interests of politics.

Cumulatively, these various shifts strengthened the notion of the professional planner as a neutral expert, well equipped with the new methodologies to help society find the 'right' answers. It can be seen as the zenith of the early trend to make planning an expert professional activity rather than a movement for social and political reform. Such was the degree of planning's incorporation within state bureaucracies after 1945, however, that it created a profoundly different conception of professional activity. The residual notion of an autonomous expert using knowledge and experience to give advice to an identifiable client had declining relevance. The truth was that planners were now mainly state functionaries serving the interests of the government agencies who employed them. In some countries, for example France, there was little attempt to disguise this. In Britain and many other countries, however, the planning profession effectively struck a 'corporatist bargain', surrendering its autonomy in return for an enhanced ability to shape what happened.

Planning in a Post-Colonial World

The wider reformulation of the world order after 1945 had huge impacts for planning. The colonial empires, if they had not collapsed already, were never going to be the same again. Often reluctantly, colonial powers recognized that indigenous desires for independence were unstoppable. Most newly independent states began on the principle of promoting their own nationals into the main political and professional roles. Yet the reality was that most lacked the technical capacity their ambitious projects to develop their countries required. There were some moves to create an organizational and training basis to build that capacity, sometimes through new planning schools but more often involving overseas training. There was also a need, certainly in the first years of independent rule, to continue to rely on foreign planning expertise.

Yet, instead of the one dimensional experience of receiving planning expertise and policy from the imperial motherland which had characterized most colonial territories, a more complex pattern appeared. The new superpowers, the United States and the Soviet Union, became significant players in the post-colonial world. Either directly or through surrogates they both sought to use planning and development initiatives to strengthen their influence in newly independent countries. For their part, the former imperial powers often tried (not always successfully) to maintain the old links. Many smaller countries also began to give planning and related technical assistance to former colonial territories.

For their part, the leaders of newly independent states recognized that they had more power under this system than under colonialism. To a large extent it was possible to choose donor countries from which they received technical assistance or to which they sent their own would-be planners for training. They could switch allegiances over time (often reflecting political regime changes), choose aid providers without a history of colonialism or deliberately avoid giving one country a monopoly. The result was a fascinating patchwork. It was possible to find American and Swiss planners playing important planning roles in 1950s India. Meanwhile a Greek planner planned the new capital of Pakistan and British planners worked in what became Bangladesh (then East Pakistan). Soviet planners worked in Vietnam and, for a time, Indonesia where there were also Danish and British planners at work. Subsequently the initial breach with the Dutch was healed so that they became major providers. In 1960s Tanzania, Canadian planners produced key plans for Dar-es-Salaam while East Germans worked in Zanzibar. In other parts of the continent, there were planners at work from South Africa, Greece and Hungary as well as former colonial powers.

Nor was the pattern a simple accumulation of bilateral aid patterns. The United Nations Organisation soon became an important force accelerating planning's wider global diffusion. A small UN planning office was established in 1951 and eventually became UN Habitat. In the early years, it directly undertook some planning roles itself but generally promoted international contacts, orchestrated the work of other international agencies, gave advice to post-colonial and developing nations about selecting foreign planners and generally promoted what it saw as good practice. It also tried to encourage local capacity building.

Culturally fascinating and important in spreading planning knowledge though these interactions were, the appropriateness of the results was often questionable. Although the stark unevenness of power between colonizer and colonized had gone, a high degree of external dependence remained. Numbers of indigenous planners rose only slowly. Most also emerged from their usually foreign planning schools excessively deferential to Western or Communist approaches. Indigenous political or professional elites might be able to challenge the proposals of foreign planners but citizen involvement was largely absent. Moreover the planning cultures of these countries remained underdeveloped, often lacking relatively autonomous reformist, pressure group and professional networks. Neither were many post-colonial governments willing to encourage the countervailing voices that could come from a more vigorous civil society.

For various reasons a few post-colonial countries quickly achieved greater affluence so that the external models seemed more relevant. But for most the presumption that they would quickly correspond to a Western or Soviet path of development was naïve. As early as the 1950s, a few more experienced and sensitive trans-national planners began to see a need for different approaches more appropriate to undeveloped post-colonial societies. They began to suggest approaches that harnessed local efforts more effectively and did not depend so much on such an expert approach. The first signs involved a shift to 'site and services' approaches to housing provision. These presaged the appearance of more radical schemes of community-based upgrading of informal settlements. Instead of highly 'scientific' plan-making with high data needs, a more pragmatic form of 'action planning' appeared, more closely matched to the limitations of indigenous technical capacity.

More generally there has been growing emphasis on local capacity building. Interestingly this has included the formation of knowledge networks and organizations that depend far less on external orchestration and are based within the post-colonial developing world rather than being primarily organizations of the affluent world. However, these have been a very recent phenomenon. The possibilities of such networking have improved dramatically with recent innovations in information technology and communications.

Modern Planning in the Post Modern World

The crisis of relevance of 'modern urban planning' which manifested itself in the post-colonial world was the earliest and most acute manifestation of a wider phenomenon. By the 1970s, even in its affluent heartlands the results of modern planning were increasingly seen as having failed to fulfil the hopes which had attended their conception. Across the West (and in more muted form even within the Soviet bloc), there were more signs of resistance. Opposition to major urban highways schemes and redevelopment projects had been growing steadily during the 1960s, especially so in North America but became increasingly common in Europe and other parts of the affluent world.

Citizens were becoming less inclined to defer to planners' expertise. They wanted planners who were more humble and less convinced of the certainties of their methodologies and the 'answers' that followed from these. They became more protective of the cities that had grown up before planners had even appeared. Historic buildings and districts and the more general texture and morphology of historic cities were increasingly valued. By the late twentieth century there was a widespread reaction against the planned functional 'compartmentalization' of cities.

Close on the heels of this popular questioning came an erosion of other certainties which had underpinned modern urban planning in its heartlands. Shifts in the world economy produced decline in the older industrial regions of Europe and North America. Cities with formerly secure niches in their national economies, now found themselves losing investment and population. The fall of Communism within Eastern Europe countries had similar results as their state-supported economies also collapsed. From the perspective of urban planners, these changes upended many assumptions on which they had based their plans. New efforts that marginalized many of planning's traditional concerns were now made to find new sources of wealth.

Planners in the United States came from a planning culture which was historically more attuned to place competition and boosterism and returned to this with a certain familiarity. But for European cities it marked a bigger change. City marketing and various strategies and instruments to privilege business became common. (See also, Nyseth, Chapter 19, this volume.) So too did the forming of very close relationships with private developers, in various forms of public-private partnerships. There were also unprecedented public investments in culture, usually reflecting multiple intentions to generate a new economy of consumption and increase the general attractiveness of the city for investment.

The third major element of change in the late twentieth century was more encouraging for planning in that it resembled in some ways the ideas of planning that had crystallized a century earlier. This was the emergence of a new commitment to sustainable forms of development. This grew from the environmental movement which had taken more active forms since the 1970s and the international development movement which recognized the need to facilitate growth while minimizing environmental damage. During the 1980s and 1990s a new global discourse of sustainable development emerged, especially so in the wake of the 1992 Earth Summit held in Rio de Janeiro.

It marked a partial rejection of one of the main foundations of modern planning and indeed modernism generally, the belief in the almost infinite possibilities of human progress fuelled by science and technology. Yet it also stressed themes that echoed some traditional elements of modern planning, such as the promotion of community and the need to balance pressures for growth against those to protect and conserve nature. There were also echoes of the faith in a new terminology to solve public problems. In the 1890s it was *Städtebau*, town planning, *urbanisme*, city planning and so on which seemed to offer the peaceful, reformist way to create a better world. A century on, the new discourses of sustainability, seeking 'compact cities', 'new urbanism' and 'smart growth', appeared to carry comparable hopes. It remains to be seen whether they can be fulfilled to a greater extent than were those of the original movement for modern urban planning.

Conclusions

This chapter has suggested something of the various cultural dimensions of urban planning history. It has emphasized the social milieux from which planning emerged, initially as a series of local and national responses to the common phenomenon of the large modern city. Yet the predominantly liberal, progressive outlook that characterized the early national movements soon drew them into international contact. From a very early stage, an international planning movement was created which both softened tendencies for overtly nationalistic expressions of planning and created a means for rapid diffusion of planning knowledge.

However, even as it became a flagship for liberal internationalist sentiments, the universal process of professionalization began to weaken the connections between planning and the reformist impulses from whence it came. Increasingly it became an expert activity, conceptualized more in technical and eventually quasi-scientific terms. As such it became potentially more detachable from the political and social values of its founders. Alongside this, the twentieth century also saw a substantial incorporation of planning as an instrument of the state and of governance.

These two processes, of professionalization and state incorporation, have been critical to understanding how planning could have become a practice deeply embedded within less than liberal political systems. From a very early stage planning became a tool first of colonialist and then of other varieties of authoritarian and totalitarian states. Even within the liberal West, professionalism and incorporation also underpinned the belief which became widespread from around the middle of the twentieth century that applying modern planning expertise was itself a good thing almost regardless of what it actually did. The truth has been that the extent of the perfectibility of planning expertise has been limited. In many demonstrable ways planners have made cultural assumptions about the way people wanted to live that were often dubious at the time and have not responded well to changed social aspirations and economic realities.

The reactions and deeper changes of recent decades have diminished popular and political faith in the kind of planning that grew up over much of the twentieth century. Shorn of its ambitious hopes of social progress, planning continues as an embedded bureaucratic function of the state, rather modestly intermediating between competing interests, usually in favour of the most powerful. A trite way to end would be to suggest some kind of cultural renewal of the planning movement, reinvigorating its reformist roots and putting professional expertise and state incorporation into their proper, subsidiary places. Instead I leave it to the reader to decide whether it will be possible or desirable to reinvent a kind of planning that might regain the near universal prominence it enjoyed during the long post-1945 boom. Sadly, the reality is that the 'peaceful path to real reform' (as Howard termed it in 1898) remains as elusive as ever.

References

Meller, H. 2007. Gender, citizenship and the making of the modern environment, in *Women and the Making of Built Space in England, 1870–1950*, edited by E. Darling and L. Whitworth. Aldershot: Ashgate, 13–32.

Ward, S.V. (ed.) 1992. *The Garden City: Past, Present and Future*. London: E & FN Spon.

—— . 2002. *Planning the Twentieth-Century City: The Advanced Capitalist World*. Chichester: Wiley.

—— . 2005. A pioneer 'global intelligence corps'? The internationalisation of planning practice 1890–1939. *Town Planning Review*, 76(2), 119–41.

—— . 2010a. Transnational planners in a postcolonial world, in *Crossing Borders: International Exchange and Planning Practices*, edited by P. Healey and R. Upton. London: Routledge, 47–72.

—— . 2010b. What did the Germans ever do for us? A century of British learning about and imagining modern planning. *Planning Perspectives*, 25(2), 117–40.

—— . 2012. Soviet communism and the British planning movement: rational learning or utopian imagining? *Planning Perspectives*, 27(4), 499–524.

Ward, S.V., Freestone, R. and Silver, C. 2011. The 'new' planning history: reflections, issues and directions. *Town Planning Review*, 82(3), 231–62.

Culture to Creativity to Environment – and Back Again[1]

Toby Miller

This chapter examines the theory and history of cultural policy, a classically Enlightenment discourse that is under threat from the new right's discourse of creative industries. It then provides a brief case study of related environmental issues to highlight the risks when industrial priorities dominate cultural ones.[2]

Culture and Policy

The word 'culture' derives from the Latin 'colere', which implied tending and developing agriculture as part of subsistence (Adorno 2009: 146). With the advent of capitalism's division of labour, culture came both to *embody* instrumentalism and to *abjure* it, via the industrialization of farming, on the one hand, and the cultivation of individual taste, on the other. Eighteenth-century German, French and Spanish dictionaries bear witness to a metaphorical shift from culinary cultivation to spiritual elevation. The spread of literacy and printing saw customs and laws passed on, governed and adjudicated through the written word. Cultural texts supplemented and supplanted physical force as guarantors of authority. With the Industrial Revolution, populations urbanized, food was imported, textual forms were exchanged and an emergent consumer society stimulated horse racing, opera, art exhibitions, masquerades and balls. The impact of this shift was indexed in cultural labour: *poligrafi* in fifteenth-century Venice and hacks in eighteenth-century London wrote popular and influential conduct books. These works of instruction on everyday urban life marked the textualization of custom and the appearance of new occupational identities (Briggs and Burke 2003). Cultural policies emerged as secular, urban alternatives to deistic, rural knowledge (Schelling 1914: 180) in an emerging capitalist age focused on *'self-realization'* (Weber 2000). These policies also informed imperial expansion through Spain's *conquista de América*,

---

[1]   The editors provided useful feedback.

[2]   As this is meant to be an omnibus survey, nations and regions are mentioned as well as cities.

Portugal's *missão civilizadora*, and France and Britain's *mission civilisatrice*, creating an anxiety that has never subsided (Mowlana 2000). The United States was a classic instance of an import culture. In 1820, Sydney Smith asked: 'In the four quarters of the globe, who reads an American book? or goes to an American play? or looks at an American picture or statue?' (1844: 141) and Herman Melville criticized the nineteenth-century US devotion to all things English as an obstacle to bringing 'Republicanism into literature' (Newcomb 1996: 94). In imperial Britain, the study of culture formed 'the core of the educational system'. It was 'believed to have peculiar virtues in producing politicians, civil servants, Imperial administrators and legislators', incarnating and indexing 'the arcane wisdom of the Establishment' (Plumb 1964: 7). Culture was expected to produce and renovate what Matthew Arnold called 'that powerful but at present somewhat narrow-toned organ, the modern Englishman' (1875: x). A century ago, US higher education was dominated in its turn by moral philosophy, Latin, and Greek in an attempt to match and transcend that organ (Ayers 2009: 25).

Kant's *Critique of Judgment* ideologized these developments. He argued that culture ensured popular 'conformity to laws without the law'. Aesthetics could generate 'morally practical precepts' by schooling the populace to transcend particular interests via the development of a *'public* sense, *i.e.* a critical faculty which in its reflective act takes account (*a priori*) of the mode of representation ... to weight its judgement with the collective reason of mankind' (1987: 151). Kant's *Political Writings* envisaged *'emergence from ... self-incurred immaturity'*, independent of religious, governmental or commercial direction and animated by the desire to lead rather than consume (Kant 1991: 54, also see Hunter 2008: 590). For Coleridge, 'the fountain heads' of culture were 'watching over' the sciences, 'cultivating and enlarging the knowledge already possessed ... in order to be citizens' (1839: 46) while Rousseau maintained that it was 'not enough to say to the citizens, be good; they must be taught to be so' (1755: 130).

Nineteenth-century capitalism generated a specialized and expansive division of person and labour, articulated through the shift from rural to urban living and the growth of European empires. Revolutionary thinkers also picked up on this issue. Marx wrote: 'it is impossible to create a moral power by paragraphs of law.' There must also be 'organic laws supplementing the Constitution' (1978: 27, 35). Gramsci (1971: 204) theorized this supplement as an 'equilibrium' between constitutional law ('political society' or a 'dictatorship or some other coercive apparatus used to control the masses in conformity with a given type of production and economy') and organic law ('civil society' or the 'hegemony of a social group over the entire nation exercised through so-called private organizations such as the church, the unions, the schools, etc.'). These organic laws and their textual manifestations represent each 'epoch's consciousness of itself' (Althusser 1969: 108); hence the extraordinary investments by audiences, creators, governments and corporations in culture.

Cultural policy today refers to institutional supports for aesthetic production and memory. It bridges the distance between art and everyday life. Governments, trade unions, colleges, social movements, community groups, foundations, charities, churches and businesses aid, fund, control, promote, teach and evaluate culture. This happens through law courts protecting exotic dancing as free speech; curricula requiring students to study texts because they are uplifting; film commissions sponsoring scripts to reflect national identity; cities working with companies to gentrify downtown districts; and foundations funding minorities to supplement dominant norms. These criteria derive from legal doctrine, citizenship education, tourism goals, urban plans or philanthropic desires.

There is a seemingly interminable struggle between consumerist and *dirigiste* approaches to cultural policy. The consumerist approach argues that culture circulates satisfactorily through the mechanics of price: whatever succeeds commercially is *ipso facto* in tune with popular taste and is an efficient, effective and just allocation of resources. The *dirigiste* approach counters that cultural improvement of the population is necessary, because markets favour pleasure over sophistication and popular taste is ephemeral. In other words, markets fail to encourage and sustain art's function of defining and developing universal human values and forms of expression. At the junction of these approaches, a contest ensues over whether 'it is more worthwhile to look at a Titian on a wall than watch a football game on television' (Dworkin 1985: 232–3). As most people supposedly prefer the latter – a preference that can be quantified through their preparedness to pay – it is deemed condescending to force them to subsidize the former as part of their generic tax burden on the grounds that timeless art can only survive if the *poloi* are required to admire it.

Cultural policy studies began in the 1970s within the positivistic social sciences as a means of managing these dilemmas. It developed through evaluations of policies and programmes through the Association of Cultural Economics; conferences on economics, social theory and the arts; and publications such as *Arts and Education Policy Review*, the *Journal of Arts Management, Law, and Society*, the *International Journal of Cultural Policy*, *Poetics* and the *Journal of Cultural Economics*, which investigated audience preferences and the ethical and technical management of the arts. Then a challenge came from the left. Habermas problematized conventional cultural policy thus:

> Cultural policy is assigned the task of operating on two fronts. On the one hand, it is supposed to discredit intellectuals as a power-crazy and non-productive social class supporting modernism; for post-material values, especially expressive needs for self-fulfillment and critical judgements of universalistic morality, are threats to the motivational resources of a functioning labor society and a depoliticized public sphere. On the other hand, conventional morality, patriotism, bourgeois religion and folk culture, is to be carefully nurtured, in order to compensate for personal burdens on one's private life and to offer some cushion against the pressures of a competitive society and its accelerated modernization. (1986: 11)

Then Stuart Cunningham suggested that applying cultural studies to policy matters might renew both areas. Cultural studies could transcend its tendency to criticize without bringing about change by drawing energy and direction from 'a social democratic view of citizenship and the trainings necessary to activate and motivate it'. (See also Stevenson, Chapter 9, this volume.) Such an 'engagement with policy' would avoid 'a politics of the status quo' as per the conventional social sciences thanks to cultural studies' concern with power (1992: 11).

A policy trend within cultural studies took off. Key figures included Tom O'Regan (1996) and Tony Bennett (1995). Working on the media and museums respectively, they drew on *dirigiste* French models of urban planning to designate cultural precincts and the like. Their methods – archival studies, survey research, policy analysis and Foucauldian theory – emphasized the foundational nature of government in the creation of the liberal individual (that is, a person open to new ideas delivered in a rational form and reasoned manner). In 1991, the Australian Academy of the Humanities responded to such developments, introducing its venerable members to cultural studies, cultural policy, feminism and multiculturalism, then adding these as fellowship categories (Morris 2005: 111–13, 116–17). In Latin America, influential engagements with cultural policy materialized in the work of Néstor García Canclini (2001) and cognate practice was underway in Canada (Robertson 2006) and Britain (Lewis 1991). But more was at stake than a new trend in cultural studies. Congruent changes were taking place in the political economy.

In the last quarter of the twentieth century, economic production in the global North shifted from a farming and manufacturing base to a cultural one. The population was harnessed to jobs in music, theatre, animation, recording, radio, TV, architecture, software, design, toys, books, heritage, tourism, advertising, fashion, crafts, photography, performance, the Internet, games, sports and cinema. In 2005, the International Intellectual Property Alliance estimated the value of the copyright industries (its term for the cultural industries) at US$1.38 trillion in the United States. That was 11.12 per cent of total Gross Domestic Product, and it accounted for 23.78 per cent of overall economic growth. Culture employed more than 11 million people across the country, over 8 per cent of the workforce. In terms of foreign sales, 2005 exports of music, software, film, television and print were US$110.82 billion (Siwek 2006).

Such developments raised the stakes of culture and policy from binding populations together to keeping them fed and housed. They also promised a way out of the opposition between choice and paternalism, which became moot when commitments to building citizens married commitments to building economies. We might say that a nineteenth-century Arnold (Matthew) endorsed a twentieth-century one (Schwarzenegger) and *vice versa*.

Creativity: The New Right of Cultural Studies

Not surprisingly, latter-day practitioners and students of cultural policy are in thrall to economism, notably the creative-industry approach of Richard Florida (2002) and his acolytes from the new right of cultural studies. (See also Sasaki, Chapter 12, this volume.) Florida speaks of a 'creative class®' that he says is revitalizing post-industrial towns in the global North devastated by the relocation of agriculture and manufacturing to places with cheaper labour pools. The revival of such cities, Florida argues, is driven by a magic elixir of tolerance, technology and talent, as measured by same-sex households, broadband connections and higher degrees. As a performative and profitable point, he has even trademarked the brand: 3298801 is Florida's registration number for 'creative class®' with the US Patent and Trademark Office (http://tess2.uspto.gov).[3]

True believers argue that cultural policy is outmoded, because post-industrialized societies have seen an efflorescence of the creative sector via new technology and small business. Kantian concerns barely apply, let alone Gramscian ones. Culture is

---

[3]   Thanks to Bill Grantham for directing me to the Office's Trademark Electronic Search System.

central to employment, rather than a mechanism for holding societies together. In the words of the lapsed-leftist cultural theorist and President of the European Bank for Reconstruction and Development Jacques Attali (2008: 31), a new 'mercantile order forms wherever a creative class masters a key innovation from navigation to accounting or, in our own time, where services are most efficiently mass produced, thus generating enormous wealth'. This is said to give rise to an 'aristocracy of talent' (Kotkin 2001: 22) as magically multiplying meritocrats luxuriate in ever-changing techniques, technologies and networks.

As they wander the globe promoting tertiary industry against agriculture and manufacturing, creative-industry consultants become branded celebrities, carpetbaggers carefully sidestepping the historic tasks laid out by the left. Prone to cybertarianism, these chorines of digital capitalism and the technological sublime pile out of business class and onto the jet way, descending on welcoming burghers eager to be made over at public expense by professors whose books appear on airport newsstands rather than cloistered scholarly shelves (Gibson and Klocker 2004). Should you wish to hear Florida at your next convention, visit the website of the 'Creative Class Group' (http://creativeclass.com). Celebrity Speakers (http://speakers.co.uk/csaWeb/speaker,JAQATT) will link you to the Global Speakers Bureau for a date with Attali.

Creative-cities chorines alight from the tarmac in three major groups. Richard Floridians hop a limousine from the airport then ride around town on bicycles, spying on ballet-loving, gay-friendly, multicultural computer geeks who have moved to deindustrialized rustbelts. Australian creationists criticize cultural policy studies as residually socialistic and textual. And Brussels bureaucrats offer blueprints to cities eager for affluence and prepared to be reinvented via culture and tolerance.

Meanwhile, the US President's Committee on Arts and the Humanities welcomes the 'creative economy' as a central focus of its activities, stressing:

> *The President's Committee focuses its leadership, with other agencies and the private sector, on the power of the arts and humanities as an economic driver, sustaining critical cultural resources and fostering civic investment in cultural assets and infrastructure. These efforts help speed innovation and expand markets and consumers, directly benefiting local economies. (http://www.pcah. gov/creative-economy [accessed 4 April 2013])*

This approach promises a pragmatic uptake of cultural difference, import substitution, export orientation, and national and regional pride and influence. Municipal, regional, state, continental and global agencies have responded by replacing older funding and administrative categories of culture with the discourse of the creative industries. The United Nations Conference on Trade and Development (UNCTAD) decrees that '[c]reativity, more than labor and capital, or even traditional technologies, is deeply embedded in every country's cultural context' (2004: 3). Nowadays, far from looking to destabilize capitalism and neo-colonialism through culture – as they once did – UNCTAD and the United Nations Development Program (UNDP) maintain that 'the interface among creativity, culture, economics and technology, as expressed in the ability to create and circulate intellectual capital, has the potential to generate income, jobs and export earnings while at the same time promoting social inclusion, cultural diversity and human development' (United Nations Conference on Trade and Development/United Nations Development Program 2008). The United Nations Educational, Scientific and Cultural Organization (UNESCO) has a Global Alliance for Cultural Diversity (2002) that heralds the creative industries as a *portmanteau* term covering the cultural sector that goes even further, beyond output and into that favourite neoliberal *canard* of process. In 2006 Rwanda convened a global conference on the 'creative economy' to take the social healing engendered by culture and commodify it. Brazil houses UNCTAD and UNDP's International Forum for Creative Industries. Even India's venerable last gasp of Nehruvianism, its Planning Commission, boasts a committee for creative industries; Singapore, Hong Kong Japan, and South Korea follow similar strategies (Cunningham 2009b, Keane 2006, Ramanathan 2006) and China is moving 'from an older, state-dominated focus on cultural industries … towards a more market-oriented pattern of creative industries' (UNCTAD 2004: 7).

The change is an intellectual one as well. This new right of cultural studies is part of an innovative interdisciplinarity that blends research and social inclusiveness (Brint et al. 2009 describe this tendency). Professors already interested in cultural policy, for reasons of cultural nationalism or in opposition to corporate culture, hope to be something more than 'the little match seller, nose pressed to the window, looking in on the grand life within' (Cunningham 2006). Many have shifted their discourse from nation-building to copyright protection, focusing on comparative advantage and competition rather than heritage and aesthetics.

Put another way, neoliberal emphases on creativity have succeeded old-school cultural patrimony. Cunningham, for instance, no longer speaks of mixing socialist ideals with reformism. His rhetoric favours 'a better matching of curriculum to career' via 'practical business challenges', such that 'non-market disciplines' generate internal markets in competition with others and forge 'an alliance with the business sector' (2009a, 2007a, 2007b).

These powerful advocates and institutions are in thrall to the idea of culture as an endlessly growing resource that can dynamize societies. The Australian Research Council, which once supported a major cultural policy initiative under the Gramscian-turned-Foucauldian Bennett, now funds a Centre of Excellence for Creative Industries and Innovation (http://cci.edu.au) run by Cunningham and, for a time, the lapsed semiotic romantic John Hartley, who preaches evolutionary economics as 'cultural science' (http://cultural-science.org). In the same vein, the Australian Academy for the Humanities calls for 'research in the humanities and creative arts' to be tax-exempt because of their contribution to research and development, and subject to the same surveys of 'employer demand' as the professions and the sciences (Cunningham 2007a).

Britain's Arts and Humanities Research Council (AHRC) and National Endowment for Science, Technology and the Arts say '[t]he arts and humanities have a particularly strong affiliation with the creative industries' and undertake research that 'helps to fuel' them and boost innovation across the economy (Bakhshi, Schneider and Walker 2008: 1). The British Academy seeks to understand and further the 'creative and cultural industries' (2004: viii) and even the very sober National Research Council of the US National Academies notes that the electronic media stimulate 'economic development' (Mitchell, Inouye and Blumenthal 2003: 1).

In other words, a neoliberal bequest of creativity has succeeded the old-school patrimony of culture. Economic transformations have comprehensively challenged the idea of culture as removed from industry because the comparatively cheap and easy access to making and distributing meaning afforded by internet media and genres is thought to have eroded the one-way hold on culture that saw a small segment of the world as producers and the larger segment as consumers. The result is said to be a democratized media, higher skill levels, more sovereign customers, powerful challenges to old patterns of expertise and institutional authority, and more liveable, sustainable and pleasurable regions, especially cities. Creativity is distributed rather than centralized and it becomes both a pleasure and a responsibility to invest in human capital and ensure a robust civil society and self.

The working assumption is that the individual talent of the creative sector is progressively overrunning the mass scale of the culture industries. (See also O'Connor, and Montgomery, Chapters 10 and 20, this volume). It's a kind of Marxist/Godardian dream, where people fish, film, fornicate and finance from morning to midnight as technology obliterates geography, sovereignty and hierarchy in an alchemy of truth and beauty. A deregulated, individuated world makes consumers into producers, frees the disabled from confinement, encourages new subjectivities, rewards intellect and competitiveness, links people across cultures, and allows billions of flowers to bloom in a post-political cornucopia. Consumption is privileged, production is discounted, and labour is forgotten (Dahlström and Hermelin 2007, Ritzer and Jurgenson 2010). This discourse avowedly stands against elitism, for populism, against subvention, for markets, against public service, for philanthropy and so on.

A Counter-Discourse

A paradox lies at the base of this movement. Thirty years ago, Foucault identified coin-operated think tanks like the American Enterprise Institute as intellectual handservants of neoliberalism, whilst recognizing that these vocalists of a 'permanent criticism of government policy' (2008: 247) actually sought permanent influence *over* such policy, using markets as their privileged 'interface of government and the individual' (2008: 253). Neoliberalism governed populations through market imperatives, invoking and training them as ratiocinative liberal actors waiting for their inner creativity to be unlocked. Consumption was turned on its head: everyone was creative, no one was simply a spectator, and we were all manufacturing pleasure while witnessing activities we had paid to watch. Internally divided – but happily so – each person was 'a consumer on the one hand, but … also a producer' (2008: 226). We can see this duality at play in the *dirigiste* mode of urban cultural planning mentioned above, which sought to do good by doing accumulation (Stevenson 2004).

Unsurprisingly, not everyone accepts the creative-industry chorines' embrace of techno-charged capitalist culture with equanimity. And cultural policy has a radical as well as a reactionary lineage. For instance, during the 1970s UNESCO sought a fundamental transformation of international cultural exchange, based on post-colonial states undoing their dependent relationships with the global North. This vibrant desire, which animated many countries and social movements, was shut down in the mid-1980s when Britain and the United States withdrew from UNESCO, crippling its funding and legitimacy (Gerbner, Mowlana and Nordenstreng 1994). Something similar happened when the Organization recognized Palestine in 2011. Again, the United States engaged in symbolic violence, refusing to pay its dues after losing a crucial vote on the matter (Nuland 2011).

Apart from these ideological concerns, creative-industries discourse is questionable on its own terms. There is minimal proof that a creative class exists or that 'creative cities' outperform their drab brethren economically following makeovers 'from the rusty coinage of "cultural industries" to the newly minted "creative industries"' (Ross 2006–7: 1). Companies certainly seek skills when deciding where to locate their businesses; but skills also seek work. City centres generally attract workers who are young and not yet breeding. And the centrality of gay culture in the Floridian calculus derives from assuming all same-sex households are queer. Even if this were accurate, many successful cities in the United States roll with reaction (consider Orlando or Phoenix). The definition of urbanism in US statistics includes the suburbs (which now hold more residents than do cities) so that, too, is suspect in terms of the importance of downtown lofts to economies. There is no evidence of an overlap of tastes, values, living arrangements and locations between artists on the one hand and accountants on the other, despite their being bundled together in the creative concept; nor is it sensible to assume other countries replicate the massive internal mobility of the US population. Finally, other surveys pour scorn on the claim that quality of life is central to selecting business campuses as opposed to low costs, good communications technology, proximity to markets, and adequate transportation systems. A European Commission evaluation of 29 Cities of Culture disclosed that their principal goal – economic growth stimulated by the public subvention of culture to renew failed cities – has failed. Glasgow, for instance, was initially hailed as a success of the programme; but despite many years of rhetoric, it has seen no sustained growth (Alanen 2007, Nathan 2005). Anyone who has spent time in Mexico City (see http://mim.promexico.gob.mx/wb/ mim/ind_perfil_del_sector for boosterism) or downtown Los Angeles (see http://www. laane.org/about-us/what-we-do for critique), where I have lived this past year, can bear witness to the impact of the creativity *ethos* as it smuggles middle-class privilege through customs and expels the homeless and the bereft in the name of renewal.

The question of social justice and redistributive democracy hovers over the new right's soaring creative rhetoric (Pratt 2011). Kultur Macht Europa issued a sterling declaration following its Fourth Federal Congress on Cultural Policy in 2007 about the necessity of ensuring that artistic infrastructure is evaluated with regard to diverse and profound textuality. We see similar concerns in the Jodhpur Initiative for Promoting Cultural Industries in the Asia-Pacific Region, adopted in 2005 as the Jodhpur Consensus by 28 countries, and the Euromayday Network (Jodhpur Initiatives 2005, http://euromayday. org). And many scholars and activists are still committed to progressive cultural policy, such as George Yúdice (2002, Miller and Yúdice 2002) in Miami and San José and Kate Oakley (2006) and David Bell (2007) in Leeds. They beaver away, weathering slings and arrows from the comfortably pure ultra-left for engaging with commerce and the state, and sending a few of their own towards those who unproblematically embrace such links. It remains to be seen whether cultural policy will be organized in accord with neo-Kantianism, profiteering or progressive politics.

Cultural Waste

In any event, before there can be a story to sell, a message to influence, a consumer to buy or a renovated city to propagate in the collective and individual use of culture, there has to be a physical medium. Books, magazines, money and other printed media rely on papermaking and printing. Radios, televisions, computers, cell phones, easels, paintings, sculptures, instruments and music players arrive in our homes, offices and studios with parts that have been assembled, packaged and transported from excavated and manufactured materials. In short, apart from the immediate surroundings of buildings, classrooms, homes, subways, cars and campuses where people engage culture, its physical foundation is machinery that is created and operated through work and leaves an impact on the Earth. This is the very dirty, very large secret of cultural policy. Unravelling it requires both philosophical thinking and social-science method.

Consider an unlikely bedfellow for Habermas – Heidegger – or at least his famous paradox of the forester, a man who participates in the destruction of the very environment that gives meaning to his life in order to supply a key resource to *bourgeois* culture, which in turn uses it to shape his opinions:

> The forester who measures the felled timber in the woods and who to all appearances walks the forest path in the same way his grandfather did is today ordered by the industry that produces commercial woods, whether he knows it or not. He is made subordinate to the orderability of cellulose, which for its part is challenged forth by the need for paper, which is then delivered to newspapers and illustrated magazines. The latter, in their turn, set public opinion to swallowing what is printed, so that a set configuration of opinion becomes available on demand. (1977: 299)

As the forester's work is subsumed into modern pulp and paper production, labour and the environment are further disarticulated from one another. Paper mills and printing presses are hailed as 'revolutionary' and newspapers, magazines, books and fine paper become signs of progress and intellectual life. They make no mention of his role or its environmental aftershocks.

For labour is acknowledged in today's brave new form of cultural policy only when it is abstracted from physical, dirty work (Mattelart 2002). The prevailing presumption is that the creative industries, especially the newer media, deliver a clean, post-industrial, capitalism. This myth has been continually reinforced by the 'virtual nature of much of the industry's content', which 'tends to obscure their responsibility for a vast proliferation of hardware, all with high levels of built-in obsolescence and decreasing levels of efficiency' (Boyce and Lewis 2009: 5).

Producing and distributing culture consumes, despoils and wastes natural resources and exploits human life at an ever-increasing rate. Cultural technologies contain toxic substances that pervade the sites and environs where they are manufactured, used and thrown away, poisoning humans, animals, vegetation, soil, air and water. Rapid cycles of innovation and planned obsolescence accelerate both the emergence of new electronic hardware and the accumulation of obsolete forms, which are transformed overnight into junk. Today's digital devices are made to break or become un-cool very rapidly indeed. This planned obsolescence reinforces consumerism and animates the ideology of growth that says technological innovation is necessary. Immediacy and interactivity induce ignorance of inter-generational effects of consumption, including long-term harm to workers and the environment. Culture's residue is poisoned waterways, sickened workers and toxic habitats, and cities grow sick because of culture and 'creativity'.

In 2007, a combination of cultural technologies and production accounted for 3 per cent of all greenhouse gas emitted around the world. Between 20 and 50 million tonnes of electronic waste (e-waste) are generated annually, much of it via cell phones, televisions and computers, which wealthy people throw out regularly in order to buy replacements. E-waste is historically produced in Australasia, Western Europe, Japan, Canada and the United States, and dumped in Latin America, Eastern Europe, Africa and Asia in the form of a thousand different, often deadly, materials for each computer. Today, of course, India and China increasingly generate their own cultural detritus. Through the combined efforts of these creative nations, e-waste has become the fastest-growing part of municipal clean-ups around the world (Herat 2007, Malmodin et al. 2010, Robinson 2009).

The accumulation of electronic hardware has caused grave environmental and health concerns that stem from the chemical and material composition of these commodities and their potential seepage into landfills, water sources and the bodies of workers. E-waste salvage yards have major implications for human health and safety wherever plastics and wires are burnt and circuit boards leached with acid or grilled then dumped in streams to minimize the volume of waste and retrieve valuable items. There are horrific implications for local and downstream land and water as well as residents. Perhaps 1 per cent of people in the global South now live as urban ragpickers, recycling the detritus of computers, printers, cell phones and the like. They amount to approximately 15 million people worldwide, often living and working in cities that have cultural policies designed to attract media technology and production – not to mention 'the creative class' – that poison land, air, water and bodies. Many ragpickers are pre-teen Chinese, Nigerian and Indian girls, picking away without protection at discarded televisions and computers in order to find precious metals and dump the remains in landfills. The metals are sold to recyclers, who rarely use landfills or labour in the global North because of environmental and industrial legislation *contra* the destruction to soil, water and workers caused by the dozens of poisonous chemicals and gases in these dangerous machines (Basel Action Network 2006, Basel Action Network and Silicon Valley Toxics Coalition 2002, Maxwell and Miller 2012, Osibanjo and Nnorom 2007, Ray et al. 2004, Tong and Wang 2004, Wong et al. 2007).[4]

Consider an environmental field very close to much cultural policy: film and television drama. The Political Economy Research Institute's 2004 *Misfortune 100: Top Corporate Air Polluters in the United States* placed media owners at numbers 1, 3, 16, 22 and 39. How can this be, given the advent of the Environmental Media Association's awards, the 2007 'Hollywood goes green' summit meeting, and *Hollywood Today*'s boast that actors give green gifts of 'vintage-inspired' camisoles and recycled jewels (Pantera 2009, Ventre 2008)?

MSNBC.com admonishes that although 'the Prius reigns supreme as the current status symbol' in Hollywood, 'trucks that carry equipment from studios to locations

---

[4]   Thanks to Rick Maxwell, many of whose ideas inform this section.

and back continue to emit exhaust from diesel engines', as do generators on-set (Pantera 2009, Ventre 2008). Research into Hollywood's environmental impact has disclosed massive use of electricity and petroleum and the release of hundreds of thousands of tons of deadly emissions each year. In fact, the motion-picture industry is the biggest producer of conventional pollutants in Los Angeles. Municipal and state-wide levels of film-related energy consumption and greenhouse-gas emissions (carbon dioxide, methane and nitrous oxide) approximate to those of the aerospace and semi-conductor industries. Film consumers are also major producers of pollutants, from auto emissions, chemical run-off from parked cars and the energy to power home-entertainment devices (Corbett and Turco 2006).

Is this the brave new world of an urban economy driven by creativity, epitomized in a city like Los Angeles that is obsessed with culture to its very core? How can we engage and criticize this urban fantasy in ways that endow foresters, their equivalents and the public with power/knowledge? Via a cultural policy that tracks the materiality of the sign from start to finish along the commodity chain that makes it, brings it to us, then disposes of it.

A growing number of artists are criticizing apolitical celebrations of digital technology. Consider *Arte Povera*'s use of found materials to rail at errant, arrogant consumption by highlighting e-waste, recycling and ragpickers, or such urban artists as Jessica Millman, Miguel Rivera, Alexdromeda, Sudhu Tewari, Natalie Jeremijenko, Nome Edonna, Chris Jordan, Erik Otto and Jane Kim. Yona Friedman and Jorge Crowe focus on artistic re-use rather than originality, while Julie Bargmann and Stacy Levy undertake creative clean-ups. Amsterdam's urbanscreens.org and Ars Electronica of Linz use electronic billboards to encourage active citizenship. Environmental art can cover both works that directly represent the environment – examples would include Monet's *London Series* or Constable's *Clouds* – and non-representational, performative works like Richard Long's *A Line Made by Walking*, James Turrell's *Skyspace* or Olafur Eliasson's *The Weather Project*. They all assume nature is occupied and shaped by humanity and *vice versa*. Cultural producers such as these can return us to the original animation of culture – in its best Enlightenment form – as we look for progressive urban policies.

References

Adorno, T.W. 2009. *Kultur* and culture (translated by M. Kalbus). *Social Text*, 99, 145–58.
Alanen, A. 2007. What's wrong with the concept of creative industries? *Framework: The Finnish Art Review*, 6. [Online]. Available at: http://www.oecd.org/dataoecd/52/52/37794008.pdf [accessed 4 April 2013].
Althusser, L. 1969. *For Marx* (translated by B. Brewster). Harmondsworth: Penguin.
Arnold, M. 1875. *Essays in Criticism*. 3rd Edition. London: Macmillan.
Attali, J. 2008. This is not America's final crisis. *New Perspectives Quarterly*, 25(2), 31–3.
Ayers, E.L. 2009. Where the humanities are. *Daedalus*, 138(1), 24–34.
Bakhshi, H., Schneider, P. and Walker, C. 2008. *Arts and Humanities Research and Innovation*. London: Arts and Humanities Research Council/National Endowment for Science, Technology and the Arts.

Basel Action Network. 2006. *JPEPA as a Step in Japan's Greater Plan to Liberalize Hazardous Waste Trade in Asia*. Seattle, WA: Basel Action Network.

Basel Action Network and Silicon Valley Toxics Coalition. 2002. *Exporting Harm: The High-Tech Trashing of Asia*. Seattle, WA: Basel Action Network.

Bell, D. 2007. Fade to grey: some reflections on policy and mundanity. *Environment and Planning A*, 39(3), 541–54.

Bennett, T. 1995. *The Birth of the Museum: History, Theory, and Politics*. London: Routledge.

Boyce, T. and Lewis, J. (eds) 2009. *Climate Change and the Media*. New York: Peter Lang.

Briggs, A. and Burke, P. 2003. *A Social History of the Media: From Gutenberg to the Internet*. Cambridge: Polity.

Brint, S.G., Turk-Bicakci, L., Proctor, K. and Murphy, S.P. 2009. Expanding the social frame of knowledge: interdisciplinary, degree-granting fields in American colleges and universities, 1975–2000. *Review of Higher Education*, 32(2), 155–83.

British Academy. 2004. *'That Full Complement of Riches': The Contributions of the Arts, Humanities and Social Sciences to the Nation's Wealth*. [Online]. Formerly available at: http://www.britac.ac.uk/policy/full-complement-riches.cfm [accessed 10 March 2012].

Coleridge, S.T. 1839. *On the Constitution of Church and State According to the Idea of Each. II: Lay Sermons* (edited by H.N. Coleridge). London: William Pickering.

Corbett, C.J. and Turco, R.P. 2006. *Sustainability in the Motion Picture Industry*. Report prepared for the Integrated Waste Management Board of the State of California by the University of California, Los Angeles, Institute of the Environment. [Online]. Available at: http://www.environment.ucla.edu/ccep/research/article.asp?parentid=753 [accessed 4 April 2013].

Cunningham, S. 1992. *Framing Culture: Criticism and Policy in Australia*. Sydney: Allen & Unwin.

—— . 2006. Business needs crash course in appreciation. *Australian Financial Review*, 14 August, 36.

—— . 2007a. Oh, the humanities! Australia's innovation system out of kilter. *Australian Universities Review*, 49(1–2), 28–30.

—— . 2007b. Taking arts into the digital era. *Courier Mail*. [Online, 22 June]. Available at: http://www.onlineopinion.com.au/view.asp?article=6031 [accessed 4 April 2013].

—— . 2009a. Creative industries as a globally contestable policy field. *Chinese Journal of Communication*, 2(1), 13–24.

—— . 2009b. Trojan horse or Rorschach blot? Creative industries discourse around the world. *International Journal of Cultural Policy*, 15(4), 375–86.

Dahlström, M. and Hermelin, B. 2007. Creative industries, spatiality and flexibility: the example of film production. *Norsk Geografisk Tiddskrift – Norwegian Journal of Geography*, 61(3), 111–21.

Dworkin, R. 1985. *A Matter of Principle*. Cambridge, MA: Harvard University Press.

Florida, R. 2002. *The Rise of the Creative Class and How it's Transforming Work, Leisure and Everyday Life*. New York: Basic Books.

Foucault, M. 2008. *The Birth of Biopolitics: Lectures at the Collège de France, 1978–79* (translated by G. Burchell and edited by M. Senellart). Houndmills: Palgrave Macmillan.

García Canclini, N. 2001. *Consumers and Citizens: Globalization and Multicultural Conflicts* (translated by G. Yúdice). Minneapolis, MN: University of Minnesota Press.

Gerbner, G., Mowlana, H. and Nordenstreng, K. (eds) 1994. *The Global Media Debate: Its Rise and Fall*. Norwood, NJ: Ablex.

Gibson, C. and Klocker, N. 2004. Academic publishing as 'creative' industry, and recent discourses of 'creative economies': some critical reflections. *Area*, 36(4), 423–34.

Gramsci, A. 1971. *Selections from the Prison Notebooks* (translated and edited by Q. Hoare and G. Nowell-Smith). New York: International Publishers.

Habermas, J. 1986. The new obscurity: the crisis of the welfare state and the exhaustion of utopian energies (translated by P. Jacobs). *Philosophy and Social Criticism*, 11(2), 1–18.

Heidegger, M. 1977. *Basic Writings from Being and Time (1927) to The Task of Thinking (1964)* (edited by D.F. Krell and translated by J. Stambaugh, J.G. Gray, D.F. Krell, J. Sallis, F.A. Capuzzi, A. Hofstadter, W.B. Barton, Jr., V. Deutsch, W. Lovitt and F.D. Wieck). New York: Harper & Row.

Herat, S. 2007. Review: sustainable management of electronic waste (e-waste). *Clean*, 35(4), 305–10.

Hunter, I. 2008. Critical response II: talking about my generation. *Critical Inquiry*, 34(3), 583–600.

Jodhpur Initiatives. 2005. *Asia-Pacific Creative Communities: A Strategy for the 21st Century*. Bangkok: UNESCO.

Kant, I. 1987. *Critique of Judgment* (translated by W.S. Pluhar). New York: Hackett Publishing.

—— . 1991. *Metaphysics of Morals* (translated by M. Gregor and edited by R. Guess). Cambridge: Cambridge University Press.

Keane, M. 2006. From made in China to created in China. *International Journal of Cultural Studies*, 9(3), 285–96.

Kotkin, J. 2001. *The New Geography: How the Digital Revolution is Reshaping the American Landscape*. New York: Random House.

Kultur Macht Europa. 2007. Culture powers Europe. [Online]. Available at: http://www.kultur-macht-europa.eu/kongress.html?L=1 [accessed 4 April 2013].

Lewis, J. 1991. *Art, Culture, and Enterprise: The Politics of Art and the Cultural Industries*. London: Routledge.

Malmodin, J., Moberg, Å., Lundén, D., Finnveden, G. and Lövehagen, N. 2010. Greenhouse gas emissions and operational electricity use in the ICT and entertainment and media sectors. *Journal of Industrial Ecology*, 14(5), 770–90.

Marx, K. 1978. *The Eighteenth Brumaire of Louis Bonaparte*. Peking: Foreign Language Press.

Mattelart, A. 2002. An archaeology of the global era: constructing a belief (translated by S. Taponier with P. Schlesinger). *Media Culture and Society*, 24(5), 591–612.

Maxwell, R. and Miller, T. 2012. *Greening the Media*. New York: Oxford University Press.

Miller, T. and Yúdice, G. 2002. *Cultural Policy*. London: Sage.

Mitchell, W.J., Inouye, A.S. and Blumenthal, M.S. (eds) 2003. *Beyond Productivity: Information Technology, Innovation, and Creativity*. Committee on Information Technology and Creativity, Computer Science and Telecommunications Board, Division on Engineering and Physical Sciences, National Research Council of the National Academies.

Morris, M. 2005. Humanities for taxpayers: some problems. *New Literary History*, 36(1), 111–29.

Mowlana, H. 2000. The renewal of the global media debate: implications for the relationship between the West and the Islamic world. In *Islam and the West in the Mass Media: Fragmented Images in a Globalizing World*, edited by K. Hafez. Cresskill: Hampton Press, 105–18.

Nathan, M. 2005. *The Wrong Stuff: Creative Class Theory, Diversity and City Performance*. Centre for Cities, Institute for Public Policy Research Discussion Paper 1.

Newcomb, H. 1996. Other people's fictions: cultural appropriation, cultural integrity, and international media strategies. In *Mass Media and Free Trade: NAFTA and the Cultural Industries*, edited by E.G. McAnany and K.T. Wilkinson. Austin, TX: University of Texas Press, 92–109.

Nuland, V. 2011. *State Department on Palestinian Admission to UNESCO*. [Online, 31 October]. Available at: http://translations.state.gov/st/english/texttrans/2011/10/201 11031170111su0.1734082.html#ixzz1p8Q4nLIn [accessed 4 April 2013].

O'Regan, Tom. 1996. *Australian National Cinema*. London: Routledge.

Oakley, K. 2006. Include us out: economic development and social policy in the creative industries. *Cultural Trends*, 15(4), 255–73.

Osibanjo, O. and Nnorom, I.C. 2007. The challenge of electronic waste (e-waste) management in developing countries. *Waste Management and Research*, 25(6), 489–501.

Pantera, G. 2009. Hollywood goes green. *Hollywoodtoday.net*, 6 May. [Online]. Available at: http://www.hollywoodtoday.net/2009/05/06/hollywood-goes-green [accessed 4 April 2013].

Plumb, J.H. 1964. Introduction. In *Crisis in the Humanities*, edited by J.H. Plumb. Harmondsworth: Penguin, 7–10.

Political Economy Research Institute. 2004. *The Misfortune 100: Top Corporate Air Polluters in the United States*. Amherst, MA: University of Massachusetts.

Pratt, A.C. 2011. The cultural contradictions of the creative city. *City, Culture and Society*, 2(3), 23–30.

Ramanathan, S. 2006. The creativity mantra. *The Hindu*, 29 October. [Online]. Available at: http://www.hindu.com/mag/2006/10/29/stories/2006102900290700.htm [accessed 4 April 2013].

Ray, M.R., Mukherjee, G., Roychowdhury, S. and Lahiri, T. 2004. Respiratory and general health impairments of ragpickers in India: a study in Delhi. *International Archives of Occupational and Environmental Health*, 77(8), 595–98.

Ritzer, G. and Jurgenson, N. 2010. Production, consumption, prosumption: the nature of capitalism in the age of the digital 'prosumer'. *Journal of Consumer Culture*, 10(1), 13–36.

Robertson, C. 2006. *Policy Matters: Administrations of Art and Culture*. Toronto: YYZBOOKS.

Robinson, B.H. 2009. E-waste: an assessment of global production and environmental impacts. *Science of the Total Environment*, 408(2), 183–91.

Ross, A. 2006–7. Nice work if you can get it: the mercurial career of creative industries policy. *Work Organisation, Labour and Globalisation*, 1(1), 1–19.

Rousseau, J. 1755. *A Discourse on Political Economy*. [Online]. Available at: http://www.constitution.org/jjr/polecon.htm [accessed 4 April 2013].

Schelling, F.E. 1914. New humanities for old. *Classical Weekly*, 7(23), 179–84.

Siwek, S.E. 2006. *Copyright Industries in the U.S. Economy*. International Intellectual Property Alliance.

Smith, S. 1844. *The Works of the Rev. Sydney Smith*. Philadelphia, PA: Carey and Hart.

Stevenson, D. 2004. 'Civic gold' rush: cultural planning and the politics of the third way. *International Journal of Cultural Policy*, 10(1), 119–31.

Tong, X. and Wang, J. 2004. Transnational flows of e-waste and spatial patterns of recycling in China. *Eurasian Geography and Economics*, 45(8), 608–21.

United Nations Conference on Trade and Development. 2004. *Creative Industries and Development*. Eleventh Session, São Paulo. TD(XI)/BP/13.

United Nations Conference on Trade and Development/United Nations Development Program. 2008. *Creative Economy Report 2008: The Challenge of Assessing the Creative Economy: Towards Informed Policy-Making*. [Online]. Available at: http://www.unctad. org/templates/webflyer.asp?docid=9750&intItemID=2068&lang=1 [accessed 10 March 2012].

United Nations Educational, Scientific and Cultural Organization. 2002. *Culture and UNESCO*. Paris: UNESCO.

Ventre, M. 2008. It's not easy being green, Hollywood discovers. *MSNBC.com*, 23 April. [Online]. Available at: http://msnbc.msn.com/id/24256817/ns/business-going_green [accessed 4 April 2013].

Weber, S. 2000. The future of the humanities: experimenting. *Culture Machine 2*. [Online]. Available at: http://www.culturemachine.net/index.php/cm/article/view/311/296 [accessed 4 April 2013].

Wong, C.S.C., Wu, S.C., Duzgoren-Aydin, N.S., Aydin, A. and Wong, M.H. 2007. Trace metal contamination of sediments in an e-waste processing village in China. *Environmental Pollution*, 145(2), 434–42.

Yúdice, G. 2002. *El recurso de la cultura: Usos de la cultura en la era global*. Barcelona: Editorial Gedisa.

Case Study Window – Global Cities: Governance Cultures and Urban Policy in New York, Paris, Tokyo and Beijing

Peter Newman and Andy Thornley

Globalization is often viewed as having a homogenizing effect. It increases competition between cities that then pursue similar strategies to capture footloose global investment. The top world cities become models for others to emulate. Similar skylines, prestige projects and policy initiatives can be found across the major cities of the world. However, in our contribution to this book we would like to show that such a perspective is exaggerated. We follow Abu-Lughod (1999) in believing that history, tradition and culture make a difference to the way in which cities respond to the pressures of economic globalization. In particular we focus on the relationship between the market, state and civil society. In our view each country has developed a national consensus around the appropriate balance between these three elements. Although this may continue to be debated and subject to shifts with changes in governments, significant core values remain and allow distinct differences in approach to be identified between countries.

One label that might be attached to such differences in approach to the market/state/civil society balance is 'political culture'. Political culture is seen as specific to a particular nation state and provides the stage on which governmental changes and the formulation of policies are played out. It differs from political ideology (for example democratic socialism or Tory corporatism) in that people will disagree on political ideology while maintaining a consensus on political culture. However, the problem in using 'political culture' is that it was first used in political science to promote the US political system, most famously in Almond and Verba's (1963) work *The Civic Culture*.

So to frame our discussion of global cities in this chapter we prefer the notion 'cultures of governance'. (See also, Duxbury and Jeannotte, and Young, Chapters 21 and 23, this volume.) In this way we hope to convey the broad political nature of our analysis while at the same time not identifying with the US focus of such work as *The Civic Culture*. 'Cultures of governance' also points towards our particular interest in the relationships between different levels of government, national, regional, city and neighbourhood. As we have said there is a certain enduring consensus on the culture of governance that distinguishes between different national approaches, however at the same time the dominant consensus may be contested. As we will see in our case studies, this can lead to modification and change and, in particular, economic globalization has presented a new pressure.

The national differences in the relationships between the market state and civil society have been analysed from different perspectives by various authors seeking to categorize nations into different approaches or 'models'. Some enter the triad from the economic direction and talk about the cultures of capitalism (for example Hampden-Turner and Trompenaars 1993, Hutton 1995, Thurow 1992). These focus particularly on the shared consensus regarding the degree and nature of state involvement in the economy, utilizing labels such as 'laissez-faire', 'social market' or 'Japanese model' (for a detailed review see Coates 1999). Others focus on how social needs are met and the way the market system is modified. For example, in a much quoted typology, Esping-Andersen (1990) refers to the three models of welfare – social democratic, conservative/corporative and liberal. The first has a strong Welfare State, the second relies on social insurance or the family and is often based on the Catholic principle of subsidiary, and the third relies on the market and is individualistic.

It is not our aim to develop or use a specific typology. The models mentioned above are ideal types and do not necessarily exactly describe particular countries. However, in the rest of this contribution we will be exploring four case studies that have been chosen to demonstrate a wide range of different cultures of governance. We start with New York that is set within a culture that is at the more liberal, individualistic and laissez-faire end of the spectrum. We then turn to Paris within the French tradition of strong state authority and centralized planning. The third example is Tokyo that operates within a Development State and our final example is Beijing – conditioned by China's socialist market economy while also retaining many cultural features from the past.

In each case we will briefly outline the dominant features of the national culture of governance and how this interacts with political forces and pressures within cities. The impact of this interaction will then be traced through to the strategic policy agendas and approaches of the cities with examples of key development projects (for a more detailed analysis of these issue in a range of world cities see Newman and Thornley 2011).

New York

In federal systems of government such as the United States we should expect governance cultures to vary as sub-national scales of government engage with and at times help shape the national negotiation of market–state relationships. Constitutionally, in the context of overriding US federal law, the states are sovereign. States take a direct role in some policy arenas and they shape the government arrangements of particular cities. In liberal capitalist societies such as the United States we expect a tension between a market liberalism and 'embedded forms' of (Keynesian Welfare State) institutional protections (Indergaard 2011). Perhaps the height of an embedded liberal phase was the New Deal of the 1930s. This progressive, interventionist style had its origins in New York. Governor Roosevelt took interventionist lessons from New York State into his

Presidency from 1932. The strong relationship between the President and New York City's mayor allowed the city itself to embark on extensive interventionist policies (Abu-Lughod 1999). This phase of history contributed to the distinctiveness of New York City government. The city has a much bigger public sector than the rather 'thin' governance of Los Angeles and public expenditure is much higher than spending in Chicago (Mollenkopf 2008: 244 and 259). In subsequent decades we could see the federal government's response to New York City's financial crisis in the 1970s as presaging the neo-liberalism of President Reagan and a city more dependent on private initiative and resources. The strong role of the private sector through the major development projects that reshaped the city in the last two decades was brought into question by the 2008 financial crisis leading to speculation about the next governance regime and new relationships with the federal scale (Indergaard 2011).

Understanding the governance culture that sets the context for planning New York City we are interested therefore in the relationships between scales of government and relationships between public and private sectors. (See also Low, Chapter 17, this volume.) In the period we cover in this chapter we can characterize a distinctive market-oriented style in the city; according to Gladstone and Fainstein a 'pro-growth, corporate led urban regime' (2001: 35). At a city region scale governance is shaped by particular interactions of public and private sectors. Kantor (2009) characterizes this institutional context as 'managed pluralism'. There is regional competition among local governments with a powerful New York City at the centre. Higher-level governments in the city region including the State of New Jersey as well as New York, address aspects of market failure and may attempt to manage intergovernmental competition. In the culture of governance that emerges, Kantor argues, public and private institutions are capable of political coordination and can come together, for example, to deliver transport infrastructure and large development projects. The socially oriented projects of earlier times are less well managed. The city's distinctive history and institutions mean that we should take Mollenkopf's (2008: 244) advice and not look at New York as a 'global template' for world cities.

Following the bankruptcy of the mid 1970s New York City evolved a particular style of planning and development. The city focused on Manhattan as its most marketable asset and on large projects. City-wide development lobbies – the New York City Partnership, the Real Estate Board of New York and the New York Chapter of the American Institute of Architects – exerted considerable influence in City Hall. But essential to the development process were the assets and powers of the public sector. The city's economic development office could offer tax breaks to attract and retain commercial development, city planning could rezone sites and offer zoning incentives, and the state's economic development corporation had the power and resources to acquire and repackage development sites to suit private investors. This proved to be an effective development machine typically led by an agency able to override the normal planning process. Deals could be struck with local interests or community groups overridden.

The ability to mobilize public powers behind private redevelopment and renewal projects marks out the success of New York in this period. Times Square was remodelled as an entertainment district but also offered a substantial amount of new office space. The railway yards on the west side of mid-town Manhattan offered another large redevelopment opportunity. In the 1990s Mayor Giuliani had imagined the Hudson Yards site as a location for a new stadium for the New York Yankees. That scheme failed but the city continued with a comprehensive rezoning plan that allows for 40m square feet of offices, as well as substantial residential development, expansion of the Jacob Javitts Convention Center and a new stadium. The stadium idea was timely as New York City considered a bid for the 2012 Olympic Games. The bid was unsuccessful but the prospect of comprehensive redevelopment made the case for extension of the No. 7 subway train to the stadium site, the city arguing that development over the long term accompanied by transport improvements starting in 2005 would 'make the areas more attractive for investment' (Doctoroff quoted in Bowles 2003). As with other large projects the state's development corporation could acquire land, and another state-controlled body, the MTA, owned the railway land. However, the city's rezoning and comprehensive development plan was initially stalled by local opposition. The opponents represented commercial interests as well as affected residential communities, and small business and manufacturing (Gross and Newman 2005). The state vetoed the Olympic stadium effectively killing off the New York bid but with improved public transport the sites continue to be attractive to development. The city's zoning powers and the state's development agencies and the capacity of 'managed pluralism' to deliver public transport projects characterize a New York style of planning and development. This alliance of public and private interest, city and state power was also evident in the process of rebuilding the sites of the World Trade Center (WTC).

The NYC Infrastructure Task Force held initial meetings with the mayor and the leaseholder of the WTC towers. To organize the development process a subsidiary of the state's development agency was created, the Lower Manhattan Development Corporation (LMDC), and its board appointed by the Governor. The state had substantial interest in redevelopment as the state-controlled Port Authority had been the original developer of the twin towers and the underground public transport links involved the MTA. Progress has not been smooth. An initial architectural competition and grander visions conflicted with the leaseholder's desire to rebuild office space as quickly as possible. An initially inclusive planning process (with 4,000 New Yorkers attending one forum) reviewed schemes drawn from the international architectural competition gave way to more pragmatic plans and wrangling between the leaseholder and the LMDC. One problem was the weakness of the office market as business preferred mid town, however any wider re-planning consideration lay outside the narrow development focus of the state-led development agency (Kantor 2002). The economic crisis in 2008 caused further negotiation over the timing of development and the amount of public subsidy being demanded from the Port Authority (Bagli 2009). Redeveloping the rail station was also delayed and the cost of the Fulton Street Transit Center doubled between 2001 and 2009.

Such project by project planning has been characteristic of the past 40 years of world city building. But returning to our earlier discussion, are there signs of a shift from the market liberal model to more interventionist ambitions? Two initiatives of Mayor Bloomberg might suggest a new approach if not replacing at least complementing the pro-growth, corporate-led planning style. An early Bloomberg initiative, the $7.5bn 'New Housing Market Place Plan', aimed to provide a range of affordable housing for low and middle incomes (NYCDHPD 2003). In 2005 the target was increased to 165,000 units by 2014 to be achieved through a range of city subsidy mechanisms. A total of 100,000 units had been delivered by 2010. However, this social investment needs to be seen in a context of the loss of 'affordable' housing through the mortgage foreclosures triggered by the 2008 crisis and the steady loss of rent controlled accommodation in the city. In 2008–9 there

Figure 4.1   New York's long term plan
Source: The authors.

were 150,000 foreclosures and these tended to be concentrated in black neighbourhoods that suffered from the aggressive marketing of sub-prime loans (Powell and Roberts 2009). Between 2003 and 2009 some 200,000 apartments affordable to low-income renters were lost (Fernandez 2009). Market forces pose a severe challenge to social planning.

A second Bloomberg initiative in 2007 was the longer term, social and environmental focus of PlaNYC 2030 (see Figure 4.1). The plan predicts substantial population growth in a context of fundamental infrastructure challenges in transportation, water supply and energy. In terms of climate change, New York may be susceptible to a 'heat island effect' with heat waves impacting on residents. This 'sustainability' plan takes a long range and comprehensive planning perspective. However, the infrastructure challenges cross the city boundary. Population growth is outside the city's control and the relative weakness of regional scale plans, and the bias of 'managed pluralism' toward project based planning expose the potential limits of the pursuit of sustainable development.

Paris

Contrasting with New York's market liberal context in which the development industry leads on major projects, Paris is located in a unitary state with a tradition of centralized planning. This tradition has been evolving. We have to acknowledge Paris as a European city in the context of the EU and the Eurozone, and some 'Europeanization' of planning as all cities adapt to EU funding streams and adoption of a general, if poorly understood (CEC 2010), notion of 'territorial cohesion' that fits a 'European model of society' (see Faludi 2007). Within France substantial reform of the state has been underway since the early 1980s. Processes of decentralization transferred planning competence from the central state to the mayors of communes and created a new, elected regional scale of government. More recently the loi Chèvenement encouraged the amalgamation of highly fragmented local governments into larger units capable of strategic planning. The outcome of this long process of reform of the state is that relationships between layers of government are often complex, with overlapping responsibilities sometimes managed through joint contracts in an institutional landscape including numerous centrally or locally created agencies.

This 're-scaling' of government has had particular implications for the nation's capital and economically dominant city-region. Historically there has been a tension within the French state between a foundational guarantee of equality across the republic and individual attachment to place as represented through the local commune. During the latter half of the twentieth century national planning sought to balance the growth and development of major cities. But, relocated within competitive European space, government took a different view and the fortunes of the country's major asset – the Paris city-region – took priority. By the turn of the century the economic competitiveness of Paris was a priority but reform of the state had created competing scales of planning. Additionally, competitive party politics impacts on regional and local responses to perceived global or European challenges.

As in the case of New York there is a dominant unit of government at the centre of the region. The city of Paris is comparatively small (with a population of about 2 million) but has a substantial tax base and comprehensive planning and development powers. Mayoral power is relatively (since 1977) recent, with the mayor taking over from state appointed prefects. But the positional power of the mayor is extensive and the city's resources allow it to initiate and achieve strategic projects. In Paris and other communes in the region the authority of the mayor rests on an embedded faith in representative government and the power to respond to citizen demands. Incumbents tend to be re-elected but over time regimes change as the social base and social attitudes change. To speak of a governance culture we therefore need to refer to a strong state but also to the complex overlayering of decision making between scales, to competitive party politics, and in the case of Paris to powerful city government and evolving tensions between governmental actors in the wider city-region. To explore these cultures of governance at work we take two different perspectives. First we examine the rivalry around strategic direction for the city-region and second we look at the style of planning behind the large-scale remodelling of north-east Paris.

Responsibility for regional planning is not clear cut. In 2007 the regional council's draft regional plan was criticized by central government for its failure to identify more opportunities for job creation (Subra 2009). It was this concern about the economic competitiveness of the capital region that led President Sarkozy's new government to put forward plans for wholesale administrative reform creating a new unit of government substantially larger than the city of Paris and incorporating the greater part of the inner core of the region (Lefèvre 2009). A new Minister for the Development of the Capital Region was charged with seeing through reform. The (socialist) city and regional governments opposed the idea. It was clear from the 2008 municipal election results that a new government at this scale might have a socialist majority and subsequently the President became less interested in administrative reform. But the competitiveness of Paris remained an issue and the President invited international teams of architects to draw up visions of regional futures looking ahead to 2050. Schemes were presented at an exhibition 'Le Grand Pari de L'agglomération Parisienne' in the spring of 2009 (Présidence de la République, 2009) and it was at the launch of the exhibition of projects that the President introduced an ambitious scheme of regional development, naming eight development zones and construction of a €35bn new automatic metro that would sweep in a 155km figure of eight around the inner part of the region. The scheme known as the 'Grand 8' provoked further political controversy. The law giving authority to the project was approved at the end of 2009 and set up a development agency Société de Grand Paris (SGP). The regional council and other sub-national actors were included as partners in SGP but control rested with central government. SGP had powers to acquire land and determine land uses for a distance of 400 metres around each of the planned 40 new stations. The agency and not local communes would receive increased tax revenues generated by development around the new stations, and SGP's plans for the station zones, to be agreed with the communes by the autumn of 2011, would override existing local plans. The three aims set out for the Grand 8 were to enhance intra-regional connectivity, exploit important economic assets within the region and to secure Paris's position as a world city, 'Ce projet intègre un objectif de croissance économique afin de soutenir la concurrence des autres métropoles mondiales' (Assemblée Nationale 2010: 1).

There was conflict over the scheme. The elected regional tier has responsibility for transport planning in the Paris region and through the regional transportation agency, STIF (Syndicat des Transports d'Ile-de-France) provides a large share of investment in lines and stock. The regional council had its own preferred rail plan, the proposed orbital 'Arc Express', a less ambitious plan but with a similar aim of improving circulation around the inner suburbs. In the autumn of 2010 both the region's Arc Express and the President's Grand 8 began four months of public scrutiny through the Débat Public process. At over 40 public meetings several thousand participants engaged in the debate about the benefits of the scheme. Conclusions were to be published in 2011 and behind this public process negotiations continued between government, region and transport authorities. The main challenge for SGP was to find the funding for its scheme. In a depressed land market it would be hard to find potential developers for the station sites and guaranteed tax revenue. Business groups in the region had made it clear that this transport infrastructure should be a state investment and opposed the idea of public-private partnership (CESR 2010). The culture of governance is that the state leads. However, within the complex and overlapping responsibilities of governments in Ile-de-France, achieving public leadership is no easy task.

In what Subra (2009) labelled a 'counter offensive' to the President's proposed reform of government in the city-region the mayor of Paris set up Paris-Métropole, an informal alliance of over 170 communes in the region (http://www.parismetropole. fr/, Lefèvre 2009). In Paris-Métropole each commune is an equal partner, leadership is handed on each year, and thus in itself the agency is politically weak but nonetheless able to engage willing neighbouring communes in joint studies and strategy. This new form of cooperation has proved effective in the strategic planning of north-east Paris (see Figure 4.2). The city of Paris cooperates with the intercommunal association Plaine Commune, a group of eight neighbouring communes.

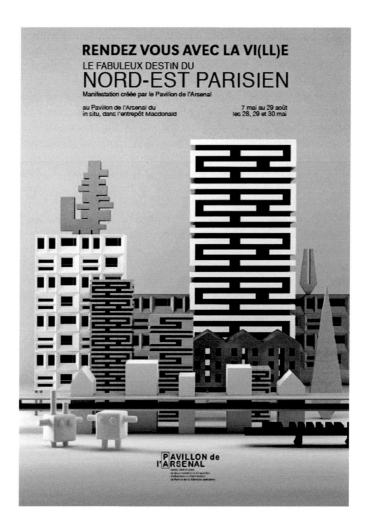

Figure 4.2   Paris presents its 'Nord-Est' projects
Source: The authors.

With the direction of a Paris deputy mayor a joint approach has been developed to urban renewal of old industrial and warehousing zones either side of the peripheral motorway. On the Paris side a €7bn project started in 2008 to develop commercial and residential areas with a range of other uses and integrated with extensions to the tramway network and regional railways (SEMAVIP 2009). The deputy mayor also represents Paris in STIF, the body allocating transport investment. It is through such political coordination that infrastructure can be integrated with ambitious development schemes and priorities shared across administrative boundaries. The development agency undertaking the larger projects within the city of Paris, SEMAVIP, includes some private partners but is 77 per cent owned by the city of Paris. A governance culture is characterized by public sector leadership able to deploy the substantial resources of the state and which defines private sector opportunities. In this case, political coordination makes the overlapping responsibilities of layers of government work.

Tokyo

Although a capitalist economy, Japan has taken a different approach to that of the United States and its alternative model of capitalism is usually termed the Developmental State or 'state development capitalism'. The approach first came to prominence in the 1960s in the light of Japan's dramatically successful post-war economic development – sometimes termed the 'post-war economic miracle'. In analysing the reasons for this the US academic Chalmers Johnson (1982) coined the term 'Developmental State'. The main characteristic of this model is the strong guiding role undertaken by the nation state. (See also Sasaki, Chapter 12, in this volume.) So although there is little government ownership of economic activity there is powerful intervention and regulation with bureaucratic elites, free from political interference, taking a leading role. This gives the state leverage over the private sector. The origin of the approach lies in the post-war crisis of Japan and the desire of the country to rebuild its economy so as to compete in the world markets. The state therefore took control over the long term economic planning of the country and used its power to support national industrial interests. Top priority across government was given to economic growth. A strong alliance grew up between political parties, the bureaucracy and economic interests.

The Japanese economy collapsed in the early 1990s and the following recession led to political confusion and a lack of clear policies. This period was labelled 'the lost decade'. This downturn coincided with the expansion of economic globalization that presented a challenge for the nationalistic and protective approach of the Development State. The period since then has been characterized by attempts to marry the approach with the demands and imperatives of a global competitive market. From 2000 onwards there has been an acceptance by the Japanese national government that there is a need for the Japanese economy to become more open and competitive and there have been various policy initiatives oriented to deregulation, particularly in urban development. However, the essential culture and aim of the developmental state remain important. Some have noted how the developmental state could become the vehicle for a particular approach to globalization and a demonstration that the US model is not the only one (for example Hill and Kim 2000). The argument for a continued role for the state has been strengthened by the 2011 earthquake and tsunami.

If the Development State prioritizes economic growth then how are social needs met? Japan does have a welfare programme, for example a universal medical system, and has less social polarization than most other developed countries. The culture of welfare provision is to pursue the Welfare State principle of social equity but with a low level of government spending. How is this achieved? Esping-Anderson provides one analysis (1997, see also Kim 2010). He concludes that Japan is a special, hybrid, case. Welfare provision occurs through three different systems with the state provision supplemented from two other directions. First, many private companies provide welfare support under the concept of life-time employment. Second, Japan still draws upon its older (Buddhist and Confucian) culture in which the family and local communities play important social welfare roles – for example a high proportion of non-working and retired women fulfil caring functions (OECD 1994). However, this approach is under a lot of pressure (Miyamoto 2003). Globalization has resulted in a less stable economic situation with decreasing life-long employment. Social attitudes towards the role of women are changing and there is a rapid growth in the proportion of elderly people in the population and signs of increasing polarization.

Thus the central state has been seen as essential to orchestrating economic growth and enabling the benefits to be enjoyed by all. The approach has been under stress with economic recession and globalization. However the central state continues to be the central element in the culture of governance, focusing on managing the economy. Other issues have been left to alternative mechanisms or even neglected. The culture of governance in relation to these non-economic issues has been rather different. Some policies, such as on housing and environment have been poorly developed (for details see Sorensen 2002). This particular combination of cultures has led to a strong central state with subordinate regional and city authorities and little public participation. Tensions, however, have arisen as the developmental state adapts to the pressures of globalization. We will now illustrate how this has affected national/city relationships and urban policy in Tokyo. In recent years there has been a gradual move to decentralize power, however this has not prevented tension between national and city governments. National government still has a strong influence on the planning of cities including Tokyo, but since his appointment as Governor of Tokyo in 1999, Ishihara has been challenging this central authority.

In 1990 the Prime Minister proposed that a new location should be found for central government functions. This proposal was made in the context of concerns about the over-concentration of activity in Tokyo and the resulting problems of congestion and high land values. However, Governor Ishihara opposed such a move and a major purpose of his Tokyo Circular Megalopolis Concept (TMG 2001) was to reverse this decentralization policy. The plan took a regional approach and contained policies to demonstrate that the region could work efficiently while still retaining the central government functions in central Tokyo. His view was that agglomeration of activity was necessary for Tokyo to operate as a competitive city. Ishihara used the Circular Megalopolis Concept to lobby central government and as many of the proposals were outside his metropolitan jurisdiction he had to get other local authorities on his side. After initial reluctance they accepted that the success of Tokyo was a necessary factor in their own economic future. The relocation of the Capital is no longer a live issue.

Another tension in the state arrangements can be seen in the growing debates over regional policy. Regional policy for Tokyo is traditionally prepared by national government, setting out major infrastructure and investment policies, and local authorities within the region have to conform to this. However, from 1979 a Capital Regional Summit of local authorities was formed although this was generally restricted to a talking shop. A step forward was taken in 2006 with the establishment of the Capital Region

Forum, which included economic interests, and greater regional co-operation developed. Meanwhile Governor Ishihara was taking a regional leadership role in the development of the Megolopolis Concept Plan and challenging central government's regional policies.

Recently, national level government has taken more interest in regional policy and passed the National Land Sustainability Act in 2005, which led to the new National Land Sustainability Plan. This included a new national plan and 'wide-area regional plans' for certain areas, including Greater Tokyo. These were seen as an important departure from the previous approach as they were responding to the new competitive challenges of globalization. Within this framework national government and the leaders of the local authorities together prepared a new Tokyo Capital Regional Plan in 2009, although consensus politics meant this lacked a strong strategic direction. Nevertheless the acceptance that globalization demands a strategy at the level of the city region has led attempts to resolve the contested issues of inter-scalar politics.

Beijing

By 2011, China had risen to become the second largest economy in the world measured by GNP. It has achieved this with its own economic approach and hence again challenges the hegemony of the US model of capitalism. The Chinese approach has undergone continual modification and adaptation over the last 30 years or so as it gradually and selectively introduces market elements into its system. Before the reforms of 1979 the country was a centrally planned economy dominated by the central state and the Communist Party. The system focused on economic planning and the process of industrialization utilizing work-place units. Such units were the basis for most services such as housing, education, recreation and medical care. The process of transition to a more market approach began after 1979 with the open-door policy – private investment was allowed in certain sectors and in certain geographical areas. A further step was taken in 1987 with the introduction of a partial land market. The selective application of market principles and decentralized authority is intended to allow the growth of a service sector and urban development projects. By 1991 national political leaders were using the term 'socialist market economy' to describe this transitional process. However, the reforms have not been universal and have therefore created a very complex situation in which both traditional socialist elements and reformist elements co-exist although the central state and the Communist Party remain in overall control. Cities have only gradually been given autonomy and there is considerable variation across the country.

Some have highlighted how this variation leads to the co-existence of two cultural approaches (Leaf 2005). One based upon past traditions of neo-Confucianism in which there is a strong fusion between the authority of the state and the role of the family and one based upon the market and the rule of law. Ding (1994) has described the resultant institutional structures as 'amphibious' where 'the lines between public and private, political and personal, formal and informal, official and nonofficial, governmental and market, legal and customary, and between procedural and substantial, are all blurred' (1994: 317). The complexity is also expressed in the existence of a dual market in land. In one market land is sold at market prices usually through tender or auction and involves outside investors and in the second land is administratively allocated, usually at low cost, through negotiation.

Another distinguishing feature of the Chinese approach has been the role of the Chinese diaspora in the process of economic globalization. Crawford (2000) has described this as leading to a distinctive 'Chinese capitalism'. Chinese business throughout the world is family-based, drawing on the principles of Confucianism. Wider networks of trust also develop between such family firms based on regional affinities (*guanxi*). With globalization these networks extend more and more across countries and overseas Chinese business networks are crucial to understanding Southeast Asia's political economy. Since China's 1979 open door policy by far the largest investor into the country has been overseas Chinese companies operating through *guanxi* and trusted local partners. We will illustrate how the complex processes involved in the socialist market economy, the combination of a strong central state and a selected decentralization, and dual cultural elements, have influenced urban policy in Beijing. We take the case of the development of the new central business district (CBD).

Municipal planners have been concerned to create a new CBD to accommodate 'world city' functions and establish Beijing as an 'international city'. The Master Plan for Beijing produced in 1993 (Beijing Institute of Urban Planning and Design 1993) included the aim to create a CBD 'as a first step to compete with Shanghai as the financial capital of China' (Ren 2008: 521). However during the 1980s the move to decentralize political

Figure 4.3   The new Beijing downtown
Source: Hyun Bang Shin.

decision-making to District governments within Beijing gave them the autonomy to develop their own schemes and plans and negotiate with multinational corporations. They also obtained the right to administer the sale of land-use rights and this gave them a major incentive to promote intense development to raise revenues. This led to competition between districts to develop the new CBD function (Gaubatz 2005).

At the centre of Beijing lies the historic Forbidden City and plans had always sought to retain the low-rise nature of this area. The CBD in the Master Plan was therefore just beyond this to the east in Chaoyong District – a location that had historically housed international businesses and embassies (see Figure 4.3). In the early 1990s this area was the location of a controversial project called Oriental Plaza that involved a complex of shopping malls, luxury flats, hotel and offices. It was promoted by the city mayor and financed from Hong Kong. Controversy developed because of the evictions involved, the attempt to bypass planning procedures, the role of outside investment, and a

Figure 4.4   Beijing downtown remodelled

Source: Hyun Bang Shin.

81

corruption scandal over the bribes received by the mayor from the main Hong Kong investor (Broudehoux 2004). The mayor was sentenced to 16 years in jail. Nevertheless Oriental Plaza was eventually opened in 1999 as the 'new address' of Beijing's CBD and the largest shopping centre in Asia (Cook 2006). In the same year, 1999, a former executive of the China Construction Bank was appointed by the central government as the new city mayor with the aim 'to build a financial district that can lure multinational firms to Beijing' (Ren 2008: 522). Many internationally-oriented high-rise offices and residential developments have subsequently been built in this area, and it was the location for new hotel developments for the 2008 Olympics.

Meanwhile on the other side of the Forbidden City, the District of Xicheng was promoting its own centre called 'Financial Street', even though this lies within the low-rise zone and contravenes the official Beijing Master Plan (Gaubatz 2005) (see Figure 4.4). The District was seeking to attract foreign firms to this area on the back of the Bank of China decision to construct its new high-rise headquarters there. In the plan this area was allocated for state-owned financial institutions and ministries but in the 1990s the district initiated a major redevelopment project relocating nearly 5,000 families and building about 3 million square metres of office space (Gaubatz 2005). The development of this alternative CBD stimulated other districts to get in on the act and, slightly further out towards the northwest, Haidian District launched its 'Financial Corridor' in 2002. Thus the political fragmentation resulting from the autonomy of local districts has led to a lack of clarity in planning a location for the new world city functions. Small local authorities without skilled staff have been able to enter into partnerships with large-scale foreign firms with experience of world city projects in other cities. Their plans tend to be used in a symbolic fashion to present a vision that can convince external investors. The application of the new approach in Beijing is proving difficult (Zhao 2010) because of the continuation of past planning traditions, the speed of urban growth and the complexities of the dual market. There are a large number of actors – both from the different levels of government and the different economic interests.

Planning and Governance Cultures

Economic competitiveness drives strategic planning in the four cities that were the focus of this chapter. But whilst concern about economic globalization may be shared we see distinctive responses due to the continuing influence of political traditions and the complex relationships between market/state/civil society and levels of government, which frame strategic planning approaches. The struggle to develop city-wide or city-regional strategy appears to be particularly acute. PlaNYC has great ambition but limited institutional support. There are competing versions of the next Paris region. In both Tokyo and Beijing national and sub-national governments compete to plan at the larger scale. We can see the desire to get the right scale of economic space as a response to economic globalization, but it is equally clear that the effectiveness of cities is firmly embedded in governance cultures. There is change, particularly as different scales of government interact with each other and with markets and civil society, but we have also shown the *longue durée* of governance traditions.

References

Abu-Lughod, J.L. 1999. *New York, Chicago, Los Angeles: America's Global Cities.* Minneapolis, MN: University of Minnesota Press.

Almond, G. and Verba, S. 1963. *The Civic Culture.* Princeton NJ: Princeton University Press.

Assemblée Nationale. 2010. Loi n.2010-597 du 3 juin 2010 relative au Grand Paris (1). [Online]. Available at: http://www.legifrance.gouv.fr/affichTexte.do?cidTexte=JORFT EXT000022308227&dateTexte=#LEGIARTI000022309428 [accessed 4 April 2013].

Bagli, C. 2009. As finance offices empty, developers rethink ground zero. *New York Times*, April 15. [Online]. Available at: http://www.nytimes.com/2009/04/15/ nyregion/15develop.html [accessed 4 April 2013].

Beijing Institute of Urban Planning and Design. 1993. *Beijing Urban Master Plan (1991–2010).* Beijing: BIUPD.

Bowles, J. 2003. *Slow Down on Far West Midtown.* [Online]. Available at: http://www. nycfuture.org/content/articles/article_view.cfm?article_id=1068&article_type=5 [accessed 4 April 2013].

Broudehoux, A-M. 2004. *The Making and Selling of Post-Mao Beijing.* London: Routledge.

CEC. 2010. *Territorial Cohesion: What Scales of Policy Intervention?.* Follow up of Green Paper on Territorial Cohesion 2nd TCUM session. Brussels, 12 March 2010. [Online]. Available at: http://ec.europa.eu/regional_policy/conferences/territorial/12032010/ index_en.cfm [accessed 10 February 2012].

CESR (Conseil Economique et Social Régional). 2010. *Quelles perspectives pour le Partenariat Public–Privé (PPP) et autres nouveaux modes de financement pour les investissements de la région Ile-De-France?* [Online]. Available at: http://www.cesr-ile-de-france. fr/documents/rapport_pdf/rapport/09_fina_ppp/rapport-comment-financer-investissements-region-ile-franc.pdf [accessed 4 April 2013].

Coates, D. 1999. Models of capitalism in the new world order: the UK case. *Political Studies*, 47(4), 643–60.

Cook, I. 2006. Beijing as an 'internationalized metropolis', in *Globalization and the Chinese City*, edited by F. Wu. London and New York: Routledge, 63–84.

Crawford, D. 2000. Chinese capitalism: cultures, the Southeast Asian region and economic globalization. *Third World Quarterly*, 21(1), 69–86.

Ding, X.L. 1994. Institutional amphibiousness and the transition from Communism: the case of China. *British Journal of Political Science*, 24(1), 293–318.

Esping-Andersen, G. 1990. *Three Worlds of Welfare Capitalism.* Cambridge: Polity Press.

——. 1997. Hybrid or unique? The Japanese welfare state between Europe and America. *Journal of European Social Policy*, 7(3), 179–89.

Faludi, A. (ed.) 2007. *Territorial Cohesion and the European Model of Society.* Cambridge, MA: Lincoln Institute.

Fernandez, M. 2009. As city adds housing for poor, market subtracts it. *New York Times*, 15 October. [Online]. Available at: http://www.nytimes.com/2009/10/15/ nyregion/15housing.html?pagewanted=all [accessed 4 April 2013].

Gaubatz, P. 2005. Globalization and the development of new central business districts in Beijing, Shanghai and Guangzhou, in *Restructuring the Chinese City*, edited by J.C. Ma and F. Wu. London and New York: Routledge, 87–108.

Gladstone, D. and Fainstein, S. 2001. Tourism in US global cities: a comparison of New York and Los Angeles. *Journal of Urban Affairs*, 23(1), 23–40.

Gross, J.S. and Newman, P. 2005. *Local Development and Political Opportunity in London and New York*. ISA Research Committee 21 Conference, Paris, June.

Hampden-Turner, C. and Trompenaars, F. 1993. *The Seven Cultures of Capitalism*. London: Piatkus.

Hill, R.C. and Kim, J.W. 2000. Global cities and developmental states: New York, Tokyo and Seoul. *Urban Studies*, 37(12), 2167–95.

Hutton, W. 1995. *The State We're In*. London: Cape.

Indergaard, M. 2011. Another Washington-New York consensus? Progressives back in contention. *Environment and Planning A*, 43(2), 286–306.

Johnson, C. 1982. *MITI and the Japanese Miracle*. Stanford, CA: Stanford University Press.

Kantor, P. 2002. Terrorism and governability in New York City: old problem, new dilemma. *Urban Affairs Review*, 38(1), 120–27.

—— . 2009. Globalisation and governance in the New York region: managed pluralism. *Progress in Planning*, 73(1), 34–8.

Kim, P.H. 2010. The East Asian welfare state debate and surrogate social policy: an exploratory study on Japan and South Korea. *Socio-Economic Review*, 8(3), 411–35.

Leaf, M. 2005. Modernity confronts tradition: the professional planner and local corporatism in the rebuilding of China's cities, in *Comparative Planning Cultures*, edited by B. Sanyal. New York and London: Routledge, 91–112.

Lefèvre, C. 2009. *Gouverner les métropoles*. Paris: Lextenso Editions.

Miyamoto, T. 2003. Dynamics of the Japanese welfare state in comparative perspective. *The Japanese Journal of Social Security Policy*, 2(2), 12–24.

Mollenkopf, J. 2008. School is out: the case of New York City. *Urban Affairs Review*, 44(2), 239–65.

New York City Department of Housing Preservation and Development (NYCDHPD). 2003. *The New Housing Market Place: Creating Housing for the Next Generation 2004–13*. New York: City of New York.

Newman, P. and Thornley, A. 2011. *Planning World Cities*. 2nd Edition. London: Palgrave.

OECD. 1994. *New Directions in Social Policy*. Paris: OECD.

Powell, M. and Roberts, J. 2009. Minorities affected most as New York foreclosures rise. *New York Times*, 15 May. [Online]. Available at: http://www.nytimes.com/2009/05/16/nyregion/16foreclose.html [accessed 4 April 2013].

Présidence de la République. 2009. *Le grand Paris. Dossier de presse*. Cité de l'Architecture et patrimoine Paris, 29 April.

Ren, X. 2008. Architecture as branding: mega project development in Beijing. *Built Environment*, 34(4), 517–31.

SEMAVIP. 2009. *SEMAVIP: For the City*. Paris: Société d'économie mixte d'aménagement de la ville de Paris.

Sorensen, A. 2002. *The Making of Urban Japan*. London and New York: Routledge.

Subra, P. 2009. *Le grand Paris*. Paris: Armand Colin.

Thurow, L. 1992. *Head to Head: The Coming Economic Battle among Japan, Europe and America*. New York: William Morrow.

TMG. 2001. *Tokyo Megalopolis Concept*. Tokyo: TMG.

Zhao, P. 2010. Implementation of the metropolitan growth management in the transition era: evidence from Beijing. *Planning Practice and Research*, 25(1), 77–93.

PART 2
Planning and Its Dimensions

Preface to Part 2

Greg Young

In the contributions to this Part the authors paint a vivid picture of resilient and 'wicked' issues such as the global North–South divide, social gender-blindness and the more recent and perplexing conundrums presented to planning by neo-liberal governmentality and insurgent politics in the United States, challenging not only the rationale of communicative theory but also perhaps democratic communication itself. The authors also develop a picture of the varied shades or degrees of the so-called 'dark side' of the *chiaroscuro* of the planning discipline. While the authors' diagnoses of the relevant issues and problems that constitute these dilemmas are varied they all in effect call for versions of what Vanessa Watson describes in her chapter as an 'enlargement of thinking'. Although these enlargements of thought are perceived variously in theoretical, analytical, therapeutic and political terms, they are construed as possible ways forward. In spite of this there is a collective mood in the chapters in which progressive solutions face the ongoing challenges of being outflanked and outmanoeuvred by factors such as the global rationalities of power, the power of privileged 'rationalities' globally or, in the case of James Throgmorton writing from the United States, the anger of an insurgent Tea Party Movement that has now firmly placed planning itself in its cross-hairs.

At a global level Vanessa Watson evidences the growing body of research challenging the hegemony of Northern epistemologies and universalized Northern theory as well as the transfer of predominantly Northern planning ideas and practices in the face of the recognition of the power and importance of local context. The role of the 'cultural turn' in its planning manifestation has been to highlight the importance and implications of local cultural variation, however, the relativism this implies for Watson raises puzzling issues of balance and judgement between competing claims, world views, cultures, identities and rationalities leading to an identified 'trap of relativism'. The solution to this conundrum is to be found through an enlargement of thought based on considering multiple perspectives and contexts such as may be engaged through cross-global comparative research offering the possibility to counter Northern predominance. Finally, she notes that there is no 'philosophical tradition' that would in principle support the escalating crime, violence and social breakdown set in motion by urban development trends producing urban marginalization and inequality.

On a significant historical scale from the late nineteenth century up until current times, Clara Greed, from a feminist perspective, traces the enduringly gendered nature of planning in the British context. This experience ranges from the birth of the modern town planning movement, the period of post-war reconstruction, neo-Marxism in the

1970s, neo-liberalism in the 1980s, the environmentalism of the 1990s and in this century contemporary planning for equality and diversity in the post-secular city. She singles out the hidden history of women in planning and the built environment professions and documents the imprint of gender relations on planning policy, city form and the built environment. However, regardless of the planning trend in fashion, the gender blindness she identifies that operates in planning now extends to a similar limitation in regard to faith, with many planners in a post-secular age needing to accommodate the value of the cultures of others and their life experiences although these may be very different to their own. She argues that if planners are to meet the needs of the majority of society in contexts such as that of the United Kingdom, they will need to address the needs of women, ethnic minorities and religious groups.

James Throgmorton, in writing from the United States, describes in his chapter a set of circumstances where the repressed needs and ambitions of a populist conservative group, 'The Tea Party', seeks to influence and co-opt national policymaking which it feels has ignored their needs and realities. The tactics employed by the Tea Party call into question the 'collaborative rationality' of the communicative turn that impacted on planning theory from the 1990s and leads Throgmorton to speculate on the role of planning theory in such a context. The originator of the theory of communicative action, Jürgen Habermas, who grew up in the shadow of the German Third Reich, developed his theory in part from Freud's psychoanalytical concept of distorted communication. Throgmorton can see that there is in the Tea Party context no genuine desire to communicate – whether in distorted or other terms. Yet while the conservative insurgency of the 'Tea Party moment' offers a seemingly deracinated discourse, Throgmorton looks sympathetically at the complex origins of the movement and in response counsels the right for groups and citizens to express anger and to be listened to. He adds to this therapeutic wisdom the recognition that planning is deeply political and needs to facilitate democratic deliberation.

In the final chapter in this Part Glen Searle scrutinizes development values from the late 1980s up until today in major harbour-side redevelopments in Sydney, Australia that confront wider community and planning values and objectives. He develops an innovative theoretical construct comprising a hierarchical framework to consider the influences of planning discourses, planning doctrines and professional planning habituses in conflicted planning contexts. 'Searle's triad' is an original approach that offers an analytical lens capable of deconstructing and illuminating the fluctuating specifics of individual redevelopment schemes.

In the case of each chapter whether from Greed, Throgmorton, Watson or Searle, planning or specific forms of planning are found wanting and inadequate to cope with the enlargement of thinking deemed by the authors in various senses to be a requirement of a more democratic, responsible and effective planning. This enlargement of thinking includes a desirable political, ethical, narrative and analytical sophistication that is able to counter the broad range of threats seeking to outflank and outmanoeuvre planning.

A Feminist Perspective on Planning Cultures: Tacit Gendered Assumptions in a Taciturn Profession

Clara Greed

Introduction: What is the Problem?

The Changing Nature of Planning

Urban planning in the United Kingdom is underpinned by nation-wide legislation, with apparently fixed rules as to what is right and wrong. But 'planning' is an elastic field of public policy, characterized by paradigm shifts and breathtaking reversals. Planning has variously focused on land-use zoning, regional economic issues, social policy and equality issues. Possibly most significant of all has been the shift to the 'green' environmental planning agenda that currently dominates. Whilst these swings appear extreme, from a feminist perspective the development of urban planning appears to have been gender-blind, retaining a tacit, substantive focus on predominantly male concerns. Gender may be defined as the cultural role ascribed to women and men in society, as against the biological sex to which they are born. Feminism describes a view of society and knowledge that is based upon the principle that men's views and interests dominate and unequal gender relations result which disadvantage women. It was in this context that the 'women and planning' movement developed from the 1970s onwards as more women entered town planning courses. It sought to change the design and planning of the built environment so that the 'different' needs of women are recognized and planned for (Booth, Darke and Yeandle 1996, Greed 1994a, Matrix 1984, Panelli 2004, Reeves 2005). In this chapter, I examine the history of urban planning from a feminist perspective beginning first by defining planning theory, culture and the planning subculture. Then I will go through the phases of planning theory and practice, drawing out key factors that belie the nature of the planning subculture and 'types' of planner in each stage (Greed and Johnson forthcoming 2013).

Definitions and uses of planning theory and culture

Whilst the definition of theory is important, its social and cultural function is probably more significant in terms of increased professional status and intellectual legitimacy. 'Theory' derives from the Greek, 'theoreo' which means 'I see' and refers to observation, speculation and conjecture (Feyerabend 1966). Within the secular, positivist scientific ethos of the modern Western world 'theory' has been exalted to 'fact' and this positioning has given many a planning theory greater veracity than it might warrant. How planners 'see' the world is a key theme throughout the chapter.

There are only two main types of planning theory (Greed 2000: 236, Taylor 1998). First, there are those that have been developed by planners either to facilitate better planning or to justify what the current generation wants planning to be, and within this category there are 'normative' theories that shape thinking about what planning should be and ought to do. Second, there are 'theories of planning', developed mainly by sociologists, which provide a critical perspective on planning (Faludi 1973). Planners have sometimes appropriated and twisted urban sociologists' theories to justify their actions, and to legitimate their position because they are 'planning for the working class'. Possessing some suitably difficult, esoteric and arcane theories is particularly important because of increasing credentialization, which is the process of limiting access to specialist professional enclaves by means of increasing the educational entrance requirements (Collins 1979: 90–91, Millerson 1964). Paradoxically the rush for academic recognition has resulted in educational inflation, so there is increased pressure on people to have a degree and belong to a profession (Greed 1991: 30). Morgan (2010: 38) notes that 'people have to go to university to get a decent job, even though most jobs do not require specialist training in higher education'.

Raymond Williams defines culture as appertaining to a whole way of life, including the material, intellectual and spiritual (Williams 1981). Culture is the *zeitgeist*, or spirit of the age, which is accepted as 'normal' and 'obvious' – 'everywhere and nowhere like the sky' (Barthes 1973). Whilst planning generally mirrors the values of the national culture, the planning profession possesses its own particular subculture, norms and ways of doing things (Greed 1994a, 2005a). At certain times 'planning' has been the epitome of the spirit of the age, for example, as the *modus operandi* of the post-Second World War reconstruction period in which it was widely accepted that 'The New Britain must be planned'. But at other times planning appears to be out of step with mainstream culture (Ogburn 1922). But the planning culture can also appear to be radical, even revolutionary. Alonso (1963: 824), for instance, notes the urban planning profession (like most adolescents perhaps) likes to revolt, to strike a pose, and to rapidly adopt and discard heroes (Greed 1994a: 30). The profession appears to have its own heroes and 'ideal types' that men can aspire to copy, follow and (subsequently) discard (Bologh 1990).

To gain some detachment and perspective on what has been going on, it is important to look behind the 'front' that the planning profession projects to the general public (Goffman 1969, Greed 1994b). I draw upon my previous research on the surveying profession and its internecine tribes (Greed 1991: 28) and upon the planners themselves

(Greed 1994a). Looking through the eyes of women (as 'other') helps to 'make the familiar strange' (Greed 1994b: 14, Delamont 1985). Planning has a number of 'blind spots' not least ignorance of women's needs although women (52 per cent of the population) are the majority of 'the planned'. 'Planning is for people' (Broady 1968) but frequently it ignores this female preponderance.

It is helpful to see the world of planning as a subculture, which is taken to mean that there are cultural traits, beliefs and lifestyles peculiar to planning. The values and attitudes of the subculture have a major influence on professional decision-making, and thus influence the nature of urban development. The need for identification with the values of the subculture blocks out the entrance of both people and alternative ideas that do not 'fit in'. The concept of 'closure' as discussed by Parkin (1979: 89–90), was developed by Weber (1964: 141–52, 236) and describes how subgroups protect their status and territory through gatekeeping, withholding of essential information, and other exclusionary mechanisms. Closure is worked out on a day to day basis at the interpersonal level, with some people being made to feel unwelcome, bullied and 'wrong'; and others being welcomed into the subculture, made to feel comfortable and encouraged to progress to decision-making levels. Gender, social class, ethnicity, disability and other so-called 'minority factors' determine who is 'suitable' to enter the planning tribe. All these processes contribute towards 'the reproduction over space of social relations' (Massey 1984: 16), especially the imprint of gender relations on the nature of planning policy, city form and the built environment.

Individual planners may not be consciously aware of the values that they subscribe to as it is tacit (unspoken) 'habitas', what is normal and obvious for them. (See also Searle and O'Connor, Chapters 8 and 10, this volume). There has been a lack of reference until relatively recently to women's needs in planning, presumably because the male is the *de facto* 'norm'. In this respect, planning was similar to other male-dominated professions, such as seafaring, where 'women' are still such an oddity that men either treat them with exaggerated courtesy or extreme hostility (Kitada 2010). If asked about 'women' the men are likely to say, 'we don't discriminate we treat everyone the same' – which is perhaps the worst kind of discrimination.

The Phases of Planning Theory

The Development of Modern Town Planning

'Modern' planning developed in Britain in the nineteenth century to deal with the effects of industrialization, urbanization, population growth, including of disease, poverty and the slum housing inhabited by the new working classes. (See also Ward, Chapter 2, this volume.) Reaction and reform was characterized by the efforts of private sector philanthropic factory owners (Greed 2000) who employed new types of architect-planners, master planners and urban designers, capable of planning entire model communities. Visionaries, utopianists and reformers, with no formal 'planning' education, all had their own ideas on how cities should be and 'every reading man has a draft of a new community in his waistcoat pocket' (Bryson 1992: 28, quoting Emerson's comment to Carlyle of 1840, popularized by Oliver Wendell Holmes' collected volume of Emerson's pithy sayings (Holmes 1892: 125)).

But where were the women in a male-dominated world of industrialists, city fathers and architectural 'geniuses'? Women had been very active in the early Garden City movement and housing reform but they were written out of the history of planning which concentrated on the great men of the movement. Although Ebenezer Howard (1898) was elevated as the Grandfather of Modern Planning (Greed 1994a: 92–5), there were many other men and women involved who were forgotten. Early plans included nurseries, communal feeding facilities and even kitchenless houses to reform 'domestic labour'. There were also many women, particularly in North America, who established their own model towns and who were architects in their own right (Hayden 1981, Pearson 1988). Educated women actively campaigned to get women's needs and 'different' perspectives incorporated into housing design, model town development, urban planning policy, architecture and welfare reform (Greed 1994a).

Early Twentieth Century

Town Planning became a profession in 1913 (Greed 1994a: 108), in that practitioners were required to take qualifying examinations in order to belong to the newly established Town Planning Institute. But formalization of qualifications and entry requirements restricted women's entry. (See also Ward, Chapter 2, this volume.) Indeed women were not permitted to be members of any of the built environment professions until the 1919 Sex (Disqualification) Removal Act. Nor were women considered 'the right type' to enter all-male gentlemen's clubs and coteries. Rather their work was likely to be despised as the voluntary efforts of mere 'Lady Bountifuls'. Nevertheless, a certain Ebenezer Howard's first wife, Elizabeth (who died in 1907), was one of the first members of the Town and Country Planning Association (established in 1903) and many other women contributed to the cause (Hayden 1981). Planning became institutionalized within the municipal government structures but few women achieved senior positions, as a marriage bar operated throughout public sector employment in Britain. Women have never quite caught up and even today women constitute less than 10 per cent of those in senior positions in the built environment professions (CIC 2009). Likewise women were unable to exert influence on the development of planning policy and theory within academia, as they were not admitted to most universities in Britain until after the Second World War. Those women who did get to college were rapidly corralled into separate academic spheres or were awarded diplomas rather than degrees (Jarvis, Kantor and Cloke 2009: 56–8). City research was split between 'masculine' abstract urban science and 'feminine' practical social work solutions. So inevitably the men created the grand theories of urban sociology and planning, excluding a woman's perspective. Women, at best, were likely to end up as research assistants or professors' wives.

Rudimentary town planning acts were introduced in 1909 and 1919 which required the production of a 'scheme' that is a plan for each local authority area showing the land-use zonings. Patrick Geddes promoted a scientific approach to planning, based on the mantra of survey, analysis, plan and the collection of detailed statistics. (See also Bianchini, Chapter 22, this volume.) He also promoted the need for 'order' for separation of land-uses in the name of functionality. Geddes' ideal rational city was divided into three zones: home, work and play (Geddes 1915). There is no place in this tripartite arrangement for a fourth factor, namely all the caring, home-making, childcare undertaken by women in the home, which does not count as either 'work' or 'play'. In Geddes' view 'the home' was a haven for men to relax after work, whereas for women it was the major locus of their work. Whilst zoning was 'justified' in the name of public health, the founding fathers of planning used it either directly or indirectly to negate women's contribution to society, to control women's place in the city and to prevent her from transgressing spatial and moral protocols and venturing into the public realm of the city of man (Lees 2004). Geddes and many other great thinkers of the time were supportive of eugenics and were keen to control the breeding instincts of the lower classes by tackling overcrowding (and assumed immorality) by reducing housing densities. Geddes saw women as inferior and in need of control and in this respect he was enamoured by Freud's association of the mother principle with 'stagnation' (Bologh 1990: 14, Geddes and Thomson 1889). Such attitudes towards women as 'dirty' undoubtedly shaped his approach to land-use zoning (Greed 1994a, Matless 1992). He was also strongly opposed to women's suffrage (Kent 1987: 35).

An investigation of the origins of the word 'zoning' reveals the murky occult roots of this apparently scientific planning principle and its association with the control of women's place in urban space by keeping them out of public life and civic space and within the domestic realm of house and home (Greed 1994a: 70–81). According to Boulding (1992: 227), in Ancient Greek the word 'zona' conveys the idea of a belt and by inference to the restriction of the 'loins' that is the control of sexuality and the production of children (the fruit of the loins), and thus is linked to the control of women's sexuality and related temptations and moral pollution. Marilyn French (1992: 76) points out the Hebrew for 'prostitute' – 'zonah' (harlot) – means 'she who goes out of doors', that is, she is in the wrong place and has lost her respectability. In Latin, 'zonam solvere' means 'to lose the virgin zone', that is to get married or lose one's virginity or, in medieval parlance, to remove the chastity belt. It is only a short etymological leap from zona, to zana, to sana, to sanitary, and to the obsession with sanitation and hygiene in the modern secular city. Social hygiene appeared to be more about controlling women than actual disease, especially the dirtiness (sexuality) of working-class women.

Zoning and separation principles were also promoted by European planning theorists, such as Le Play, who wrote of *lieu, travail* and *famille* (place, work and family) (Greed 1994a). Le Corbusier drew his inspiration from ancient occult and Masonic sources, in seeking to codify and control space and women's place both within the city of man and in the layout of the home (Birkstead 2009). Although Le Corbusier stated that a house is a machine for living in neither he nor his fellow members of the Modern Movement had much idea how ordinary women used their 'machine' and the problems they encountered in the city of man (Greed 1994a: 121–4).

The archaeology of the past weighs heavily upon the modern planning profession (Foucault 1972). What passed as scientific principles in the new secular town planning profession were replete with deep-seated religious beliefs about the place of women in society and the city (Greed 1994a: 116–17). Dualistic, Gnostic influences were at work in planning which stressed the philosophical concept of there being a split between spirit and body, between pure and impure (Douglas 1966). One of the characteristics of religion is to make a clear division between sacred and profane, between right and wrong, to impose order on the world and give meaning to existence (Eliade 1959)

A reality is created in which 'man' is at the centre and 'right' and woman is usually wrong, impure and 'other', This is presumably because of women's association with the inferior realms of body, nature, sexuality, and thus mortality, sin and filth. Women were seen as dragging men down earthwards from the higher, purer, spiritual realms (Jagger 1983). The trinity of mind, body and soul promoted by Descartes harks back to Aristotle's division of knowing, doing and feeling, and strongly parallels the home, work, play zoning mentality of the planners (Boardman 1978, Doxiadis, 1968: 22, Greed 1994a: 117).

Post-War Reconstruction

Modern town planning in Britain developed from these murky beginnings into a major arm of government after the Second World War as the nation undertook reconstruction and sought, literally, to 'Build a Better Britain' (Abercrombie 1945). Emphasis was put upon the importance of designating new zones specifically for (male) employment which were often decentralized in new industrial estates on the outskirts of the town. Separation of work and home might have been ideal for the male breadwinner as it enabled him to escape the confines and stagnation of the domestic realm every day, but it was not necessarily very practical for women. New Towns increased women's isolation in the domestic zone as each was divided into neighbourhood units where all the necessities of life would be provided without any need to stray further. It was assumed women were happy to be housewives who were supported by their breadwinner husbands (Wilson 1980). Growing suburbanization increased the sense of confinement and limited horizons – 'the problem without a name' (Greed 1994a). Even when well-intentioned men undertook urban sociological study they brought with them assumptions about women's place in society. Young and Willmott's study of the East End of London (1957) 'naturally' assumes that women residents were primarily 'Mums' and care-givers with no life or work of their own out of the home (Rose 1993).

The Gendered Inheritance of City Form

Research has demonstrated that women suffer disadvantage within a built environment that is developed by men, primarily for other men (Greed 2011a, 2005a, 2005b, 1994a). Women's concerns are not limited to matters of house and home, including childcare although it is very important with 68 per cent of women and 78 per cent of men of working age in employment (ONS 2010). Women use and experience the built environment differently from men, and therefore have distinct needs and expectations in terms of the nature of urban structure and planning policy. As well as being workers, women are more likely to be the ones responsible for childcare, shopping and a range of other caring roles, all of which generate different usage of urban space. Fewer women than men have access to a car and so they comprise the majority of public transport users in many areas, and their journey patterns are different from those of men. A woman's daily journeys might be as follows; home > school > work > shops > school > and back home again. Thus women tend to trip-chain their journeys, and often undertake complicated, intermittent, lateral journeys, rather than radial journeys straight to and from the city centre. Such journeys are often undertaken outside the rush hour if they work part-time, and by public transport or on foot (Uteng and Cresswell 2008). As an alternative solution, women planners have recommended the creation of 'the city of everyday life', of short distances, multiple centres, mixed land-uses and adequate social facilities (Eurofem 1998).

The Sixties and Systems Planning

Planning education in Britain underwent a period of expansion (albeit of mainly male students) in the 1960s. There was a need to make planning more intellectually respectable by increasing the academic and theoretical basis of the subject – ideally one that would embody the 'white heat of technology' as favoured by Harold Wilson's Labour government of the time which put great emphasis upon state planning. Many planners were enamoured by the new urban planning ideas coming out of North America and new computer-based methodologies, especially the systems view of planning was very enticing (McLoughlin 1969). The city was seen as a system in which everything was linked to everything else, and the land-uses, roads and development were seen as the spatial end-product of the economic, social and political 'aspatial' forces at work in the city (Foley 1964: 37). Planners could control and guide the city 'like the helmsman of a ship' by analysing data using the new computer technology. Although a systems approach to planning ostensibly was about feeding 'everything' into the computer, in reality it worked much better on quantitative data, and especially on counting and predicting traffic flows, and appeared completely 'peopleless' and very male in its priorities. I first encountered systems planning as a young planning student in the late 1960s and when I questioned its quantitative bias I was informed by my tutor, 'If you can't count it, it doesn't count' (that is 'if you can't measure it, you can't manage it'). When I naively asked, 'but what about women? a woman's work is never done', I was told by the male lecturer, 'don't be stupid, we're not talking about that!' Systems theory, and related methodologies, have subsequently been much criticized for being a-political, socially unaware and dominated by the male obsession with the motor car. Retail Gravity Models purported to predict shopping demand according to time and distance to shopping centres by car. But most shoppers were women, and bus timetables, public toilets, accessibility and child-friendly environments were far more important (Greed 1994a).

The Seventies and Neo Marxism

In the 1970s, planners increasingly were becoming frustrated in their role of advisors especially to local councillors, and because they lacked any direct power. As a result many planners embraced Marxist urban theory which paradoxically was extremely anti-spatial and yet offered greater power to the planners. According to Marx the economic material base of society determines its legal and political superstructure (Taylor 1998: 104–7). The built environment was merely an outward manifestation of this superstructure, rather like the icing on top of a fruit cake (Greed 2000: 229–30). Therefore trying to improve society by changing the spatial nature of cities was like rearranging the deckchairs on the Titanic. There was a need to change the economic basis of society, namely the capitalist system, and so there was no point in doing anything until that was achieved – 'after the Revolution'. So, all the ongoing urban problems, design criteria, housing overcrowding and development pressures that had previously been the focus of planners' work strangely faded into the background. To achieve revolution it was necessary for an elite intellectual vanguard (the planners in this instance) to mobilize the proletariat. In Britain, the 'worker' was generally 'seen' to be male, Northern, working in heavy industry or the mines, redolent with homoerotic, Soviet images of the noble, heroic worker. In contrast women were portrayed as selfish, lazy, bourgeois housewives who did not work at all, or did a little bit of work for 'pin money', and were only fit to make the tea whilst the men ran the revolution (Hartmann 1981). It took a very long time for feminists to convince both academia and policy makers that women were workers too, and that their domestic labour created the workforce in the first place (McDowell 1991). Like all radical theories, neo-Marxism served the purpose of providing some within the profession with the theoretical tools to rebel against their elders and eventually to take over.

1980s and Non-Planning or Some-Planning

By the 1980s, the pendulum had swung again and neo-Marxism had gone out of fashion. In Britain, the Conservative Party led by Prime Minister Margaret Thatcher came to power and as a result neoliberalism and market economics became hegemonic. Planning was seen as slowing down the system and preventing property development from taking place. The planning process was to be made market-led and many local government planning departments were cut back, whilst entrepreneurial, bright, young graduates increasingly entered the private sector of planning consultancy and property development.

This was a period in which women began to make inroads in erstwhile male-dominated professions and management structures, bringing a new perspective on planning policy and priorities. But did more women mean better planning? As more women had entered planning schools from the early 1970s, a feminist view of planning became more widespread and the 'women and planning' movement was born. Much of the foundational literature was North American (Hayden 1984, 1981), but gradually British women planners rediscovered their heritage and developed their own analysis and policy recommendations (Booth, Darke and Yeandle 1996, Greed 1994a, Little 1994). There was a short period when 'women and planning' peaked in the early 1990s, and many reports, books, academic articles and conference papers were produced (Greed 2005b). But few men attended 'women and planning' conferences and the cause was not strong enough to create a paradigm shift. Indeed, women's issues were sidelined, first by neo-Marxism, then by market-led planning and subsequently by environmentalism. The dominant planning culture remained unchanged. Attempts to gain planning permission for developments that met women's needs such as for child care provision, public toilets, disability provision and community facilities were frequently refused. They were seen as *ultra vires*, that is 'not a land use matter'; they were 'social' matters outside the remit of urban planning (Greed 2005a, 2005b, 1994a). In contrast, planning proposals for playing fields mainly used for men's sports are seen as 'physical land-use matters' and not as social uses. The Sports Council and football organizations are listed as official consultees in the planning process, and such 'public' open space is shown on all statutory plans.

In the 1980s 'planning theory' and critical discourse were not in fashion but there was a resurgence of more traditional, spatial planning concerns. The urban conservation movement grew in strength in Britain as people became concerned about the loss of historic architectural heritage. There was a desire to create a better-designed townscape and to apply human-scale design principles to residential areas (Prince of Wales 1989). A range of strange alliances were formed across the political spectrum to question the nature of planning, including those concerned with community planning, crime and design, accessibility, women and planning, vernacular architecture, urban design and cultural heritage.

1990s and Environmentalism

The planning discourse was being reshaped as new forces were at work at an international level that were to dominate 'planning' for many years to come. Concern was expressed about the future of the planet, global warming and the possible destruction of planet earth by its own people. The United Nation's Rio Declaration (UN 1992) required all signatory member-states to integrate environmental controls and policies within their planning systems. Emphasis was placed upon sustainability – with leaving the planet in good shape, with adequate natural resources, for future generations. Sustainability originally had three dimensions: environmental sustainability, economic viability and social equality – the three 'Ps' of planet, prosperity, people (Bruntland 1987). When these principles reached the UK planning system, given the cultural values of the planning tribe, greatest emphasis was given to the physical environmental aspects, and little emphasis was given to either the economic or social equity dimensions, let alone to the relationship between gender and sustainability. In some European countries, especially in Scandinavia, sustainability policy has been more strongly linked with equality considerations (Skjerve 1993).

Environmentalism and 'green issues' became the new 'gospel of redemption' for planners. It provided the impetus for a shift in attitudes away from planning for the motorcar, to restrictions on car use. At first this seemed to be against the planning culture's traditional values, as the tribe always promoted the use of the car and tore down entire city centres to accommodate it. The policies have been reversed but the level of power wielded by planners remains unchanged. Prioritizing environmental concerns about individual carbon footprints without linking this to the social, economic and spatial realities encountered by ordinary people leads to spatial oppression (Greed 2011b). Many ordinary suburban dwellers in Britain lack access to reliable bus routes or local railway stations (Beeching 1963) and so have to use their car regardless of the condemnation of the green planners (Uteng and Cresswell 2008).

Environmental planning, like systems planning, is better at dealing with quantifiable data, using EIA (Environmental Impact Assessment) methods for evaluating new development proposals. But little has been done about SIA (Social Impact Assessment) – in the UK at least – in spite of the existence of a range of qualitative methodologies (Greed 2005b, RTPI 2003). Likewise scant attention is given to women's 'different' travel needs and thus to the challenges the sustainable city present to women, such as the very tight time-budgets, and the complex trip-chaining patterns, discussed above, that women have in order to carry out all their home and work tasks. But if one dares to ask 'what about women?' they are likely to be told that 'we've done women; you should be concerned with the environment'.

Planning for Equality and Diversity in the New Millennium

The Labour Government under Tony Blair (1997–2007) and Gordon Brown (2007–10) put considerable emphasis upon social inclusion, diversity and the community, and so planning appeared 'softer' and socially aware. In spite of its references to 'joined up thinking', gender considerations were not meshed into environmental policy and were marginalized. The equalities and diversity movement appeared good in that it raised the importance of 'people issues', but gender was low on the pecking order, beneath ethnicity and sexuality. Trying to include all the different diversity issues became increasingly complex and difficult to administer. Equalities monitoring techniques simply required planners to fill up a tick-box form to confirm they had taken a whole list of equality issues into account with reference to specific policy outcomes (37 in some London boroughs) (Greed 2005b). Much of this monitoring appeared 'disembodied' and 'space less'. It did not relate to the impact of specific planning policies on different categories of people within society. Whilst 'gender' had been ignored in the past, it was now outflanked by a mass of other competing equality issues.

Planning theory tried to encompass this all-inclusive approach by stressing collaborative and communicative approaches to planning which downplayed the differences and potential conflicts and varying levels of power amongst different members of the community (Healey 1997, see also Throgmorton, Chapter 6, this volume). Remarkably 'the working class', for whose supposed benefit planning ostensibly existed, the great talisman of planners' success, was seldom mentioned in this new conciliatory urban social theory and was not included at all in the 2010 Equality Act.

Meanwhile in academia there were attempts to develop the theory of intersectionality theory (Bagilhole 2009), which sought to take into account the fact that different diversity characteristics overlap and interact within an individual's being, whilst stressing that some issues, such as gender are likely to be more overarching in their importance than others. The diversity agenda of post-structuralist sociology has become so complicated, abstract, philosophical and relativistic, that few understand it and even 'gender' is now questioned as a valid category (Jones 2011).

The Twenty-First Century and the Post-Secular City

The widening diversity agenda has allowed for a greater range of social and cultural issues to be considered, including religion, which is of particular concern to women who comprise the majority of many congregations and faith groups (Greed 2011b). By the end of the twentieth century, according to some society was entering a post-secular phase (AlSayyad and Massoumi 2010, Baker and Beaumont 2011, Gorringe 2002), manifested in renewed interest in spiritual matters and a concern for faith issues in civil society and the city (Beaumont 2008) after the predominance of a secular, humanistic culture during the twentieth century (Cox 1965). This is a new phase of society that contains many paradoxes. Post-secularism is characterized by a contest between religious lobbyists (including new fundamentalists) and secular pressure groups who promote diversity and equality (Baines 2009). Nevertheless, post-secularism has allowed some urban theorists to ask a new set of questions about the city, related to social justice, morality and whether the city is 'good' (Amin 2006). Planners have always prided themselves on being neutral and not there to make moral judgments (Fewings 2009, Howe 1994, Sandercock 1997). When women have complained that granting planning permission for 'lap dancing' clubs and 24-hour bars may destabilize a neighbourhood and create personal safety issues for women wishing to walk through the area unmolested, they are more likely to be dismissed as 'religious fundamentalists' or men haters (Greed 2011b, Lewisham 2008, WDS 2009).

Planners' lack of religious awareness puts large sectors of the minority population at a disadvantage (Engwicht 2010, Sandercock 2006, see also Watson, Chapter 1, this volume). Whilst there has been a decline in religious affiliation and church attendance within the traditional (white) denominations across European Christendom, there has been a growth in church attendance in the developing world especially in Africa and South America. As a result of increased immigration and globalization (Fenster 2004) some of the largest churches in the United Kingdom comprise Pentecostal congregations which draw much of their membership from ethnic minority groups. Women play a prominent role both as the majority of these congregations and in pastoral roles and in community development (Onuoha and Greed 2003). Many such faith communities have had great difficulty obtaining planning permission for church building (CAG Consultants 2008). Kingsway International Christian Centre (KICC), which is a mega-church with over 12,000 members of 46 different nationalities, and is one of the largest churches in Western Europe, has been the subject of a long-running planning saga because it failed to get planning permission for a large new church building on an industrial estate in Dagenham, East London. Its previous premises in Hackney were requisitioned as part of the 2012 Olympic Games site development and KICC leaders imagined the planners would deal favourably with their application. But they chose not to do so (DCLG 2009). Perhaps both God and gender (faith and feminism) share the misfortune of not being recognized as valid land-use issues within the UK planning system (Greed 2011b). Much needs to change, not least the planners' understanding of the value of others' cultures, and life experiences so different from their own. In the post-secular age planners need to understand the importance of faith to many sectors of the community today. Otherwise they will remain unaware and unable to meet the needs of the majority of society, including women, ethnic minority and religious groups within society.

References

Abercrombie, P. 1945. *The Greater London Development Plan*. London: HMSO.
Alonso, W. 1963. Cities and city planners. *Daedalus*, 92(4), 824–939.

AlSayyad, N. and Massoumi, M. 2010. *The Fundamentalist City? Religiosity and the Making of Space*. London: Routledge.

Amin, A. 2006. The good city. *Urban Studies*, 43(5–6), 109–232.

Bagilhole, B. 2009. *Understanding Equal Opportunities and Diversity: The Social Differentiations and Intersections of Inequality*. Bristol: Policy Press.

Baines, B. 2009. Must feminists identify as secular citizens? Lessons from Ontario, in *Gender Equality: Dimensions of Women's Equal Citizenship*, edited by L.C. McCain and J.L. Grossman. Cambridge: Cambridge University Press, 83–106.

Baker, C. and Beaumont, J. (eds) 2011. *Post-Secular Cities: Space, Theory and Practice*. London: Continuum.

Barthes, R. 1973. *Mythologies*. London: Paladin.

Beaumont, J. 2008. Introduction: faith-based organisations and urban social issues. *Urban Studies*, 45(10), 2019–34.

Beeching, R. 1963. *The Beeching Report: The Reshaping of British Railways*. London: HMSO.

Birkstead, J.K. 2009. *Le Corbusier and the Occult*. London: MIT Press.

Boardman, P. 1978. *The World of Patrick Geddes*. London: Routledge.

Bologh, R. 1990. *Love or Greatness: Max Weber and Masculine Thinking: A Feminist Inquiry*. London: Unwin Hyman.

Booth, C., Darke, J. and Yeandle, S. (eds) 1996. *Changing Places: Women's Lives in the City*. London: Paul Chapman.

Boulding, E. 1992. *The Underside of History*. Volume I. London: Sage.

Broady, M. 1968. *Planning for People*. London: NCSS/Bedford Square Press.

Bruntland Report. 1987. *Our Common Future: World Commission on Environment and Development*. Oxford: Oxford University Press.

Bryson, V. 1992. *Feminist Political Thought: An Introduction*. London: Macmillan.

CAG Consultants. 2008. *Responding to the Needs of Faith Communities: Places of Worship: Final Report*. London: CAG, Cooperative Advisory Group Planning Consultants.

CIC. 2009. *Gathering and Reviewing Data on Diversity within the Construction Professions*. London: Construction Industry Council (produced at the University of the West of England, Bristol by A. De Graft-Johnson, R. Sara, F. Gleed and N. Brjlak).

Collins, R. 1979. *The Credential Society*. New York: Academic Press.

Cox, H. 1965. *The Secular City: Secularisation and Urbanisation in Theological Retrospect*. Harmondsworth: Penguin.

Delamont, S. 1985. Fighting familiarity. *Strategies of Qualitative Research in Education*. Warwick: ESRC Summer School.

Department for Communities and Local Government (DCLG). 2009. *Appeal by Kingsway International Christian Centre and the London Development Agency*. Ref: U0006.7/LBGH, London: Department for Communities and Local Government.

Douglas, M. 2002. *Purity and Danger: An Analysis of the Concepts of Pollution and Taboo*. London: Routledge.

Doxiadis, C. 1968. *Ekistics: An Introduction to the Science of Human Settlements*. London: Hutchinson.

Eliade, M. 1959. *The Sacred and Profane: The Nature of Religion*. New York: Harvest, Brace and World.

Engwicht, D. 2010. *The Connection between Religion and Urban Planning*. [Online, Research paper]. Available at: http://www.creative-communities.com/wp-content/uploads/downloads/2010/08/Religion.pdf [accessed 2007].

Eurofem. 1998. *Gender and Human Settlements: Conference Report on Local and Regional Sustainable Human Development from a Gender Perspective*. Hameenlina: Eurofem.

Faludi, M. 1973. *A Reader in Planning Theory*. Oxford: Pergamon.

Fenster, T. 2004. *The Global City and the Holy City: Narratives in Knowledge, Planning and Diversity*. Harlow: Pearson.

Fewings, P. 2009. *Ethics for the Built Environment*. London: Taylor and Francis.

Feyerabend, K. 1966. *Langenscheidt's Pocket Greek Dictionary*. London: Hodder and Stoughton.

Foley, D. 1964. An approach to urban metropolitan structure, in *Explorations into Urban Structure*, edited by M. Webber et al. Philadelphia, PA: University of Pennsylvania Press, 1–71.

Foucault, M. 1972. *The Archaeology of Knowledge*. New York: Pantheon.

French, M. 1992. *The War Against Women*. London: Hamish Hamilton.

Geddes, P. 1915. *Cities in Evolution: An Introduction to the Town Planning Movement and to the Study of Civics*. London: Architectural Press.

Geddes, P. and Thomson, J.A. 1889. *The Evolution of Sex*. London: Scott.

Goffman, E. 1969. *Presentation of Self in Everyday Life*. Harmondsworth: Penguin.

Gorringe, T.J. 2002. *A Theology of the Built Environment: Justice, Empowerment and Redemption*. Cambridge: Cambridge University Press.

Greed, C. 1991. *Surveying Sisters: Women in a Traditional Male Profession*. London: Routledge.

—— . 1994a. *Women and Planning: Creating Gendered Realities*. London: Routledge.

—— . 1994b. The place of ethnography in planning. *Planning Practice and Research*, 9(2): 119–27.

—— . 2000. *Introducing Planning*. London: Continuum.

—— . 2005a. Overcoming the factors inhibiting the mainstreaming of gender into spatial planning policy in the United Kingdom. *Urban Studies*, 42(4), 1–31.

—— . 2005b. An investigation into the effectiveness of gender mainstreaming as a means of integrating the needs of women and men into spatial planning in the United Kingdom. *Progress in Planning*, 64(4), 239–321.

—— . 2011a. Planning for sustainable urban areas or everyday life and inclusion? *Proceedings of the ICE – Urban Design and Planning*, 164(2), 107–19.

—— . 2011b. A feminist critique of the post secular city: god and gender, in *Post-Secular Cities: Space, Theory and Practice*, edited by C. Baker and J. Beaumont. London: Continuum, 104–19.

Greed, C. and Johnson, D. 2013 (forthcoming). *Planning in the United Kingdom: An Introduction*. London: Palgrave Macmillan.

Hartmann, H. 1981. The unhappy marriage of Marxism and feminism: towards a more progressive union, in *Women and Revolution*, edited by L. Sargent. London: Pluto, 1– 42.

Hayden, D. 1981. *The Grand Domestic Revolution: Feminist Designs for Homes, Neighbourhoods and Cities*. Cambridge, MA: MIT Press.

—— . 1984. *Redesigning the American Dream*. London: Norton.

Healey, P. 1997. *Collaborative Planning: Shaping Places in Fragmented Societies*. London: Macmillan.

Holmes, O.W. 1892. *Ralph Waldo Emerson*. New York: Services Corporation.

Howard, E. 1898. *Garden Cities of Tomorrow*. London: Faber and Faber.

Howe, E. 1994. *Acting on Ethics in City Planning*. Rutgers, NJ: Center for Urban Research.

Jagger, A. 1983. *Feminist Politics and Human Nature*. Brighton: Harvester.

Jarvis, H., Kantor, P. and Cloke, J. 2009. *Cities and Gender*. London: Routledge.

Jones, R. 2011. *Irigaray: Towards a Sexuate Philosophy*. Bristol: Polity.

Kent, S.K. 1987. *Sex and Suffrage in Britain 1860–1924*. London: Routledge.

Kitada, M. 2010. *Women Seafarers and Their Identities*, unpublished PhD thesis. Cardiff: University of Cardiff.

Lees, L. 2004. *The Emancipatory City?: Paradoxes and Possibilities*. London: Sage.

Lewisham. 2008. *Report from Planning Committee (A) on Elizabeth Industrial Estate Change of Use of Ground Floor*. Case File DE/237/C/TP, DCS Number 100-054-444. London: London Borough of Lewisham. Available at: www.planningresource.co.uk [accessed 10 June 2012].

Little, J. 1994. *Gender, Planning and the Policy Process*. Oxford: Elsevier.

Massey, D. 1984. *Spatial Divisions of Labour: Social Structures and the Geography of Production*. London: Macmillan.

Matless, D. 1992. Regional surveys and local knowledges: the geographical imagination of Britain, 1918–39. *Transactions*, 17(4), 464–80.

Matrix. 1984. *Making Space, Women and the Man Made Environment*. London: Pluto.

McDowell, L. 1991. Restructuring production and reproduction: some theoretical and empirical issues relating to gender, or women in Britain, in *Urban Life in Transition*, edited by M. Gottdiener and C. Pickvance. London: Sage, 77–105.

McLoughlin, J.B. 1969. *Urban and Regional Planning: A Systems Approach*. London: Faber.

Millerson, G. 1964. *The Qualifying Associations*. London: Routledge and Kegan Paul.

Morgan, J. 2010. Appetite for education. *Times Higher Education Supplement*, December: 32–3.

Office for National Statistics (ONS). 2010. *Social Trends*. London: Office of National Statistics, London. [Online]. Available at: http://www.statistics.gov.uk/statbase/product.asp?vlnk=13675 [accessed 10 June 2012].

Ogburn, W. 1922. *Social Change: With Respect to Cultural and Original Nature*. Oxford: Delta Books.

Onuoha, C. and Greed, C. 2003. *Racial Discrimination in Local Planning Authority Development Control Procedures in London Boroughs*. Occasional Paper 15. Bristol: Faculty of the Built Environment, University of the West of England, Bristol.

Panelli, R. 2004. *Social Geographies: From Difference to Action*. London: Sage.

Parkin, F. 1979. *Marxism and Class Theory: A Bourgeois Critique*. London: Tavistock.

Pearson, L. 1988. *The Architectural and Social History of Cooperative Living*. London: Macmillan.

Prince of Wales. 1989. *A Vision of Britain*. London: Doubleday.

Reeves, D. 2005. *Planning for Diversity: Planning and Policy in a World of Difference*. London: Routledge.

Rose, G. 1993. *Feminism and Geography: The Limits of Geographical Knowledge*. Cambridge: Polity Press.

Royal Town Planning Institute (RTPI). 2003. *The Gender Mainstreaming Toolkit*. London: Royal Town Planning Institute.

Sandercock, L. 1997. *Towards Cosmopolis: Planning for Post-secular Cities*. London: Wiley.

——. 2006. Spirituality and the urban professions: the paradox at the heart of urban planning. *Planning Theory and Practice*, 7(1), 69–75 (Interface section).

Skjerve, R. (ed.) 1993. *Manual for Alternative Municipal Planning*. Oslo: Ministry of the Environment.

Taylor, N. 1998. *Urban Planning Theory since 1945*. London: Sage.

United Nations (UN). 1992. *The Rio Declaration: United Conference on the Environment at Rio De Janiero: Conference Proceedings*. New York: United Nations.

Uteng, T. and Cresswell, T. 2008. *Gendered Mobilities*. Aldershot: Ashgate.

Weber, M. 1964. *The Theory of Social and Economic Organisation (Wirtschaft und Gesellschaft)*. New York: Free Press.

Williams, R. 1981. *The Sociology of Culture*. Chicago, IL: University of Chicago Press.

Wilson, E. 1980. *Only Half Way to Paradise*. London: Tavistock.

Women's Design Service (WDS) 2009. *GenderSite*. [Online]. Available at: www.gendersite.org searchable database on 'women and planning' [accessed 10 June 2012].

Young, M. and Willmott, P. 1957. *Family and Kinship in East London*. Harmondsworth: Penguin.

What Can Planning Theory Be Now? Storytelling and Community Identity in a Tea Party Moment

James A. Throgmorton

Earlier chapters in this edited collection have discussed shifts in planning and planning theory over time. As those chapters have revealed, planning theory took a turn in the 1990s, which – although challenged by rational-technical scientists and proponents of a Foucauldian conception of power – resulted in interactive-communicative planning becoming preeminent.

The communicative turn is rooted in discursive democracy, and it presumes the goodness and effectiveness of 'collaborative rationality'. Consequently, it has stressed the value of interpretation, argumentation, negotiation, collaboration, consensus-building, reflective and deliberative practice, and attending to the contexts in which practice takes place.

Events since the middle of 2008 have, however, witnessed the rise of a 'Tea Party' movement that sharply challenges most of these values. Thinking of themselves as 'real Americans' who want to 'take America back', a variety of people began objecting to proposals emanating from the new Obama administration. In the summer of 2009 they displayed outrage during 'town hall' meetings that elected members of Congress held concerning proposed national health care reforms. Soon the public discourse was full of talk about how 'Obamacare' sought to take away our freedoms and to 'pull the plug on Grandma'. At the same time, accusations were hurled that efforts to slow the rate of global climate change were part of a radical environmentalist agenda. So too did fears of immigration by 'illegal aliens' cause Arizona and other states to adopt extraordinarily restrictive and possibly unconstitutional new laws. By November 2010 this new Tea Party movement had succeeded in influencing enough voters to give control of the US House of Representatives and many states to Republicans who had embraced the Tea Party's core ideas.

How can planning theory adjust to this new context, where real dialogue appears profoundly difficult to accomplish or facilitate? What can planning theory be now?

A Brief Overview of the Communicative Turn in Planning

Thirty or more years ago, a relatively small set of scholars in urban and regional planning began advocating a 'communicative turn' in planning theory and practice. Although they drew upon an eclectic array of scholarly sources, all their work was rooted in a careful study of planning practice. By the mid-1990s these scholars had succeeded in shifting the attention of most planning theorists, if not practitioners, away from older ideas rooted in technical rationality (Innes 1995). Additional research over the past decade and a half has provided further justification for making this turn.[1]

These scholars understood the communicative turn to be part of a post-positive, post-modernist or post-structural intellectual wave that swept across many disciplines and fields in the 1980s. Instead of presuming a radical separation between knowledge and emotions, and between science and politics, they argued that these spheres are deeply intertwined. For them, planning should not be defined as a purely technical process through which elite experts find the best way to achieve pre-determined objectives. They argued (and demonstrated) instead that planning is best conceived as a collaborative process in which values, knowledge and action are all co-produced through the interaction of diverse actors. Consequently communicative planning theorists focused on the social dynamics of practice in the context of particular places, institutional structures and processes of governance. For most of these theorists, the communicative turn also offered a way to include a broader range of stakeholders (including traditionally marginalized groups) into planning processes.

John Forester has played an especially important role in developing these ideas. In *Planning in the Face of Power* (1989) he argued that it's better to think of planning as 'attention-shaping, communicative action' rather than as 'instrumental action' which seeks to achieve particular ends. He claimed that planning has a strongly ethical component as well: to help create the possibility of democratic argumentation 'free from domination'. And he suggested a number of actions that planners could take to complement their technical work and foster more genuine political participation. In *The Deliberative Practitioner* (1999), he argued that planners are 'reflective practitioners' (who learn from one another through 'practice stories') and 'deliberative practitioners' (who learn through engagement with others). In his view these practice stories 'do work by organizing attention, practically and politically, not only to the facts at hand but to why the facts at hand matter' (1999: 29). The 'messiness' of such stories also 'teaches us that before problems are solved, they have to be constructed', and that the rationality of problem solving depends 'on the prior practical rationality of attending to what "the problem" really is' (1999: 37) and avoiding 'the rush to interpretation'.

---

[1]   Key contributions to this turn include: Healey (2010, 2009, 1997), Innes and Booher (2010), Harper and Stein (2006), Throgmorton (1996) and precursor articles, Verma (1996), Mandelbaum, Mazza and Burchell (1995), Forester (1993), Fischer and Forester (1993), Forester (1989) and precursor articles, Hoch (1984), Schön (1983) and Friedmann (1973).

By the mid-1990s the communicative turn had largely been accomplished, at least among planning theorists in the West. This induced a round of intellectual, practice-based and political critiques from Hillier (2007), Yiftachel (2006), Fainstein (2005), Sandercock (2003), Flyvbjerg (2002) and other planning scholars (Healey 2012). In Healey's view, these critiques have enabled proponents of the communicative turn to sharpen their ideas about collaborative practices. The proponents' ideas have continued to attract attention, moreover, because they help practitioners deal with the uncertainty, complexity and conflict that characterize the contemporary context.

Persuasive Argumentation and Storytelling as Part of the Communicative Turn

My own contribution to the communicative turn has been to emphasize the importance of rhetoric (persuasive argumentation directed at an audience) and persuasive storytelling about the future. This emphasis emerged from a dialogue between my work in the practical world and my engagement with other scholars. The origins of this dialogue go back at least to the 1970s when I worked, first, for a local air pollution control agency implementing the 1970 Clean Air Act, and, second, with a private consulting firm doing highly rational-technical research for the US Environmental Protection Agency.

In what follows I will offer a synthesis of my contributions. In brief, this synthesis will lead me to conclude that national policy-making and the places in which ordinary people live have both become far more complex than non-experts can understand.

The Complexity of National Policy Implementation

Scientists frequently express dismay about the gap between what they know needs to be done (for example, with regard to health care or global climate change) and what actually gets accomplished in the political arena. They often respond to this gap by emphasizing the need to educate the public. Proponents of the communicative turn argue instead that it is a mistake to construe public communication merely as a means for disseminating knowledge because doing so provides no space for meaningful input from diverse citizens who lack *technical* expertise or have alternate frameworks of understanding, and it forecloses consideration of other important value questions.

The scientists' dismay makes sense only if one assumes a sharp divide between science and politics; that is, between facts and values. For major public problems, however, especially 'wicked problems' (Rittel and Weber 1973) that involve national policy implementation over a relatively long period of time, science and politics interact 'all the way down' – that is, to states, localities and the level of 'public talk'. In other words, policy-making continues during implementation.

To explain, let me briefly recall some research I conducted 30 years ago when the United States was experiencing the 'energy crisis' (Throgmorton 1987, 1984, 1983).

Several interrelated factors had created the possibility that the electric power industry might change significantly. Wanting to know in what direction the industry was most likely to change, I focused on public policy toward the interconnection of electric utilities with 'independent power producers'. Within the context of mid-1970s efforts to analyse and solve the energy crisis and the historical evolution of the electric power industry as a regulated natural monopoly, I devised a 'national policy implementation' framework to discern how federal regulation was changing the industry.

Drawing upon the policy implementation literature available at that moment, and focusing on national policies implemented through a complex intergovernmental system, this framework articulated a series of expectations, including the following six. First, participants in the policy-making process would interpret the meaning of the new policy in diverse ways. Second, although the initial policy proposal and responses to it would be supported by technical plans and policy analyses, formal policy makers in Congress would adopt legislation that was 'good enough' rather than try to rationally calculate all likely consequences of all major policy options. Third, bargaining and negotiating among Congressional policy-makers would distort the original theoretical coherence of a policy initiative, and the resulting law or policy would be vague on several crucial points. Fourth, Congress would delegate rulemaking to an agency in the executive branch of the federal government, which in turn would delegate further implementation to state agencies that had their own sources of funding and authority. Given considerable discretionary power, and influenced by factors that varied from state to state, these state agencies would produce more detailed policies that likewise varied considerably. Fifth, unexpected (or unanticipated) events would occur during the process of implementation, and therefore cause participants to reconsider the meaning and necessity of the new policy. And sixth, parties that felt harmed by policy decisions would appeal those decisions to state and federal courts. Shaped more by political interaction than by rational analysis, policy outcomes would differ significantly from ones intended by the policy's initiators (new groups might, for example, be formed in response to the new policy's implementation), and the implementation process would take considerably longer than initially expected.

In brief, I found very strong support for this framework. I also observed a profound clash between the people who devised policies, laws and regulations, and the ordinary people who experienced consequences on the ground. For the latter, the effects of the new legislation came 'like a bolt of lightning' out of a clear blue sky (Throgmorton 1987: 359). The new legislation emerged from a process they neither understood nor could influence.

The Importance of Persuasive Argumentation and Storytelling

For major public problems, especially 'wicked' ones that involve national policy implementation over a relatively long period of time, the interrelated activities of *persuasive argumentation* and *storytelling* play crucial roles.

Scientists, policy analysts, planners and other practitioners help form and implement national policies, but they can do so in a variety of ways. Wherever they act, they find themselves embedded in a complex rhetorical situation created by the interaction of three primary audiences (scientists, politicians and laypeople), each of which has its own normal discourse and agreed-upon conventions of persuasion. Given this rhetorical situation, they find themselves enacting one of several possible roles: scientist, politician and lay advocate, plus three others formed through their interaction: the policy analyst, the advocacy planner and the political entrepreneur. In the centre of it all, conceptually at least, stands the active mediator (Throgmorton 1996).

As John Forester emphasized, a crucial aspect of persuasive argumentation is that these practitioners have to construct 'the problem' before they can articulate and analyse alternative solutions. This activity is deeply interpretative, rhetorical and constitutive.

It is *interpretative* because practitioners have to translate the messiness of ordinary life into 'problems' that make sense and can be acted upon. It is *rhetorical* in the sense that it involves persuasive argumentation directed at specific audiences under specific conditions. This kind of argumentation involves a social process of utterance, reply and counter-reply, which can be highly emotional. It takes place, moreover, not just at the federal or Congressional level but 'all the way down'. To argue persuasively, therefore, skilled practitioners have to pay attention to contextual features, especially 'the oral, the particular, the local, and the timely' (Toulmin 1990: 186). They must take their audience into account, be aware of differing or opposing views, and – since the meaning of an utterance always goes beyond the conscious control of the practitioner – think about how audiences construct the meanings of utterances (Throgmorton 1996). Lastly, how one defines a problem and articulates possible solutions is also *constitutive*; one's rhetoric has the power to include or exclude stakeholders who can influence action. Consequently, successful participation in processes of argumentation also entails skilful relationship building, negotiating and so on.

The complexity of process and multiplicity of roles combine to produce considerable uncertainty and confusion. Whereas practitioners working within the confines of a national policy implementation process might frame 'the problem' in technical ways, and have detailed ideas about how the process should be carried out, ordinary people would interpret it differently and find the whole process to be mightily opaque. Consequently, if one looks carefully at problem framing 'all the way down', one finds that storytelling plays a role that is far from trivial.

First, when something happens, people tell stories about it. The stories they tell are unavoidably selective and purposeful, with the purposes being tightly connected to the teller's emotions. Believing that policies should be based on facts and good science, scientifically-inclined practitioners tend to dismiss such stories and their telling. Stories, however, enable people to make sense out of facts (whether true or alleged), reframe how they think about events, and hence decide how they should respond to them (Simons 2006). When people share stories with one another, moreover, they build a sense of community and culture.

Where do such stories come from? In Hannah Arendt's (1958) view they emerge to an important degree from action itself. As she puts it: '[i]t is in the nature of beginning that something new is started which cannot be expected from whatever may have happened before' (1958: 157). But action is riddled with frustrations, primarily because action and speech take place within a 'web of relationships'. Every action, therefore, stimulates a chain reaction and every process becomes the cause of new processes (1958: 168).

In this context it is important to emphasize that stories (for example Obama's 'socialist' health care policy will force us to 'pull the plug on Grandma'), which help people make sense out of particular facts and events, circulate and interact through *webs of relationships*. These webs involve both face-to-face interactions (which are

109

deeply influenced by the spatial distribution of people by race, class and other key socioeconomic markers) and virtual interactions via communication technologies and media.

Second, although storytelling is ubiquitous, many tales coalesce into and express locally-grounded versions of 'common urban narratives' (Finnegan 1998). For people who live in American cities, these narratives include what I have elsewhere called the Founders' Tale, the City as Nightmare, the City of Oppression, the City of Boiling Frogs, the Modernist Planners' Tale, the City of Ghosts, the Immigrants' Tale, and Sex and the Creative City (Throgmorton 2007, 2005). These narratives are in turn related to the diverse ways in which people feel connected to places, a point that I will return to later in this section.

Third, stories (or case studies) can also be crafted about an unfolding sequence of historical events pertaining to a topic or a place. Consequently the flow of utterances, replies and counter-replies included in any argumentative process can be emplotted as a flow of action and hence as a narrative (Throgmorton 1996). Here it is important to distinguish between listing facts chronologically and weaving facts together into a potentially persuasive story. The latter explicitly seeks to persuade a target audience to adopt preferred beliefs and actions, and its persuasiveness largely depends on the author's skill at the storytelling craft.

When applied to cities, city-regions and nations, persuasive storytelling about the future can be used to convey a vision of how entire places can be transformed in a preferred direction. This involves crafting texts that seek to turn the flow of action in the preferred direction through the use of particular tropes by particular characters at particular times and places. Success in altering the flow requires skilful attention to literary factors such as point of view, texts, plots, characters, conflict resolution, settings and tropes (Throgmorton 1996, 1992). It also requires awareness of differing or opposing views, and hence understanding that plans and analyses can be interpreted in diverse and often conflicting ways.

A crucial question arises at this point: persuasive to whom? One can construct a story that is designed to persuade a limited set of powerful actors. Conversely one can, by weaving diverse tales together, craft a future-oriented story that enables diverse people to 'make sense together' and hence to imagine jointly and create sustainable places. This kind of story would be persuasive to, and hence perceived as trustworthy by, a wide range of readers (Throgmorton 2003). It might also stimulate the diverse residents of a place to ask: Who are *we*? How do we want to live with one another? Whose story, what culture, what sense of community and what collective identity does our planning help sustain?

Ideally all these observations suggest the need for policymaking to be an interactive collaborative process. Storytelling can play a significant role in such processes (Innes and Booher 2010, LeBaron 2002). But not all potential actors are willing to engage with others collaboratively. Instead they often strive to exclude or marginalize other actors. Moreover, policy-making at the national scale might include interactive collaborative processes, but that does not necessarily mean that the collaborations will extend 'all the way down'. In this context, planners and analysts need to recognize the power of cable television news and radio talk show hosts to frame issues and shape how their audiences think. In terms of planning theory, this is a woefully understudied force.

In every type of storytelling it can often be difficult to discern the difference between bald-faced lying and weighing/configuring facts in ways that are consistent with one's purposes and interests. The difference matters. People make mistakes all the time when telling stories. Upon hearing such a mistake, a listener might quickly condemn the teller's story as a lie. Even when a storyteller gets the facts wrong, however, it can be wise to acknowledge the *rationality of emotions* (Nussbaum 1990) and to look for the *emotional truth* and *telling errors* contained in the teller's story (Eckstein 2003, Portelli 1991). This emotional truth can often be found by listening carefully for ways in which people misremember past events and tell the story to serve their purposes; that is, by listening carefully for a story's 'telling errors'.

Lastly, crafting a story is one thing, but telling it is another. In the end it is the *telling* that really matters. Why? Because the interaction between the teller and the listener has the *potential* to exert a powerful democratic force. When analysing storytelling, therefore, it is important to ask questions such as: Who is speaking to whom? What kind of person does the teller's story invite the listener to be? What kind of community does it seek to create, both among its listeners and between the listener and the storyteller? To achieve its democratic potential, therefore, persuasive storytelling must also make space for the diverse ways in which other storytellers have already woven facts into their stories, and hence to craft a more capacious tale that is more likely to persuade a larger number of people.

Storytelling and the Construction of Place

In the previous section I noted that locally-grounded common urban narratives are related to the diverse ways in which people feel connected to places. *Place* can loosely be defined as a space that people have made meaningful – that is, a 'space invested with meaning in the context of power' (Cresswell 2004: 12). It has a geographic location, constitutes a material setting for social relations and evokes a 'sense of place' among its diverse users.

According to Doreen Massey (2005), geography scholars have identified two competing ways of conceptualizing *space*. One is a world of separate bounded places; it is a surface containing fixed, closed entities easily represented on a map, each of which has its own essentialized identity. The other is a world of flows, 'a depthless horizontality of immediate connections' (Massey 2005: 76) enabled by transportation and communication technologies. Massey rejects this dualism and argues that space can better be understood as 'a simultaneity of stories-so-far' (2005: 9) or 'plurality of trajectories' (2005: 12) in which 'space unfolds as interaction' (2005: 61). *Places* can in turn more precisely be understood as the unfolding of 'a power-geometry of intersecting trajectories' (2005: 64).

To comprehend how such an unfolding occurs, one might begin by evoking the idea of *home*. The house, neighbourhood and city in which one lives has no personal meaning until one begins inhabiting it, developing an emotional attachment to it and transforming it into a home. Thinking of a place as home often generates stories that elaborate on what I've termed 'the Founders Tale'. Such nostalgic tales typically focus on the families, institutions and buildings that have long been familiar parts of the place. At their best such tales can help people inhabit places with care, affection and a sense of belonging.

But they can also be profoundly exclusionary. For instance, they often omit the people who live in the place for a few years and then move on. They disregard the multiple ways in which the place is (and has always been) connected with the external world. And they tend to gloss over the aspects of the place's history that old-timers would like to forget. At the extreme, such stories presume that residents of the place share a common identity authentically rooted in history. In brief, such stories essentially claim, 'this is our place' and around here 'we' have always done things this way.

According to this 'essentialist' way of thinking about 'platial' identity, authentic places are under threat from a variety of forces, especially 'dangerous outsiders', mobile workers and tourists, and the homogenizing tendencies of global capital. In response to this threat, some people celebrate the place's unique features and traditions as an act of resistance, whereas others strive to exclude or marginalize unwanted newcomers. Still others don't see a problem; they advertise the place's unique qualities as a way of attracting new visitors and investment. Contrary to the essentialist view, therefore, one can argue that the meaning of a place is never finished but always becoming and always being performed. Such performances are, however, always constrained by structures (material landscapes, laws, rules, cultural and social expectations) that users did not create.

In addition to thinking of a place as home, one can feel connected to a place in at least four other ways. The five dimensions combine to form complex places and senses of place connection or, in Massey's terms, sites for the meeting up of 'intersecting trajectories'.

First, all places are, to one degree or another, embedded in complex technosystems, environmental pathways and intergovernmental linkages that tie distant places to one another (Buell 2001, Throgmorton 2005). Consequently our homes and neighbourhoods are tied to other locales via 'tentacular radiations' or 'paths out of town'. At the risk of belabouring the obvious, large urbanized areas such as the ones in which a large majority of Americans live could not exist without having the ability to import key goods, services and resources, and to export market products and contaminants.

Second, places have histories and are constantly changing. These changes superimpose upon the visible surface an unseen layer of usage, memory and significance, and hence places can be saturated with the histories of previous inhabitants and the events that have occurred in specific locations. If one has lived in a place for a long time, one is likely to be acutely conscious of that unseen layer of usage and memory. Even newcomers will gradually become aware of the extent to which powerful emotions such as joy, anger, love, fear and hope, circulate through the place via stories that people tell. As these emotionally-resonant stories circulate, they help construct a psychogeography of place (Coverly 2006).

A third additional type of connection derives from the fact that people are constantly moving into or departing from places. Thus any one place contains its residents' accumulated or composite memories of all places that have been significant to them over time. Having moved into one's place from somewhere else, one brings memories of those other places and the pathways leading away from them.

Fourth, fictive or virtual places can also matter. In some cases, as with architectural renderings of a possible development, such imagined worlds can have direct and immediate effects on the physical features of a place. But even novels, poetry, sculpture and scientifically-grounded projections and scenarios can exert a powerful influence on people both consciously and subconsciously and thereby affect expectations, hopes, fears and choices about how people invest their time, energy and resources in the here and now.

To sum up: the extraordinary complexity of national policy implementation, the variety of rhetorics practitioners use, the ubiquity and power of storytelling, and the increasing complexity of places and place connection have created a situation that leaves many ordinary people feeling completely bewildered and often outraged. The situation has become ripe for trouble in ways that we planning theorists have not yet fully grasped.

Enter the Tea Party.

The Rise of the 'Tea Party' Movement

In the months following Barack Obama's election in November, 2008, a loose coalition of people who think of themselves as 'real Americans' began forming a 'Tea Party' movement that sought to 'take America back'. Two years later this movement had succeeded in transforming public discourse and the political direction of the country. One might be tempted to condemn these Tea Partiers as neo-fascist ethno-nationalists, but that is not how they characterize themselves. Who are they? And how does their movement relate to the communicative turn in planning?

According to Kate Zernike (2010), the Tea Party movement emerged out of a powerful emotional response to the financial meltdown of 2008, the federal government's 'bailout' of large financial firms and automobile manufacturers, and the subsequent 'Great Recession'. Large numbers of people felt afraid of what lay in store for them, angry at those who had caused the meltdown and bailed out the big firms, and betrayed by mainstream institutions. Consequently, the Tea Party movement emerged from 'the grassroots' and attracted a large number of supporters. According to Zernike,

> *Tea Partiers tended to believe that they had done all the right things in life ...*
> *They had earned their place in the middle class, and they were out to protect what*
> *they saw as theirs. They distrusted people they regarded as elites, most notably*
> *the Obama administration ... And, above all, they had a visceral belief that*
> *government had taken control of their lives – and they wanted it back. (2010: 10)*

Zernike suggests that their motivations were multiple and complex. Many, however, drew language and ideology from earlier conservative uprisings and from a long-standing anti-governmental current within American culture. Although there are differences of opinion within the movement, Tea Partiers believed (far more than other Americans) that illegal immigration was a very real threat, that global warming would have no serious impact, that gay marriage should not be legally recognized and that the US Supreme Court's legalization of abortion had been a very bad decision. And they felt that people in power neither respected nor listened to them.

The earliest manifestation of the movement was an 'Anti-Porkulus Protest' held in mid-February 2009. Three days after that protest, a CNBC financial news commentator ranted on the floor of the Chicago Mercantile Exchange that the Obama administration's proposed mortgage assistance plan was 'promoting bad behavior' and rewarding 'the losers' at the expense of people who had followed the rules. He told the cheering commodities traders, 'We're thinking of having a Chicago Tea Party in July' (Zernike 2010: 13). The commentator's 'rant' 'went viral' and about a million people watched it on YouTube within the next few days. Fox Cable News began promoting Tax Day Tea Parties on its news programmes and commercials and dedicated an hour-long special to 'The 9/12 Project'. The 9/12 groups that formed in response soon joined a growing loose federation of Tea Party groups around the country, facilitated by use of the new social media. FreedomWorks and other conservative groups that had been around for years contributed ideological guidance and practical political/organizational skills. Tea Party Patriots formed a broad nationwide coalition with local affiliates in every state. Tea Party groups held Tax Day rallies in mid-April, and soon thereafter they began organizing to confront their Congressional representatives about the proposed health care legislation.

To help people prepare for the 'town hall' meetings that representatives would be holding during the summer recess of 2009, a Tea Party Patriots organizer encouraged them to 'use the [progressive organizer Saul] Alinsky playbook of which the left is so fond: freeze it, attack it, personalize it, and polarize it'. Pack the halls, he advised, and 'watch for any opportunity to yell out and challenge the Rep[resentative]'s statements early' (quoted in Zernike 2010: 83).

Tea Partiers followed the organizer's advice. In a large number of very hostile confrontations, they accused the lawmakers of promoting socialism, trampling on the Constitution and trying to kill elderly people. More than 70,000 protesters descended on Washington, DC, for the FreedomWorks 9/12 march, and in January 2010 their favoured candidate shocked the political world by winning a special election to fill a Senate seat that had been considered safely Democratic. A month later the first National Tea Party Convention convened in Nashville. Supporters came to organize and to learn how to win elections, but they had difficulty agreeing on what the Tea Party stood for. Libertarians wanted to focus on economic issues and reducing the size of government, whereas social conservatives wanted to attack 'multiculturalism', accuse President Obama of being a Muslim in disguise who had not been born in the United States, take on a range of controversial social issues, and get rid of all big government programmes while simultaneously protecting their coverage under Medicare and Social Security.

On the day the House of Representatives was scheduled to vote on the health care bill, activists from the Tea Party and other groups swarmed the Capitol. Iowa Representative Steve King told Tea Partiers, 'Let's beat the other side to a pulp!' 'Let's chase them down! There's going to be a reckoning' (quoted in Zernike 2010: 138). Someone called Democratic Party Representative John Lewis a 'nigger' as he was leaving the House Office Building, and someone else called Rep. Barney Frank a 'homo'. When coupled with the many virulent signs that had appeared at rallies over the preceding months, this provided substantial evidence that, at a minimum, the Tea Party contained a significant contingent of racists and anti-gays.

As the months passed, Tea Party activists concluded that to defeat the Democratic Party they first had to remake the Republican Party in the Tea Party's image. In part this conclusion derived from their belief that Republicans had contributed to the expansion of the federal government and that 'Republicans In Name Only' were not serious about cutting spending. This led to many successes and some notable failures in the

2010 primaries and general election. Most notably they helped the Republican Party (including a large contingent who considered themselves close to the Tea Party) to win control of the US House of Representatives, to gain seven seats in the US Senate and to win control of many state legislatures.

As noted above, conservative groups intervened at an early stage to influence how Tea Partiers defined the problematic situation that had emerged late in 2008. According to Zernike, 'young Turks' who were well versed in Twitter, Facebook and YouTube, 'provided the movement with an ideology, largely libertarian and marked by a purist and "originalist" view of the Constitution' (2010: 8). A 26-year-old conservative organization named FreedomWorks sought to 'channel outrage into action', while simultaneously giving people something to do with their anger: to organize. Led primarily by a former Republican Congressman and a young conservative organizer, it provided ideological guidance and practical political/organizational skills. FreedomWorks was underwritten by the Koch family, which had a track record of supporting libertarian causes, and think tanks like the Cato Institute, which in turn had been founded on the theories of public choice and the work of Austrian economists Ludwig von Mises and Friedrich Hayek. Guided by these Libertarian economists, FreedomWorks 'believed that the market ... should be left free from regulation, with consumers and price signals determining the flow of money ... [and they] prized strict fidelity to the original words of the Constitution' (Zernike 2010: 38).

So too did Jared Taylor of the National Center for Constitutional Studies who not only advocated returning to the founder' original words but also praised W. Cleon Skousen's 1981 book *The 5000 Year Leap*. He reemphasized Skousen's claim that the Constitution had been inspired by the Christian faith, and that the framers had never intended for the government to take taxes from one group and spend it on another. According to Skousen, the founders' first principle was a belief in God's Law. 'Without religion', he proclaimed, 'the government of a people cannot be maintained' (quoted in Zernike 2010: 75).

According to Zernike, the Tea Partiers saw their connection with the American Revolutionary War as more than a gimmick or a metaphor. For them, '[i]t was a frame of mind. They saw themselves the way they saw the founders, as liberty-loving people, rebelling against a distant and increasingly overbearing government. By getting back to what the founders intended, they believed they could right what was wrong with the country' (2010: 66). For many of these Tea Partiers, their reading of the Constitution constituted the only legitimate assessment of its meaning; all other readings were mere interpretations.

In their collective view, the Constitution did not give Congress the power to establish the Federal Reserve, to establish the Social Security system, or to adopt federal policies on education, energy, health or any other issue. So too they argued that the interstate commerce clause only authorized Congress to regulate trade between *the states* and to prevent the states from erecting barriers to free trade. Many also believed that Congress had violated the Constitution's Tenth Amendment, and insisted that individual states should decide how and whether to regulate marriage, abortion and other controversial issues.

Lastly, Zernike reports that participation in the movement instilled a sense of community among Tea Party activists. 'The Tea Party, they're just everybody', said an older woman Republican voter in Kentucky (Zernike 2010: 152). They saw themselves as the nation's true silent majority and they trusted information from fellow supporters more than any other source. They saw themselves in a battle 'between righteousness and evil, between freedom and slavery ... To the Tea Partiers, it was an all-or-nothing struggle' (Zernike 2010: 127).

What Can Planning Theory Be Now?

The tale just recounted highlights five features of the Tea Party movement's rise. First, it emerged in the context of a novel and extremely complex problematic situation. Second, its activists were driven by fear, anger and a sense of betrayal. These emotions and the means by which they were expressed provided activists with great energy and a strong sense of shared purpose and community. Third, already extant conservative organizations played a powerful role in shaping how Tea Party activists transformed the problematic situation into a problem focused on government failure. Fourth, their way of perceiving the world and defining the problem relied very heavily on essentialist conceptions of American identity. Their essentialist conception of who rightly can be considered a 'true American' and hence deserving to participate in American politics was rooted in legal, economic and religious fundamentalisms. And last, Fox Cable News, radio talk shows and new social media greatly amplified the Tea Party movement's voice.

How should proponents of the communicative turn interpret and respond to this movement?

For reasons articulated earlier in this chapter, my sense is that places and policy-making have both become so complicated as to defy understanding. The economic collapse of 2008 brought that complexity and incomprehension into sharp relief, and people who now support the Tea Party movement responded by retreating into idealized images of a nostalgic past. In making this retreat, they are being guided by an originalist interpretation of the US Constitution, a fundamentalist reading of the Christian Bible, and libertarian idealization of free market economies in a context of economic globalization; that is, by a very naïve, unstable and internally incoherent version of the Founders' Tale. Tea Party activists are deeply committed to these fundamentalisms and seem to be completely unwilling to engage in interaction, dialogue and collaboration with people who do not share their beliefs, and they seem determined to minimize the ability of 'traditionally marginalized groups' to participate in public deliberation.

At a theoretical level, the Tea Party movement and debates relating to it support claims by planning theorists that we no longer live in a moment when purely technical means can be relied upon to find the one best way to achieve pre-determined objectives. Instead, the Tea Party emerged in response to a messy problematic situation in which the primary challenge was to convert the situation into a problem that made sense and could be acted upon. Given this situation, Forester's emphasis on 'attention-shaping, communicative action' seems especially apt. Likewise, I believe it is ethically right to argue that the interactive-communicative processes through which problems are defined should be free from domination, and hence that plans and public policies should be devised through interaction, persuasion, negotiation and coalition building.

That said, the Tea Party movement presents a sharp challenge to proponents of the communicative turn. In part this challenge derives from the planning theorists' tendency to focus on planners and the practice of planning. This practice tends to be local in scale and limited in scope, whereas the Tea Party has a larger agenda focused on limiting the role of government and restoring an essentialist conception of community identity. In light of that agenda, communicative planning theorists are challenged to collaborate with other like-minded theorists and practitioners to devise processes that facilitate democratic argumentation about who we (US Americans) are and who we want to become.

In this sense, one might say that the rise of the Tea Party exemplifies the positive role of conflict: it will force planning theorists to pay attention to the movement's particular issues and interests, and the implications thereof for planning theory and practice. It will force us to devise practices that better accommodate extremely sharp differences in passionately held views. Those practices will have to acknowledge that people of whatever political persuasion have a right to be angry at times and to recognize that real public discourse – especially when people who don't normally have a voice also get to be heard – is usually passionate and messy, takes time, and requires actively listening to people's frustrations and trying to learn from them. In brief, it takes conflict.

The Tea Party's challenge also reveals the limits of defining interactive-communicative processes as an exercise in *rational* argumentation. The Tea Party's Founders Tale and argumentative claims that derive from stories related to it are ultimately driven by fear, anger and the sense of betrayal that Tea Partiers' feel. Planning theorists need to acknowledge those emotions, treat them as legitimate and make space for them to be expressed (Sandercock 2004).

What then shall be done with the erroneous facts, indeed total fabrications, that one often finds embedded in the Tea Partiers' stories and claims? They need to be challenged, no less than other factual errors should be. But theorists need to distinguish between blatant lying and weighing/configuring facts. Erroneous facts that circulate in webs of relationships should be treated as 'telling errors' that reveal what really matters to the Tea Partiers.

My own sense is that Tea Party activists feel fearful, angry and betrayed largely because they have been ignored and left behind by the transnational corporations that are shaping the economic transformation of the global economy. Strikingly, however, conservative organizations have managed to define the problem as a consequence of government failure. These organizations have proven to be very effective at persuasive argumentation and storytelling. This too is a lesson that planning theorists should learn.

Leaders of these organizations understand that they are authors who are articulating and enacting a story that is rooted in and, to a degree, consistent with what I've termed the Founders Tale of the United States. They know other stories are being told, and they are trying to turn the interaction of the stories in the direction they prefer by using particular tropes at particular times and places (that is, through persuasive storytelling). Consequently, the rise of the Tea Party movement can be understood as part of a struggle to define and control the spatial boundaries of places (at every scale from the nation to the neighbourhood) and the composition and identity of the community of people who live within them.

This poses an especially sharp challenge to progressive proponents/critics of the communicative turn. Sandercock and Attili's (2010) argument on behalf of multimedia-based storytelling as a means of facilitating interaction, dialogue, collaboration and participation, especially on the part of traditionally marginalized groups, brings the difference into sharp relief. Tea Partiers embrace an essentialist conception of community identity, and believe that they have been marginalized and ignored, and hence that justice is on *their* side. Proponents of the communicative turn, on the other hand, believe that identity is socially constructed and emerges from interactive communicative processes. Given this divide, I see no obvious way for proponents of the communicative turn to facilitate dialogue between essentialists and traditionally marginalized groups (such as Latinos, African Americans, gays and lesbians, environmentalists and others). Instead of productively engaging in public conversations, Tea Party fundamentalists condemn efforts to promote dialogue and inclusiveness with people they consider illegitimate and lacking any right to be in America or to be considered a 'true American'. Believing there are 'real Americans' with 'real American values', and provided financial support from wealthy interests, they skilfully use multimedia-based storytelling to churn up fear and anger toward others.

The core of the challenge lies in a fact that most communicative planning theorists recognize but do not really incorporate into their practices; namely, that planning is deeply political as well as technical. This unavoidably political aspect of planning means that planning theorists need to articulate a coherent and persuasive *political* rationale for rejecting essentialist conceptions while also providing democratic space for both fundamentalist and multicultural ideas to be articulated. This rationale must be articulated, not just to fellow scholars in journals, but to ordinary people in the public arena. This rationale must include a process for addressing the two key questions posed by the Tea Party movement: first, who should be included within and excluded from this community and, second, whose stories should be told, be heard and legitimately influence the construction of place? In other words, we planning theorists need to devise a viable process for facilitating democratic deliberation about who we (US Americans) are and who we want to become.

References

Arendt, H. 1958. *The Human Condition*. Garden City, NY: Doubleday Anchor.

Buell, L. 2001. *Writing for an Endangered World*. Cambridge, MA: Belknap Press of Harvard University Press.

Coverly, M. 2006. *Psychogeography*. Harpenden: Pocket Essentials.

Cresswell, T. 2004. *Place*. Malden, MA: Blackwell.

Eckstein, B. 2003. Making space, in *Story and Sustainability*, edited by B. Eckstein and J.A. Throgmorton. Cambridge, MA: MIT Press, 13–36.

Fainstein, S. 2005. Planning theory and the city. *Journal of Planning Education and Research*, 25(2), 121–30.

Finnegan, R. 1998. *Tales of the City*. Cambridge: Cambridge University Press.

Fischer, F. and Forester, J. (eds) 1993. *The Argumentative Turn in Planning and Policy Analysis*. Durham, NC: Duke University Press.

Flyvbjerg, B. 2002. Bringing power to planning research. *Journal of Planning Education and Research*, 21(4), 353–66.

Forester, J. 1989. *Planning in the Face of Power*. Berkeley, CA: University of California Press.

——. 1993. *Critical Theory, Public Policy, and Planning Practice*. Albany, NY: SUNY Press.

——. 1999. *The Deliberative Practitioner*. Cambridge, MA: MIT Press.

Friedmann, J. 1973. *Re-tracking America*. New York: Anchor Press.

Harper, T. and S. Stein. 2006. *Dialogical Planning in a Fragmented Society*. New Brunswick, NJ: CUPR Press.

Healey, P. 1997. *Collaborative Planning: Shaping Places in Fragmented Societies*. London: Macmillan.

——. 2009. The pragmatic tradition in planning thought. *Journal of Planning Education and Research*, 28(3), 277–92.

——. 2010. *Making Better Places*. London: Palgrave Macmillan.

——. 2012. Communicative planning: practices, concepts and rhetorics, in *Planning Ideas that Matter*, edited by B. Sanyal, L. Vale and C. Rosan. Cambridge, MA: MIT Press, 333–357.

Hillier, J. 2007. *Stretching Beyond the Horizon*. Burlington, VT: Ashgate.

Innes, J.E. 1995. Planning theory's emerging paradigm. *Journal of Planning Education and Research*, 14(3), 183–9.

Hoch, C. 1984. Doing good and being right. *Journal of the American Planning Association* 50: 335-345.

Innes, J.E. and Booher, D.E. 2010. *Planning with Complexity*. London: Routledge.

LeBaron, M. 2002. *Bridging Troubled Waters*. San Francisco, CA: Jossey-Bass.

Mandelbaum, S., Mazza, L. and Burchell, R. (eds) 1995. *Explorations in Planning Theory*. New Brunswick, NJ: CUPR Press.

Massey, D. 2005. *For Space*. Los Angeles, CA: Sage.

Nussbaum, M. 1990. *Love's Knowledge*. Oxford: Oxford University Press.

Portelli, A. 1991. *The Death of Luigi Trastulli and Other Stories*. Albany, NY: SUNY Press.

Rittel, H. and Webber, M. 1973. Dilemmas in a general theory of planning. *Policy Sciences*, 4(2), 155–69.

Sandercock, L. 2003. *Cosmopolis II: Mongrel Cities in the 21st Century*. New York: Continuum.

——. 2004. Towards a planning imagination for the 21st century. *Journal of the American Planning Association*, 70(2), 133–41.

Sandercock, L. and Attili, G. (eds) 2010. *Multimedia Explorations in Urban Policy and Planning*. New York: Springer.

Schön, D. 1983. *The Reflective Practitioner*. New York: Basic Books.

Simons, A. 2006. *The Story Factor*. 2nd Edition Revision. New York: Basic Books.

Skousen, W.C. 1981. *The 5000 Year Leap*. Malta, ID: National Center for Constitutional Studies.

Throgmorton, J.A. 1983. *A Bridge to a Distant Shore: Implementing Section 210 of the Public Utility Regulatory Policies Act of 1978*. PhD thesis. Los Angeles, CA: Graduate School of Architecture and Urban Planning, University of California.

——. 1984. Section 210 of PURPA as seen from a national policy implementation perspective, in *Proceedings of the Fourth NARUC Biennial Regulatory Information Conference*, edited by R.W. Lawton. Columbus, OH: National Regulatory Research Institute, 445–67.

—— . 1987. Community energy planning: winds of change from the San Gorgonio Pass. *Journal of the American Planning Association*, 53(3), 358–67.

—— . 1992. Planning as persuasive storytelling about the future: negotiating an electric power rate settlement in Illinois. *Journal of Planning Education and Research*, 12(1), 17–31.

—— . 1996. *Planning as Persuasive Storytelling: The Rhetorical Construction of Chicago's Electric Future*. Chicago, IL: University of Chicago Press.

—— . 2003. Planning as persuasive storytelling in a global-scale web of relationships. *Planning Theory*, 2(2), 125–51.

—— . 2005. Planning as persuasive storytelling in the context of 'the Network Society', in *The Network Society* edited by L. Albrecht and S.J. Mandelbaum. London: Routledge, 125–45.

—— . 2007. Inventing 'the Greatest'. *Planning Theory*, 6(3), 237–62.

Toulmin, S. 1990. *Cosmopolis*. New York: Free Press.

Verma, N. 1996. Pragmatic rationality and planning theory. *Journal of Planning Education and Research*, 16(1), 5–14.

Yiftachel, O. 2006. *Ethnocracy*. Philadelphia, PA: University of Pennsylvania Press.

Zernike, K. 2010. *Boiling Mad*. New York: Times Books.

Planning Theory and Practice in a Global Context

Vanessa Watson

Extract 1: On Kinshasa

So that other city, that peripheral city that is the real city, has developed according to its own notion of what capital might mean, or what forms of accumulation might mean. In order to exist socially in a city like Kinshasa, expenditure, circulation and conspicuous consumption are far more important than accumulation or maximalization of profit. Accumulation requires a directionality, a teleology, a specific temporality which is not the temporality of the city today. The city, on the contrary, is a space of the sudden, the unforeseen, the unexpected and fleeting moment. In order to survive it, one has to know how to capture that moment. It is this practice of capture and seizure that determines life and survival in the city, which itself is often compared to the space of the forest. As such the city does not function according to a standard capitalist logic as we know it. That also means that the urbanscape is not so much shaped by the dynamics of modernity but rather that it is constantly infused with all kinds of notions and moralities that often have longstanding, rural roots. (De Boeck 2010: 36–7)

Extract 2: Essay Topics for Urban Planning Students

- What kinds of dwellings do you build for people who have a tendency to rape or be raped?
- What public leisure facilities should you design for people who are likely to be murdered?
- What are the civic participation processes most suitable for hijackers and drunks?
- What governance structures will best serve the interests of people addicted to consumer goods?

- What should the layout of commercial zones be in a city where most people are unemployed?
- How many paintings should each child be able to see on the way home from school? (Press 2010: 67)

Introduction

The intention of the above quotes is to unsettle taken-for-granted assumptions about the production of space in urban environments and the role that urban planning plays in this. Most urban and planning theory is produced by authors writing in English or various European languages and through publishing companies for a market readership located in the global North. Although increasingly there are exceptions it is usually the case that these Northern authors assume that the socio-political and cultural characteristics with which they are familiar (primarily, variations of Western liberal democracy and relatively well resourced although unequal societies) are also true for most other parts of the world. Many such urban and planning theories and substantive ideas do not even specify the contextual assumptions upon which they are based, but simply put forward propositions and recommendations in a generalized manner as if they were true in all contexts. The central argument of this chapter is that the globe is characterized by deep economic, social and cultural divides and differences, and that assumptions underpinning many mainstream theories often do not hold outside their region of origin. This is not to suggest that all planning ideas, models or practices are only applicable in the place they were conceived. There are examples around the globe of situations where ideas have been successfully transferred across contexts; there are also far more examples of where this kind of transfer has been inappropriate, has taken place for the wrong reasons and has had disastrous consequences for those affected by it.

Three important dynamics shape planning ideas as they 'travel' from their regions of origin to other parts of an increasingly interconnected world: postcolonial relationships, the nature of state–society relations, and cultural variation and change. These factors interact with each other in complex and locally specific ways and influence how ideas from 'elsewhere' are received and appropriated into local contexts. Much hangs on the way in which 'culture' and 'difference' are interpreted. This chapter takes a position on culture which holds that it is dynamic, socially produced and shaped by power. It can be strategically mobilized, often in relation to processes of marginalization or domination.

The chapter begins with a discussion on culture and difference in relation to the production of space. It then considers why and how these factors are important in terms of shaping and moving planning ideas, drawing on examples. Finally it considers some methodological and conceptual ideas on how to move forward in a culture-rich world.

Culture and Difference – a Position

Some ten years ago, Storper (2001) noted that the 'cultural turn' in the social sciences was already underway and political economy and structuralist approaches had been replaced with an agency-driven interest in postmodernism, post-colonialism and cultural politics. (See also Evans, Chapter 13, this volume.) The dislodging of the possibility of meta-narratives and universalist analyses and solutions was a welcome shift to many, but the cultural turn brought with it the problem that if all knowledge and practice is culturally determined then how is any comparison or judgement possible between or across different cultural groups. This is the problem of relativism. The cultural turn has been criticized – by Storper (2001) – for its misuse as a celebration of difference for its own good, for its depoliticizing tendencies and for its localism, and by Fraser (2000) for its privileging of recognition over redistribution. But the problem of relativism remains, particularly in planning. Given that planning is concerned with ethical judgement, frequently involving the allocation of resources or other advantages between places and groups, the question of how to balance one claim above another is a serious one (see Watson 2006).

Philosopher Alastair MacIntyre's (1988) position on cultural relativism is a persuasive one. He conceives of different world-views which arise from different philosophical traditions. These traditions (of which liberalism is one) are fundamentally shaped by the time and place in which they emerged, by contingent circumstances, by particular societal concerns and disagreements of the time, and are articulated in terms of the particular language and culture of that order. Each of these traditions has distinctive conceptualizations of practical rationality and justice which are not necessarily compatible with each other. These world-views are also capable of mutating and hybridizing. Traditions change and evolve as a result of new situations which are encountered, or through contact with other communities and traditions (through migration or warfare and invasion, and as well through internet and TV), which means that 'internal' texts, beliefs or authorities are challenged (experience epistemological crises) and have to be reformulated. If more appropriate or attractive theoretical resources are to be found in another tradition, MacIntyre (1988) argues, they will be adopted and will come to be shared by traditions.

This dynamic and evolving notion of cultures is usefully complemented through the work of Jacobs and Fincher (1998: 2) who consider how identity is constituted and negotiated and the ways in which 'empowerment, oppression, and exclusion work through regimes of difference'. Culture and identity are therefore not static or fixed but rather socially produced and multiply located. What this points to, they argue, is the multiplicity of differences that may cohere around any one person: 'social distinctions are constituted in specific contexts through multiple and interpenetrating axes of difference … and at any one time we may be fixed into or strategically mobilize different aspects of the array of differences through which our embodied selves are known' (Jacobs and Fincher 1998: 9). Which aspect dominates is not haphazard – often the attribute to be emphasized is that which contributes most significantly to a subject's marginalization or empowerment. Concepts of difference are thus inextricably linked to the issue of power. Jacobs and Fincher's argument should not be taken to mean that nothing is real, that all aspects of identity are entirely contingent. Individual and group values, or world-views, (which are always present to some degree) ultimately circumscribe the range of possible aspects of difference which any individual may be prepared to mobilize.

In many cities of the global South, where rapid urbanization in a context of severe levels of poverty and inequality, and weak states, produces highly fluid and contested social environments, culture and identity can be opportunistically foregrounded as bases for claims against the state or other social groupings.[1] In such contexts, identity/culture is often a product of hybridization, fusion and cultural innovation. It is frequently self-generated and self-constructed, sometimes with a renewed stress on ethnic identity or 'retribalization', sometimes intertwined with global identities (De Boeck 1996, Simone 2004). De Boek's (2010) description of the production of urbanscapes in the context of Kinshasa (Extract 1, above) is relevant here. Arce and Long (2000) refer to a fusion of the institutions and practices of Western modernity with local ways of coping in a situation of rapid change and economic crisis. The modernizing efforts of planning and urban regeneration may be absorbed selectively by target communities and are mutated within local traditions and ways of doing things. Frequently a situation of 'conflicting rationalities' (Watson 2009, 2003) arises between incompatible world views or logics: the outcomes may be direct conflict and resistance, passive acceptance or the fusion and hybridization of practices referred to above.

MacIntyre's notion of hybridizing and evolving world views, and Jacobs and Fincher's idea of the opportunistic use of culture/identity, provide useful starting points for thinking about planning and culture in a global context. These ideas certainly challenge planning notions of 'the public good', universal ethics, standard models of the good city and the unquestioned transfer of 'best practice' solutions from one context to another. How, then, do planners decide what is good or bad, just or unjust, when dealing with competing world views, different cultures and identities, and conflicting rationalities? MacIntyre's (1988) view is that traditions are contextually informed and situated with their own ways of thinking about practical rationality and justice. One implication of this position is that no one tradition can assert its principles of practical rationality and justice as universal, or as being of a higher or better order than that contained in any other tradition. There is no 'neutral space' outside of traditions from which one can judge different and competing claims. Rational judgements can be made relative to the standards of a particular tradition, but they cannot be rational as such. One set of rationalities is as good as any other and we cannot pass judgement on those of a differing tradition (MacIntyre 1988: 352). It is this position that those within the liberal tradition find particularly hard to grasp, given that the central characteristic of liberalism has been an assumption of its own universality.

One alternative way to think about this problem is to take an inductive approach to arriving at a judgement in situations of competing world views or culture/identities. MacIntyre argues that the answer is to be found in understanding the ways in which traditions change over time and the possibility that shared values *can* emerge (from the inside out, rather than from a 'neutral' outside in). Hence 'finding the common ground is not subsequent to understanding, but a condition of it ...' (Donald Davidson, quoted

---

[1]   For example, in the region of Israel/Palestine, where conflict is fuelled both from the local and the global, Yiftachel and Yacobi (2003) describe 'ethnocratic cities' as territories where an exclusionary Israeli-Jewish identity has worked to essentialize and segregate Arabs and Jews.

in MacIntyre 1988: 3). In another slant to this idea, Iris Marion Young (2000) draws on the feminist epistemology of writers such as Donna Haraway (1991) and her conception of 'situated knowledges'. Young interprets this as 'a conception of objectivity as constructed from the partial and situated perspectives of differently positioned social actors' (Young 2000: 2). She suggests that the goal should be to arrive at judgements rather than principles or technical solutions. The concept of judgement adopted here is not one which assumes that it is possible to bring particular (situated) positions under a universal, or aims to construct a general standpoint outside and above particular views, but rather one which involves an 'enlargement of thought' that comes from considering the perspectives of many differently situated people.

The position on culture/identity taken here is that it is an expression of evolving and mutating world views, logics or rationalities which may in some respects be fundamentally different from each other, or in other respects may be the outcome of shared or intertwined understandings. This in turn has implications for how planners think about planning values and judgement on better or worse planning outcomes. It also recognizes that cultural positioning may be, and often is, tactical and strategic and can be closely linked to processes of exclusion or domination. The next section considers two examples of how planning interventions have engaged (negatively in both cases) with local contextual material and social specificities.

Post-colonialism, Globalization and Circulating Development Models

The 'cultural turn' in the social sciences referred to by Storper (2001) entered planning discourse in the latter part of the twentieth century as well, building on the shift away from earlier rational technocratic approaches to planning (see Greed, Chapter 5 in this volume). At the same time increasing global interconnectedness and the use of the Internet ushered the rapid circulation[2] of 'best practice' planning ideas between very different parts of the world, but not necessarily a clear understanding of what cultural difference might mean. This section of the chapter will argue that earlier (modernist) assumptions of the hegemonic nature of Western democratic liberalism persisted alongside the post-modern cultural turn in planning, providing an intellectual frame which shaped and constrained much of mainstream planning thought. There is also a strong argument to suggest that the rise of the global neo-liberal order during this period (with the discovery of 'agency' and neo-liberalism's individualistic conception of society mutually reinforcing each other) opened the way for a commodification of culture in planning which several chapters in the volume illustrate well.

The two examples which follow illustrate how planning can take place with complete disregard for culture, inequality and difference. The first example is the planning of the city of Abuja, the new capital of Nigeria, but similar stories play themselves out in Latin America, the rest of Africa and now in most of the Middle and Far East. Here postcoloniality makes itself felt in subtle ways, expressed in an undervaluing of

---

2   The international transfer of planning ideas had of course been occurring for a very long time (see Healey and Upton 2010) but has been facilitated by more recent technologies of communication.

local histories, sense of place and environmental factors, a disregard for deep income inequalities, poverty and unemployment, as well as cultural practices, in an ongoing obeisance to Western urban modernism, and in the continuing power of globally circulating urban experts and 'best practice' ideas.

Abuja

In 1975 Nigerian leaders revived an older, colonial, idea that a new capital city outside of Lagos was needed to unite the many different ethnic groups into one 'modern' nation within its colonially defined boundaries.[3] They hired Thomas Todd of the Philadelphia firm Wallace Roberts McHarg and Todd, and members of the International Congress of Modern Architecture (CIAM) such as Kenzo Tange and Doxiades. The particular concept promoted by these architects was 'monumental urban modernism' which found favour with Nigeria's emerging elites in a climate in which petro-dollars were fuelling a construction boom.

The original plan for Abuja drawn up by Todd was very similar to that of Washington DC, with Todd at the time also working on the Capitol Hill complex in Washington. Seemingly the fact that Nigeria was moving towards democracy at the time was justification for adopting ideas from Capitol Hill – as the epicentre of Western democracy. There was also strong influence from Le Corbusier's Chandigarh (India), dividing the city into sectors, districts and neighbourhoods (following the US neighbourhood unit idea) separated by a road hierarchy designed for a car-based movement system. The city was designed around functional zones, separating out residences, business, government and so on, controlled by land use zoning, with the monumental scale and layout of the central area designed to impress foreign visitors and government officials with vistas, axes and harmonious flows of space.[4] It appears that there was a strong belief amongst the architects that spatially copying physical symbols of democracy and 'enlightenment' from the United States to Nigeria would foster these political characteristics in Nigeria as well. This way of thinking has sometimes been referred to as spatial determinism.

The layout of the new town of Milton Keynes (UK) also had a marked influence on Abuja, through the transfer of professionals from the Milton Keynes Development Corporation to the Abuja project. Residential district designs are based on the Milton Keynes layout with a similar hierarchy of movement routes based on car travel, a concentration of facilities in the district's centre and provision for middle class housing. Scattered indigenous villages previously on the site were moved and their inhabitants incorporated into new satellite towns around Abuja. When asked if there had been a discussion about reflecting Nigerian architectural heritage in Abuja the answer was that the foreign professionals had been instructed (by the Nigerian government) to produce a modern twenty-first-century city of which all Nigerians could be proud.

The result of imposed urban modernism is a city which is decidedly at odds with the realities of life in urban Nigeria. Unemployment, poverty, slums and growing

---

[3] The full story of the building of Abuja is in N. Elleh (2001). *Abuja: The single Most Ambitious Urban Design Project of the 20th Century.*

[4] Elleh (2001) describes how somewhat different ideas by Todd were later changed by Tange.

informality have all found their way into the city of Abuja. Attempts by the state to return Abuja to the vision of the Master Plan have been draconian. By 2006, 800,000 people had been evicted from land that was 'zoned for other purposes under the Master Plan', supposedly to achieve the beautification of the city, privatization and 'cleaning up' criminals (COHRE 2006). The demolitions included villages which had existed for decades (but were suddenly declared informal and illegal) and had solidly constructed houses and community facilities. Ancestral graves and landholdings were destroyed. Evictions were often accompanied by violence and destruction of private property as little warning was given of removals (COHRE 2008).

In other respects the modernist planned city fails to accommodate the needs of a largely poor population. Monofunctional land-use zoning prevents use of the home as an economic unit – an essential survival strategy under conditions of high unemployment. Street-trading, an integral part of most African cities and often the dominant income-generating activity, does not fit into the vision of a modern Western city and is strictly controlled. Planners and officials are determined that Abuja will remain faithful to the intentions of the original Master Plan and '… must be preserved from the processes of change, informality and complexity that dominate Lagos' (COHRE 2008: 76). Movement systems are designed to cater for a middle-class car-owning population as in cities such as Milton Keynes, not for poor and public transport dependent commuters. This is particularly serious for the newly declared 'squatters' (occupants of existing villages considered obstacles to the Master Plan) who faced removal to satellite towns such as Pegi located some 33 km away from Abuja.

Shatkin (2011: 79) draws attention to the different 'worlding' efforts of planners as they '… endeavor to reshape urban social, political, and cultural life and spatial relations to conform to an ideal of a globalized, cosmopolitan, economically integrated, and competitive city'. In the case of Abuja the worlding efforts of planners (and politicians) appear to remain focused on an imagined need to mimic the cities of older colonial masters rather than on global economic competitiveness, but the outcomes remain just as destructive for those who get in the way of this imperative. Shatkin (2011: 79–80) argues instead that planners should be concerned with '… *actually existing urbanisms*, that are rooted in alternative social dynamics (informality, violence, alternative cultural, and social visions, vote-bank politics), and that resist worlding practice'.

Metro Manila

The second example illustrates a similar process of imposition to that which took place in Abuja, but in this case it was driven by a somewhat different coalition of actors. Shatkin (2008) draws on the case of Metro Manila to ask if there is a growing convergence between recent urban interventions in the burgeoning mega-cities of the global South and large cities in countries such as the United States, which are polarizing between sprawling suburbs of 'rampant consumerism' and blighted inner cities. Is there, therefore, a proliferating American urban model which will in future characterize global cities in all parts of the world?

Metro Manila is a city of some ten million people, which together with five surrounding provinces, produces some 50 per cent of the Philippines' annual GDP. Here globalization has supported the emergence of a small but powerful elite made up of a professional and merchant class, along with foreign investors and ethnic Chinese business families. Interestingly, a significant part of the new demand for urban space comes from Filipino workers abroad who remit large sums annually to their home city. Urban space is also in demand from poorer workers and recent arrivals in the city from rural Philippines: occupying older buildings and informal settlements, they consistently pose a challenge and a threat, for private developers and their government partners, to what Shatkin (2008) calls the 'public city'.

The privatization of urban planning and development (as a new model of urban governance) has taken an extensive and all-encompassing form in Metro Manila that goes well beyond individual shopping mall and gated village developments. Here it has taken the form of private investment in cross-city transport schemes the purpose of which is to unlock the development potential of parcels of land serviced by the new movement infrastructure. In Metro Manila and other large cities of the East, these new developments are massive: Shatkin refers to projected populations of 750,000 to a million. Moreover they are scattered across the urban region and do not necessarily take on the 'edge city' form found in many US cities. These new systems of transport and residential/commercial development are overlaid onto the existing congested and decayed urban form but are detached from it, creating entirely separate pathways of circulation and land use. Shatkin (2008) refers to this as 'bypass-implant urbanism'. These schemes attract large government subsidies and this in turn drains public finance available for public space or to address the needs of poorer communities. The consequence is the steady decline of the existing older city, as well as the threatened and actual removal of communities who find themselves in the way of the new development projects and transit routes.

Both Abuja and Metro Manila demonstrate a process of foreign (but particularly US) adoption of urban forms that, firstly, is not particularly new (in Abuja it was evident in the 1970s) and secondly, is driven as much by local actors as it is by Western experts and professionals. However, there are also broader structural factors – post-colonial relationships in the case of Abuja and globalization in the case of Metro Manila – which interact with and shape the imperatives of local actors. Shatkin (2008) questions what may be a simplistic interpretation of Philippines' new urban developments as purely a copying of an American model of urbanism and argues that instead influential local actors (in this case primarily property developers and their clients) have sought to selectively adapt international models of planning and design to the context of Southeast Asian urbanization. This echoes MacIntyre's (1988) concept of cultural change as mutating world views shaped by shared or intertwined understandings. It is significant however that the sharing appears to be primarily in one direction, from West to East, and there is little evidence of American cities adopting urban forms from the East. It is also significant that what is often referred to as international is in fact highly parochial: it is the urban forms demanded by the middle and wealthy classes in the United States which are desired elsewhere, even if they are subsequently adapted to local conditions. The power relations which underlie post-colonialism and globalization, which extend in subtle ways to influence taste and preferences, and which insidiously reinforce older global relationships of domination and obeisance, cannot be ignored.

Both Abuja and Metro Manila also show how the adoption and adaption of Western urban models impacts on the lives of the poorer and marginalized groups. In Abuja it was the previously existing villagers that lost their livelihoods and their heritage as they were moved from long-established settlements to newly designed 'satellite towns', as well as more recent construction workers who were shifted out of the city once their labour was no longer required. In Metro Manila, communities in the way of privatized developments have faced forced removal and older areas generally have fallen into decline as government attention and resources are pulled into the wake of developer-driven projects. One way of interpreting this is to see a newer metropolitan culture and lifestyle replacing an older indigenous way of life – an inevitable if unfortunate cost of progress and development. But the longer term impacts may be far-reaching. Urban processes of these kinds contribute to, and materialize, the growing income inequalities which are part and parcel of a globalized economy. The income gap between rich and poor finds direct expression in the glitzy shopping malls, office blocks and leisure centres – some of which express a crude form of heritage appropriation – in juxtaposition with shacks and decayed inner city buildings. Income inequalities and poverty are exacerbated by skewed allocations of public resources which further undermine the viability of working-class areas and the life-prospects of their inhabitants. Highly unequal cities, as many parts of Latin America have shown, can quickly become violent and crime-ridden cities (Holston 2009) controlled by drug-related and criminal gangs. Ultimately, in aspiring to the urban models of Los Angeles or Las Vegas, cities such as Abuja and Metro Manila may find themselves closer to Rio or São Paulo.

What then of MacIntyre's (1988) assertion that one set of rationalities is as good as another and there is no 'neutral space' from which one can pass judgement? In both Abuja and Metro Manila the form of urban modernization as American appropriation has been instigated by local actors and perhaps they could have equally chosen to develop cities in ways that respect and extend local histories, identities and built culture. The transplant of Washington and Milton Keynes to Abuja and 'bypass-implant urbanism' in Metro Manila is no more than an expression of locally mutating culture borrowing from other traditions and cannot be judged negatively by anyone located within other and different traditions. At one level this may be correct, but at another level urban research and precedent tell us that this path of urban modernization is likely to lead to worsening inequality and marginalization of poorer urban communities, and ultimately to violence, crime and socio-political breakdown. The urban elite may be able to isolate themselves in gated villages and segregated transit systems for a while but these artificial arrangements are not sustainable, at least politically, as 2011 events in the Middle East have shown. From any perspective this cannot be a good thing.

Young's (2000) notion of reaching judgement, not through constructing a general standpoint outside and above particular views, but rather as an 'enlargement of thought' from the perspectives of many differently situated people, may be one way to deal with this conundrum. The starting point, however, would be to recognize that a general or universal standpoint on the nature of urban modernization does already exist: it is the notion (or mind-set) that the American (perhaps more broadly Anglo-American) urban model is globally the best on offer to adopt and adapt to meet the tastes of emerging urban elites and politicians seeking national and international status. If urban and planning research has any role to play, then it is in starting to shift this taken-for-granted assumption. Thereafter, there may be room for the emergence of more situated, diverse and representative views of how rapidly growing cities should be planned and managed to meet the requirements of all its citizens, as well as responding to other current imperatives absent from American urbanism, such as climate change and resource depletion. The last section turns briefly to a methodological approach.

Challenging Urban Hegemonies

There is a growing body of work, particularly in the urban field, which argues for a challenge to the hegemony of Northern epistemologies and suggests research methodologies to take forward this agenda. The inspiration for some of this thinking comes from social theorist Raewyn Connell (2007) who argues that the hegemony of Northern knowledge (in sociology) obliterates knowledge from all other sources. (See also Pieterse, Chapter 14, this volume.) It relegates ideas which do originate in the South to 'traditions' of ethnographic interest and will not allow them to be considered alongside 'metropolitan' knowledge.

Connell (2007) calls for 'Southern theory' which draws attention to relationships (of authority, exclusion and inclusion, hegemony, partnership and so on) between global North and South rather than to sharply bounded categories of states or societies. To counter the universalization of ideas which have originated in just one part of the world, she argues for a deeper understanding of the specific contexts from which generalization grows, and for theory which will '… illuminate a situation in its concreteness' (Connell 2007: 207). She also argues for a 'multiplying of local sources of thinking', implying, as is the case in her book, the drawing of theoretical ideas from sources outside of the metropole. One way to begin to build ideas or theories about the global South, which can enter into a dialogue with Northern theory, is through comparative case research which tracks common issues across very different contexts. For example, this could involve comparing forces underlying urban spatial change across a large Northern city and a large Southern city, as a starting point for understanding the production of space from a global perspective. This would be very different to theory development which draws on global North patterns and trends and attempts to universalize these ideas to the rest of the world, either as theories or 'best practice' urban models. Or more specifically, comparing the manifestation of, and responses to, informality across global North and South cases will give a far richer and more generalizable set of ideas about this phenomenon than research which has been carried out in one kind of context only.

Significantly, Connell's (2007) research method proposals have been echoed in the urban studies field (see McFarlane 2010, Robinson 2011). Following Connell (2007) these authors also take this position not just as a method but as a political strategy. The aim of these writings and their advocated research approach is to counter Northern epistemological hegemony and restructure the geopolitics of knowledge production. McFarlane (2010: 726) argues that cross-global comparative case research is a strategy for revealing the limitations of the assumptions on which much current (Northern-origin) urban theory is based as well as for formulating new lines of enquiry from more situated accounts.

This is a useful starting point for thinking about urban forms and processes which will be new (in fact need to be, given the changed circumstances of the twenty-first century) through an approach which aligns closely to Young's (2000) idea of an 'enlargement of thought' from diverse sources. It is not a position which is necessarily 'value-neutral': it is possible to argue that any urban strategy which results in a widening gap between rich and poor, which leads to violence and societal breakdown, which destroys the very city which it claims to improve, is to be avoided.

Developing new urban ideas through comparative, global urban research also does not easily cope with the problem of urban elites' and politicians' tastes for Disneyworld shopping malls, glass tower office blocks or gated communities, with the unfortunate trend towards the commodification of culture in the built environment in 'heritage districts' and urban regeneration projects, or for the simplistic celebration of culture for its own good. But the unsettling of taken-for-granted assumptions about the built environment and the shifting of mind-sets about what inevitably represents progress and modernization in urban forms is certainly a starting point for a more diversified and context-informed set of ideas about how cities in all parts of the world can and should take shape in the future.

Conclusion

This chapter examined the geopolitical production of urban planning knowledge and ideas on urban development and how this engages with local specificities, and social identities and culture. As Extracts 1 and 2 (above) illustrate, it is possible to argue that in many cities of the global South there is a major disconnect between the visions and practices of urban modernization and the lived reality of urban inhabitants. Clearly social difference needs to be acknowledged in planning, and this in turn requires a challenge to what may be considered mainstream planning approaches – shaped by the global circulation of 'best practice' ideas, the tastes of a global elite urban market, and the built form fashions promoted by international and local architects and property developers.

Acknowledging social difference immediately raises the question of how this should be understood and how it should be responded to by planned interventions. I take the position here that difference is frequently 'deep' and can usually be traced to roots in different world philosophical traditions or cultures, but at the same time is constantly mutating and hybridizing, and responding to local contingencies. It can be used opportunistically to further certain ends. Avoiding the trap of relativism in relation to deep difference, I draw on the work of Iris Marion Young (2000) who argues for an 'enlargement of thought' that comes from considering the perspectives of many differently situated people. In the case of current patterns and trends of urban development (Abuja and Metro Manila illustrate common outcomes in response to global development forces at different points in time) it is evident that the marginalization and inequality which these development processes set in motion have likely consequences in escalating crime, violence and social breakdown. There is no philosophical tradition which would in principle support this outcome, and there are undoubtedly many 'differently situated people' who would not support it either. This is not to say, of course, that individuals may (and do) use reference to cultural traditions in an opportunistic way to forward their own particular ends.

The chapter therefore argues that it is possible to judge the outcomes of dominant urban development processes (even while accepting that fundamentally different and culturally embedded perspectives on them may exist) and to find them wanting. That said, the task is then to challenge the source of ideas and beliefs in current urban modernization processes, and to seek for alternative sources. This requires tackling the geopolitical hegemony of (urban) knowledge production, lying largely in the global North. Drawing on the work of both social science and urban theorists concerned with the same issue, it is possible to suggest a research method involving the comparative case study approach in which common issues are compared across very different parts of the world. This approach can raise questions about the assumptions on which much Northern theory is based and the validity of their universalization. It can also open the door to new understandings from sites outside the global North which will hopefully provide inspiration for new, locally responsive, inclusionary and environmentally appropriate ways of dealing with urban development.

References

Arce, A. and Long, N. (eds) 2000. *Anthropology, Development and Modernities*. London: Routledge.

COHRE. 2006. *Forced Evictions: Violations of Human Rights (10)*. Geneva: COHRE.

—— . 2008. *The Myth of the Master Plan: Forced Evictions as Urban Planning in Abuja, Nigeria*. Geneva: COHRE.

Connell, R. 2007. *Southern Theory: The Global Dynamics of Knowledge in Social Science*. Sydney: Allen and Unwin.

De Boeck, F. 1996. Postcolonialism, power and identity: local and global perspectives from Zaire, in *Postcolonial Identities in Africa*, edited by R. Werbner and T. Ranger. London: Zed Books, 75–106.

—— . 2010. Urbanism beyond architecture: African cities as infrastructure, in *African Cities Reader*, edited by E. Pieterse. Cape Town: Chimurenga and the African Centre for Cities, 32–45.

Elleh, N. 2001 *Abuja: The Single Most Ambitious Urban Design Project of the Twentieth Century*. Weimar: VDG Verlag Datenbank fur Geisteswissenschaften.

Fraser, N. 2000. Rethinking recognition. *New Left Review*, 3, 107–20.

Haraway, D. 1991. *Simians, Cyborgs and Women: The Reinvention of Nature*. London: Free Association Books.

Healey, P. and Upton, R. (eds) 2010. *Crossing Borders: International Exchange and Planning Practices*. London and New York: Routledge.

Holston, J. 2009. Dangerous spaces of citizenship: gang talk, rights and rule of law in Brazil. *Planning Theory*, 8(1), 12–31.

Jacobs, J. and Fincher, R. 1998. Introduction, in *Cities of Difference*, edited by R. Fincher and J. Jacobs. New York and London: Guilford Press, 1–25.

MacIntyre, A. 1988. *Whose Justice? Which Rationality?* London: Duckworth and Co.

McFarlane, C. 2010. The comparative city: knowledge, learning and urbanism. *International Journal of Urban and Regional Research*, 34(4), 725–42.

Press, K. 2010. Closer than this: extracts from an open source book for urban planners, in *African Cities Reader*, edited by E. Pieterse. Cape Town: Chimurenga and the African Centre for Cities, 60–77.

Robinson, J. 2011. Cities in a world of cities: the comparative gesture. *International Journal of Urban and Regional Research*, 35(1), 1–23.

Shatkin, G. 2008. The city and the bottom line: urban megaprojects and the privatization of planning in Southeast Asia. *Environment and Planning A*, 40(2), 383–401.

——. 2011. Coping with actually existing urbanisms: the real politics of planning in the global era. *Planning Theory*, 10(1), 79–87.

Simone, A. 2004. *For the City yet to Come: Changing African Life in Four Cities*. Durham, NC and London: Duke University Press.

Storper, M. 2001. The poverty of radical theory today: from the false promises of Marxism to the mirage of the cultural turn. *International Journal of Urban and Regional Research*, 25(1), 155–79.

Watson, V. 2003. Conflicting rationalities: implications for planning theory and ethics. *Planning Theory and Practice*, 4(4), 395–408.

——. 2006. Deep difference: diversity, planning and ethics. *Planning Theory*, 5(1), 31–50.

——. 2009. Seeing from the South: refocusing urban planning on the globe's central urban issues. *Urban Studies*, 46(11), 2259–75.

Yiftachel, O. and Yacobi, H. 2003. Control, resistance and informality: urban ethnocracy in Beer-Sheva, Israel, in *Urban Informality: Transnational Perspectives from the Middle East, Latin America and South Asia*, edited by N. Al-Sayyad and A. Roy. Boulder, CO: Lexington Books, 673–93.

Young, I.M. 2000. *Situated Knowledges and Democratic Discussions*. Paper presented at the conference: New Challenges to Gender, Democracy, Welfare States: Politics of Empowerment and Inclusion, Aalborg University, Denmark, 18–20 August.

Case Study Window – Discourse, Doctrine and Habitus: Redevelopment Contestation on Sydney's Harbour-Edge

Glen Searle

Planning can be seen as an essentially political activity that involves contestation of outcomes. At its simplest, this involves a contest between property rights and the public interest, but at a more complex and nuanced level it incorporates competition between different planning discourses and different motives underlying support for each discourse. This chapter argues that an understanding of these motives, in the form of the doctrines and habitus of protagonists, can provide deeper insights into the basic forces shaping planning conflicts.

The chapter uses a hierarchical framework to consider the various influences of planning discourse, doctrine and habitus in conflicted planning situations. Support for particular discourses is viewed as being frequently driven by certain doctrinal beliefs and, more deeply, by the specific work habitus of many protagonists. This framework is argued in the next section. It is then applied to two case studies involving the redevelopment of old port areas alongside Sydney Harbour: inner west Pyrmont-Ultimo and the Darling Harbour precinct of Barangaroo.

Theoretical Framework

The framework employed here of different levels of influences on planning outcomes is not new. Allmendinger notes several applications of this approach to understanding public policy change (Giddens 1984, Healey 2007, 2004, Schön and Rein 1994; see Allmendinger 2011: 46). This chapter uses a hierarchy of influences that is similar to, but conceived independently of, that of Schmidt (2008). Her conceptualization uses three levels of influences. At the first level, policy solutions are generated by policy-makers. Paradigms comprise the second level, encompassing the principles and assumptions underlying policy. The third level concerns public philosophies, denoting the values and principles of knowledge and society behind policies and paradigms. In this chapter, the first level consists of discourses used to argue the specific nature of redevelopment in harbourside Sydney. The second level comprises prevailing doctrines of the key protagonists that guide the underlying nature of development that is acceptable to them. Finally, the different habitus or ways in which planners, developers and bureaucrats work and think, forms a third level that is imbued distinctively within the practices of each group. (See also Greed, Chapter 5, this volume.) The habitus of each is formed and reinforced by the ongoing practices, rules and philosophies used in daily working life, and sets the basic approach that each group brings to achieving its development aspirations. The role of actor power in shaping development outcomes operates through these various lenses, which I discuss further below.

The most immediate set of influences shaping the emergence of types of urban development is that relating to the *discourses* used to justify the nature of the development. Here I apply the notion of discourse to debates and arguments that are played out within the professional and institutional contexts that pre-exist the development proposal in question. This contrasts with Schmidt's (2008) approach in which professional codes, planning doctrines, and local cultures and preferences embody 'background' discourses that influence planning policies and paradigms (Allmendinger 2011: 47). I argue that the doctrines and habitus of protagonists that incorporate such background discourses need to be considered separately from 'foreground' discourses because they underpin the way in which the latter are interpreted and used to achieve desired development outcomes.

The role of *doctrines*, the second level of influences used here, has come to be recognized within planning as setting particular spatial arrangements that keep on being applied. The notion of planning doctrine has been most fully developed by Faludi as describing embedded modes of thought within planning (Faludi 1997, Faludi and van der Valk 1994). In particular, planning doctrine refers to a body of thought about a planning policy area that involves consensus and shared coordination between professionals and planning-minded administrators and politicians (Faludi 1997: 83). The use of doctrine by planners to support certain spatial forms and developments, as well as the underlying legitimacy of planning, has been noted by Allmendinger (2011: 43). The on-going preservation and reinforcement of the Green Heart of the Netherlands, for example, has been argued by Faludi and van der Valk (1994) to be a long term planning doctrine. The inverse of the Green Heart, the green belt, is seen by Allmendinger (2011: 46) as akin to a paradigm in Schmidt's schema. It could as easily be conceived as a planning doctrine, which perhaps better conveys the fixedness and widespread acceptance of that planning concept, rather than the strict 'example or pattern' meaning of paradigm.

The concept of planning doctrine has, as Allmendinger (2011: 55) notes, become a mainstay in planning studies (Alexander and Faludi 1996, Murdoch and Abram 2002, Needham 1996). In this chapter I extend the application of the concept of doctrine to other key actors in a particular development. Thus understanding of the relevant implicit doctrine(s) of the state, a central player in the two case studies, is used to illuminate the predisposition of the state to its preferred outcomes. Allmendinger (2011: 55) observes that planning doctrine has recently been relabelled 'planning culture', with attempts to embed planning doctrines within overarching local and national cultures (Sanyal 2005). In this, planning culture incorporates attitudes, beliefs and values, cognitive frames, and rules and norms, *inter alia* (Allmendinger 2011: 55). This chapter sees such general and embedded influences as better analysed separately as part of habitus, representing a more foundational explanatory level than doctrine, here used to convey acceptance of particular types of planning and development outcomes.

The concept of *habitus* was developed by Bourdieu to understand how the lived world of groups with shared expectations and experiences (such as planners) shapes their experiences, interests, capacities and actions (Hillier 2002: 13). Thus habitus is '… a system of schemes of perception and appreciation of practices, cognitive and evaluative structures which are acquired through the lasting experience of a social position' (Bourdieu 1989: 19). Hence 'habituses predispose agents to act in particular ways without reducing them to cultural dopes or inhibiting their strategic capacities' (Crossley 2001: 84). Or as Painter (2000: 242) puts it, habitus 'refers to the embodiment in individual actors of systems of social norms, understandings and patterns of behaviour, which, while not wholly determining action … do ensure that individuals are more disposed to act in some ways than others'. Painter (1997: 138) has suggested the habituses of different groups centrally involved in the case studies in this chapter. Public sector professionals such as urban planners have a habitus grounded in procedures influenced by norms generated through professional training, and in common sense based on detachment, objectivity and public service. The habitus of local politicians is grounded, *inter alia*, on 'common sense based on … grassroots support and legitimacy' (Painter 1997: 138).

The understandings and patterns of behaviour embodied within habituses will produce outcomes/practices within a particular field such as planning or property development through interaction with different kinds of capital (Huxley 2002, see also Dovey, Chapter 15, this volume). Four types of capital are identified by Bourdieu (1977: 177–83): economic (money, property and so on), social (acquaintances and social networks), cultural (cultural property, including education) and symbolic. The last incorporates the notion of legitimation, the ability to have certain dominant conceptions accepted as appropriate (Swartz 1997: 74). Such conceptions in turn relate closely to the idea of doctrine; thus practice within a particular field is produced by the way that an underlying habitus interacts with particular doctrines. In a particular development context, the specific detail of practice will in turn be shaped by the discourses relating to that development.

Nevertheless the limitations on the possibility of new practices that are imposed by the concept of habitus, and indeed by concepts of prevailing discourse and doctrine, should be recognized. Here the distinction of Deleuze and Guattari (1987) between striated space, where spatial practices are stabilized, and smooth space, where spatial practices show instability and have the possibility of new directions, is potentially useful. Smooth space lets rhizomes – underground migrating forms of life (Dovey 2008: 23) – emerge unpredictably, though they may in turn lead to new, locally stable systems (Bateson 1972). This conceptualization allows new, unchoreographed events beyond the discourse-doctrine-habitus hierarchy to come forth. It is particularly appropriate as a means of understanding the evolution of urban design in the Barangaroo case study analysed here.

The Redevelopment of Pyrmont-Ultimo

The Sydney suburb of Pyrmont is situated on an elevated harbourside peninsula immediately west of Sydney's central business district CBD, separated from it by Darling Harbour. Its neighbouring suburb of Ultimo lies adjacent to the south. Darling Harbour was Sydney's main seaport precinct until the late 1970s, when most of its port functions were shifted to new docks at Port Botany to the south. The former port role had generated a range of long-standing industrial developments in Pyrmont and Ultimo, including sugar refining, wool storage and electricity generation, creating an ageing industrial fabric that extended into the old rail lines, wharves, roads and sheds of the port itself. The loss of shipping activity meant that the New South Wales state government owned a substantial area of waterfront land near the CBD that could now be used for other purposes. This provided the main incentive for the intensive redevelopment of the peninsula under state auspices over the next two decades.

The redevelopment of Pyrmont-Ultimo got under way in 1988 when a Central Sydney Strategy, produced by the state government and Sydney City Council, identified the peninsula's development potential. The drive for (re)development was impelled by several related discourses. The first argued the need to support Sydney's role as Australia's main centre of finance and corporate control, and its role as a major Asia Pacific financial centre, a discourse supported by the state government (Department of Environment and Planning 1988: 60) and the property industry. While the latter dimension grew in strength during the 1990s (Searle 1996), Sydney's global role had already emerged following the deregulation of Australia's financial sector after 1983 and by the late 1980s had started to generate pressure on CBD office space. Pyrmont-Ultimo was seen as providing potential for further expansion of central Sydney office activities. This discourse was supported by a variant that derived from the then-prevailing discourse supporting high technology development. This variant envisaged 'an advanced technology-based community with world class telecommunications infrastructure' for living and working (Travis Partners 1991: 4).

A related discourse supporting redevelopment was that of the need for urban intensification in areas such as Pyrmont-Ultimo with good access to jobs and services. Redevelopment plans for the suburbs proposed mixed commercial-high density residential development that would provide developer flexibility and contribute to government intensification goals for Sydney. This intensification discourse had been increasingly pushed by the state government after 1980 in order to save on state infrastructure costs and meet the needs of smaller households (Searle 2007). Again, this discourse had been supported by major apartment developers, who were able to garner sizeable profits from rezoning of old industrial sites and from a local planning culture of discretionary decision-making that favoured developers (Punter 2005).

The state government drew on these discourses to produce a regional plan for Pyrmont-Ultimo to increase the population from 2,800 in 1986 to an initial target of 16,500, and to increase employment from 14,000 to 54,000 (Department of Planning 1991, 1990). To fast track new development, the regional plan had several key elements (Searle and Byrne 2002: 13–14). It rezoned Pyrmont-Ultimo and set out development principles for the proposed new development, bypassing normal local council powers to do this under the Environmental Planning and Assessment Act. The regional plan also provided for an Urban Development Plan containing detailed controls over development that would also normally be prepared by Sydney City Council. In addition, the regional plan provided for master plans on key sites, principally the harbourside sugar refinery area and the large holdings of the state port authority. On master plan sites, landowners or lessees could prepare plans providing details of proposed development, which only needed the consent of the Minister for Planning. This gave owners/developers a high level of control over their development. To facilitate development, the state government also set up the City West Development Corporation to administer government-owned sites, provide major infrastructure and implement the redevelopment plans.

A variant on the pro-development discourse was added by Greenpeace Australia (Bell 1993). In 1991, it won a national design ideas competition for its car-free redevelopment design of Pyrmont. Car parks were to be confined to the perimeter, and a light rail line would link Pyrmont to the city. Most new residential buildings would be a maximum of four storeys high, while existing industrial buildings would be reused to the maximum extent possible.

The most conservative variant of the redevelopment discourse was that of the surviving local community. Its vision was one of constraining development so that the existing close-knit sense of community and its physical associations could be preserved (King and Cadavini 1994). The community expressed its alarm at the prospect of development bringing hordes of 'yuppies' into the area (O'Brien 1992), after the government's social impact assessment stated that new residents would 'predominantly be drawn from high and middle income groups' (Brian Elton and Associates 1991: 43–4). Residents also expressed fears about the impact of tall buildings on the area's character and the associated loss of views, overshadowing and loss of privacy (Brian Elton and Associates 1991: 38). The proposals for a number of high buildings in the draft regional plan were seen by a leading community member as 'vandalism' and an 'absolute denial of Pyrmont's uniqueness' (King and Cadavini 1994). For the community, keeping the character of Pyrmont meant existing housing should be preserved and the heritage character of the area maintained (Brian Elton and Associates 1991: 35). This community discourse was successful in saving some key nineteenth-century cottages from demolition, but not others (Searle and Byrne 2002: 17–18) (see Figure 8.1). The main concentration of existing housing was in the central part of Pyrmont peninsula, which was slated for redevelopment with high-rise buildings in early planning. A heritage architect and community activist developed an alternative plan in which tall 'monumental' buildings would be built along the harbour front instead of scaling down from the centre to the harbour, thus preserving the main group of old houses (King and Cadavini 1994). The activist claimed his knowledge of government decision-making enabled him to get his proposal incorporated into the final regional plan (Hillier and Searle 1995). Nevertheless, this merely changed the location of new development under the plan, and left its total scale unchanged. But the strongest element of community discourse related to the construction of Sydney's first casino on the eastern side of the peninsula, following state government authorization in 1993. Residents feared it would block views and bring even more high-density development (King and Cavadini 1994). Their stand was supported by the City Council, which reported that the casino's size would detrimentally affect the peninsula's residential amenity, contrary to the objectives of the regional plan (Searle and Bounds 1999). To avoid a legal challenge to its development approval, the government promulgated a state planning policy that allowed inconsistency with the regional plan and enabled the casino to be built.

Figure 8.1   Traditional housing preserved after community protest, Pyrmont
Source: Glen Searle.

While the community's anti-development discourse was founded on a desire to retain as much of the existing environment as possible, the discourses of the main agents sponsoring development – state government bureaucrats and planners, and developers – had deeper doctrinal roots. Government planners saw the redevelopment of the peninsula as the best opportunity in Sydney for significant urban consolidation (urban densification) (New South Wales Government 1995), with its target population of 15,000. The government's urban consolidation policy had originally been premised on infrastructure costs savings through reduced urban expansion on greenfield sites, and these savings had been confirmed in a commissioned consultant study in 1990 (Searle 2007). The trend to smaller household sizes was a second compelling rationale for the policy. The advantages of urban consolidation in encouraging public transport and lessening motor vehicle travel were also used to justify the policy in the 1980s (Department of Environment and Planning 1988), a relationship that was confirmed in Newman and Kenworthy's (1989) influential *Cities and Automobile Dependence*. This relationship then buttressed the perceived role of urban consolidation in addressing the emerging global problem of greenhouse gas emissions. The planning case for urban consolidation thus seemed ironclad, and by the mid-1980s it had the status of a planning doctrine for government planners and politicians, with both main political parties supporting it.

Nevertheless the urban consolidation doctrine on the scale applied in Pyrmont was challenged by non-government professionals, notably by city council planners and by architects. The City Council commissioned a report from an urban designer on development outcomes to 1996, which said that the density, bulk, concentration and proximity of new residential development was 'producing an undesirable urban form inconsistent with the objectives of high quality design in the ... regional ... plan' with none of the desirable qualities of relaxation, calm, interaction, stimulation, pleasure and delight (Hogarth 1996). Architecture professor Winston Barnett had earlier predicted that a built form would emerge that was 'totally unlike [Pyrmont's] historic form and more akin to high-density development in other parts of Asia' (O'Brien 1993: 5A). The Royal Australian Institute of Architects considered that the master plan for Pyrmont Point would generate residential development that 'has the character of resort accommodation and does not retain the principles and therefore the memories and traditions of the built character remaining in the Ultimo-Pyrmont area' (Maher 1993) (see Figures 8.2 and 8.3).

Figure 8.2   High-rise apartment development, Pyrmont
Source: Glen Searle.

For state government bureaucrats and politicians, a doctrine of neo-liberalism had taken root from the late 1980s. This favoured reductions in government spending and of controls that limited investor freedom, and the promotion of increased opportunities for investment. Neo-liberalism had emerged as a hegemonic ideology within the Australian federal government bureaucracy during the 1980s (Pusey 1991) and supported a major redirection of economic policy by the Australian Government after 1983, following the examples of Reagan in the United States and Thatcher in the UK. A neo-liberalist policy bent emerged within New South Wales with the election of a Liberal-National

Figure 8.3   High-rise apartment development, Pyrmont
Source: Glen Searle.

state government in 1988 (Laffin and Painter 1995). The redevelopment of Pyrmont-Ultimo embodied the doctrine of the new government by reducing the government's onerous capital liabilities and outgoings through sales of surplus state properties, and by accommodating new dwellings at lower state infrastructure costs than in greenfield locations.

In turn, support by the various redevelopment protagonists for particular discourses and doctrines can be argued as having roots in the respective habitus. For urban designers and in large part for local planners and architects, the habitus is grounded in an understanding of an appropriate composition and texture of the local public realm and an appropriate relationship between building form and scale and the scale of the street. The precise content of this understanding might be contested and be split among the precepts of different schools, such as those of contextualism qua Sitte and rationalism qua Wagner (Cuthbert 2006: 179–86). Even so, most urban designers and architects could be argued to have shared enough common habitus elements to respond in similar (negative) fashion to development on the scale of that in Pyrmont and Ultimo, as indicated by the previous quotations from individual professionals. For a City Council that in the 1990s was controlled by community-based councillors headed by a strong Lord Mayor rather than by those representing major political parties, it was this locally-scaled habitus centring on the design amenity of new development that drove its opposition to the scheme. In this, the concordance of the councillors' support base, which came mainly from Pyrmont-Ultimo, with the ideals of the urban design professionals was a contributing factor.

Conversely, the habitus of state planners supporting intensive redevelopment of Pyrmont-Ultimo was derived from an instrumental rationality that prioritized metropolitan strategic objectives. The 1988 metropolitan strategy targeted a 50 per cent increase in multi-unit dwelling construction in the existing urban area based on concomitant state infrastructure savings (Department of Environment and Planning 1988), and this was significantly increased again by planners for the 1995 strategy (New South Wales Government 1995). The importance of supporting the development of central Sydney's emerging global city functions was also becoming apparent. The large harbourside area of obsolete port and industrial land in Pyrmont provided a major opportunity to implement these strategic goals. The state planners had little concern for design issues associated with strategy implementation. It was not until the end of the decade that the government turned its attention to measures to improve higher density residential design quality as the legitimacy of the urban consolidation programme became threatened by widespread community protests.

Across the government as a whole, a bureaucratic habitus had evolved from the late 1970s that incorporated a corporatist form of management managerial approach typical of large private companies (Considine and Painter 1997). Within the New South Wales government, a Ministerial Advisory Unit staffed by economists was set up in 1976 as the main source of advice to the new Labor Premier, focusing on departmental financial accountability (Hill 2006). The new habitus deified government efficiency and the fiscal bottom line. The Liberal-National state government of 1988–95 strengthened this modus operandi with full-blown neoliberal governance incorporating privatization, deficit reduction and the sale of redundant public assets. The redevelopment of Pyrmont-Ultimo using revenue from the sale of surplus government land and the fast-tracking of private investment that would save state greenfield infrastructure costs was very much in accord with the emergent corporatist-neoliberal mode of bureaucratic thinking and doing.

The alignment of state planner and state bureaucrat doctrines and habitus with conditions favouring private developer profit-making meant that a strong set of factors was in place for redevelopment. With actual and potential conflicts between the wider range of discourses, doctrines and habitus in play, however, the issue becomes one of understanding why particular examples prevailed to generate the redevelopment that actually happened. One way of approaching this is to view such projects in terms of their part in meeting society's expectations about what the prevailing socio-economic system should deliver. In this, the state's role is to participate in a way that provides legitimation for its existence (Habermas 1976). A crisis of capital accumulation in the West in the 1970s challenged the prevailing social democratic mode of government and its legitimacy, and generated a turn to neo-liberalism (Harvey 2005). By the late 1980s this had taken root in the New South Wales state. It was played out in strengthened urban consolidation policies (Searle 2007) and the sale of state assets that were encapsulated by Pyrmont-Ultimo's redevelopment under a public–private partnership. While the instrumental rationality of the state planners in metropolitan planning was essentially a modernist approach from earlier in the century that had by the 1990s induced a reaction against it in Europe as the role of government was questioned (Allmendinger 2009: 78), it remained influential in Australia because of a unique level of state influence over urban development (Searle and Bunker 2010). This state control meant that the City Council would be stopped by state legal powers from setting the development agenda in Pyrmont-Ultimo if that could conflict with the government's own agenda. The pre-existing local community lacked the numbers and resources for its discourse to threaten the government's legitimacy, unlike proposed redevelopment of old industrial areas in nearby gentrified Balmain several years later (Searle 2005). Finally, the role of Australia's largest developer, Lend Lease, in ratcheting up the scale of building on the old sugar refinery site beyond the regional plan's original controls was a significant factor in shaping redevelopment outcomes. This up-scaling was accepted by the government as it enabled redevelopment targets to be more quickly met, but it was merely the latest example of developer windfalls from Sydney's discretionary planning system (Punter 2005).

The Redevelopment of Barangaroo, Darling Harbour

The Barangaroo project involves the redevelopment of the last shipping wharf and associated state dockside land in Darling Harbour, totalling 22 hectares, on the western side of Sydney's CBD. As in Pyrmont-Ultimo, value capture of redeveloped state land would be used to fund necessary infrastructure and a large harbourside park. This followed years of slow growth and stagnating construction in Sydney following the 2000 Olympics. The project was endorsed by the state government as offering a major stimulus to construction in the short term and boosting longer term economic development by 'secur[ing] Sydney's role as a global financial services hub in the Asia Pacific region' (Barangaroo Delivery Authority 2010a).

An international design competition for the site, with the state setting target office space levels and a requirement for half the site to be public parkland ('Headland Park'), was held in 2005. A consortium led by a local firm won the competition. A concept plan based on the winning entry was approved by the state government in early 2007, using new Part 3A powers under the Environmental Planning and Assessment Act that allowed the minister to approve major projects instead of the local council. The government rezoned the land in late 2007. The maximum approved floor space was 388,000 square metres. Lend Lease was awarded development rights and a 99-year lease, in return for providing waterfront plaza infrastructure and funding development of Headland Park, building a pedestrian tunnel from the nearest rail station, and relocating the passenger cruise terminal, at a total cost of $521m (Barangaroo Delivery Authority 2010b).

At this point contestation over the concept plan began, leading to significant alterations. The master plan was revised to increase the maximum floor space to 508,000 square metres. The Pyrmont developer ploy of negotiating floor space increases after initial approval was played out again, a process that was rampant across Sydney: 'By a process of attrition, community endorsed Development Control Plans or Master Plans are amended again and again until developers get the building heights they want' (Moore 2002). Soon after, the design excellence review panel, led by former Labor Prime Minister Paul Keating, revised the concept plan, drawing on the scheme of the design competition runner-up led by Lord Richard Rogers. The new plan 'reinstated' the Headland Park and Northern Cove elements into a more 'natural' physical form in which the shoreline and profile would closely resemble the landscape as it existed in 1836. While this was traded off by increased maximum heights for the office towers, still more development was required at the southern end to offset Lend Lease's costs of reproducing a more natural foreshore, now totalling $150m (Lenaghan 2010). In 2010 Lend Lease applied to increase floor space by a further 60,000 square metres and to increase the height of the towers to a maximum of 209 metres, creating and valorizing better corporate views across the harbour. Much of the extra floor space would be located in a new high-rise hotel to be built on a new pier. After intense opposition from community groups (Friends of Barangaroo, Barangaroo Action Group and Australians for Sustainable Development), the government approved a modified application that reduced the commercial towers from four to three (but kept total floor space the same), reduced the hotel height to a still significant 170 metres, and reduced the pier length to 85 metres. In response, in early 2011 almost 60 architects and planners produced an alternative plan that reinstated the straight edge of the existing wharf, kept the international cruise ship terminal, eliminated the hotel and its pier, aligned roads with the city grid and instituted stepped towers (Munro 2011). After a community petition against the government's revised plan with 11,000 signatories was presented to parliament, the new Liberal-National state government (elected in March 2011) appointed an independent review panel. The panel's report broadly supported the revised plan though it criticized the bulk of the three office towers. The government rejected the recommendation to redesign the towers, but bowed to community and professional opposition and deleted the proposed pier with its high-rise hotel.

The principal discourse in this case was that of global city competition, with a sub-text of Sydney's need for stronger economic growth. The new state Treasurer claimed the project 'would enable us to realise our ambition of becoming a global financial centre' (Salusinszky 2011). The more immediate need for major development activity involving 9,000 jobs was the rationale for the new Premier to reject redesigning of the towers: '… we simply cannot afford to delay Barangaroo' (Moore 2011). This discourse was supported by the property and development industry, which saw the project as meeting almost half the underlying office demand in the CBD over the following decade, and providing the large floor plates now sought by large corporations (Hopkins 2010).

The discourse of the anti-development community groups focused on the excessive height and bulk of the project, the commercial arrangements with Lend Lease, and on the perceived threat to the character of Sydney Harbour from the hotel in particular. While one group mainly comprised disaffected local residents concerned with threats to their amenity, another group had wider planning concerns. A City Councillor who helped set up the group said it was about 'a real caring for our harbour' (Tovey 2010). This strand of the discourse reflected the iconic status which the wider Sydney community had long attributed to the harbour. Its opposition to the project thus took on doctrinal overtones.

The global city discourse can also be viewed as having the status of a doctrine within the state government by the time redevelopment at Barangaroo began to be planned. McGuirk (2004) has argued that from the mid-1990s Sydney's governance was hegemonically organized and practised to produce a competitive global city. By 2005, the priority of securing Sydney's global role was very evident within the government, with the new metropolitan strategy claiming to be a 'broad framework to secure Sydney's place in the global economy' (New South Wales Government 2005: 6). The strategy showed the CBD and North Sydney as the site of 'Global Sydney', within which Barangaroo offered the greatest development opportunity. In turn, while Barangaroo's planning can be seen as aligned with an instrumental rationality prioritizing metropolitan strategic objectives such as global Sydney, this has accommodated the project rather than been its active product as at Pyrmont-Ultimo. This perhaps reflects a weakening of a technocratic paradigm/habitus within state planning (Madanipour 2010). Barangaroo is not mentioned in the 2005 metropolitan strategy, for example.

State neo-liberal doctrine also underpinned this project, as at Pyrmont. The government has proceeded with the project on the basis that it is 'cost-neutral'. This has required the scale of office development to be massive in order to offset the extensive infrastructure and site improvement costs. Again, as at Pyrmont-Ultimo, this reflects the contemporary bureaucratic habitus of state governance emphasizing a corporatist 'new public management' approach involving, *inter alia*, flexibility and entrepreneurialism that maximize private funding to meet government objectives (see Madanipour 2010: 361).

The architectural habitus underpinning the project was split into two basic camps. A minority philosophy was that the scheme needed to make an iconic 'statement', such as the red-tinted high-rise hotel jutting into the harbour proposed by the project's lead architect Richard Rogers. Prominent architectural critic Elizabeth Farrelly considered that the government's commitment to a given bulk (floor space) meant that it has relinquished 'all real control over design' (Farrelly 2011). For her, Barangaroo is thereby 'too boring' rather than too big, epitomized by the axing of the hotel-in-the-harbour. The majority philosophy was similar to that in Pyrmont-Ultimo, centring on scale and urban grain. A series of articles in the May/June 2010 issue of *Architecture Australia* set out the majority's concerns: lack of fine-grained urbanism exacerbated by having only one developer, lack of an inclusive public realm, the failure to extend the city grid, and restricted potential for spontaneity and change. The most telling critique of the revised plan has come from Sydney City Council. The two residential towers were seen as too tall and incompatible with the adjacent proposed innovation centre. The Council also considered the heights of the three major commercial towers should be adjusted to reduce excessive bulk and to provide height differentiation. Similarly, the building floor plates were considered too large above lower levels because of their bulk and negative visual impact. The proposed street widths were considered to be too narrow and lacking in 'refinement', while streets connecting with the city grid were not properly aligned with existing streets in the grid. The high-rise hotel proposed beyond the existing shore line in Darling Harbour was seen as weakening the 'distinctive western edge of the compact high-rise city' (City of Sydney 2010). In a wider sense, the two architectural philosophies here represent the opposition of older rationalist/ contextualist schools and newer, more post-modern stances.

A further frame is needed to complete the analysis of Barangaroo. The intervention of Keating to reinstate the early nineteenth-century foreshore is a powerful example of agency exerting an influence beyond prevailing discourses, doctrines and habitus. Though his was a political appointment, his intervention turned out to have a rhizomatic quality, a subterranean influence emerging unpredictably. Keating's long-standing desire to return Sydney Harbour to a less urbanized state had emerged publicly over a decade earlier when he sought demolition of a historic finger wharf further east at Woolloomooloo.

The Barangaroo project, and the government's exercise of its power to bring it about, can – like Pyrmont-Ultimo – also be seen as helping to meet the government's need for legitimation. Towards the end of its rule (1995–2011), the state Labor government was experiencing a legitimation crisis (Habermas 1976) caused by widespread failures of state governance (Sartor 2011). Barangaroo thus can be argued to represent 'a complex dialectic whereby overt expressions of power in space tend to be commensurate with the vulnerability of that power' (Dovey 2008: 17, see also Dovey, Chapter 15, this volume). Discourses and doctrines of globalization, economic development and neoliberalism supported the way in which the project provided legitimation in the face of counter discourses and habitus.

Conclusion

Successive phases of state-led harbourside redevelopment in inner Sydney at Pyrmont and Barangaroo have involved similar competing discourses, doctrines and Bourdiean habitus at multiple scales, with similar outcomes in terms of excessiveness of scale and concomitant subordination of community and architect concerns. Yet the two cases also show how variations in the nature and strength of these analytical lenses can illuminate understanding of the specifics of each redevelopment. A state legitimation crisis had emerged by the time planning for Barangaroo had started, and the project was able to address this by recourse to globalization and economic development discourses in particular, underpinned by neoliberalist and new public management doctrines. Pyrmont-Ultimo redevelopment, by contrast, emerged primarily from a planning doctrine of urban consolidation, supported by a combination of instrumental rationality habitus and neoliberal doctrine. The Barangaroo project evolved in a more opportunistic fashion that can be understood by adding a Deleuzian-Guattarian rhizomatic perspective.

This chapter has attempted to show that a simple reading of contested planning outcomes in terms of the power of the state and major developers in determining prevailing discourses requires a deeper and less contextually-driven understanding of the underlying imperatives of doctrine and habitus. These elements have been conceptualized here as comprising successively embedded shapers of planning and development outcomes from specific development contexts to underlying modes of thinking and practice. This perspective can allow a fuller understanding of the deeper roots of planning conflict and the possibilities of conflict resolution.

The basic elements of this framework – discourse, doctrine and habitus – and their hierarchical integration are less distinct in practice than in theory. Doctrines can emerge out of long-accepted discourses, and the distinction between doctrines and their application as habitus is not always apparent. And the role of habitus was less apparent in the Barangaroo case than in Pyrmont-Ultimo. Even so, each element has potentially distinctive insights to contribute to understanding planning contestation, as this chapter has postulated.

References

Alexander, E.R. and Faludi, A. 1996. Planning doctrine: its uses and implications. *Planning Theory*, 14(10), 1487–500.

Allmendinger, P. 2009. *Planning Theory*. 2nd Edition. Basingstoke: Palgrave.

—— . 2011. *New Labour and Planning: From New Right to New Left*. London: Routledge.

Barangaroo Delivery Authority. 2010a. *Barangaroo Newsletter*, August.

—— . 2010b. *GIPAA: Supplementary Information Report*. Sydney: Barangaroo Delivery Authority.

Bateson, G. 1972. *Steps to an Ecology of Mind*. New York: Ballantine Books.

Bell, K. 1993. *Strategy for a Sustainable Sydney*. Surry Hills, NSW: Greenpeace Australia.

Bourdieu, P. 1977. *Outline of a Theory of Practice*. Cambridge: Cambridge University Press.

—— . 1989. Social space and symbolic power. *Sociological Theory*, 7(1), 14–25.

Brian Elton and Associates. 1991. *City West Urban Strategy Social Impact Assessment*. Sydney: Department of Planning.

City of Sydney. 2010. *Barangaroo Concept Plan Modification MP06_0162 Mod 4. Submission to the NSW Department of Planning*. Sydney: Council of the City of Sydney.

Considine, M. and Painter, M. 1997. *Managerialism: The Great Debate*. Melbourne: Melbourne University Press.

Crossley, N. 2001. The phenomenological habitus and its construction. *Theory and Society*, 30(1), 81–120.

Cuthbert, A.R. 2006. *The Form of Cities*. Malden, MA: Blackwell.

Deleuze, G. and Guattari, F. 1987. *A Thousand Plateaus*. Minneapolis, MN: University of Minnesota Press.

Department of Environment and Planning. 1988. *Sydney into its Third Century*. Sydney: DEP.

Department of Planning. 1990. *City West Urban Strategy: Planning Opportunities*. Sydney: Department of Planning.

—— . 1991. *City West Draft Regional Environmental Plan No. 26 and Draft Urban Development Plan*. Sydney: Department of Planning.

Dovey, K. 2008. *Framing Places: Mediating Power in Built Form*. 2nd Edition. London: Routledge.

Faludi, A. 1997. A planning doctrine for Jerusalem? *International Planning Studies*, 2(1), 83–102.

Faludi, A. and van der Valk, A. 1994. *Rule and Order: Dutch Planning Doctrine in the Twentieth Century*. Dordrecht: Kluwer.

Farrelly, E. 2011. Think big but make it exciting. *Sydney Morning Herald*, 12 May, 11.

Giddens, A. 1984. *The Constitution of Society*. Cambridge: Polity.

Habermas, J. 1976. *Legitimation Crisis*. London: Heinemann.

Harvey, D. 2005. *A Brief History of Neoliberalism*. Oxford: Oxford University Press.

Healey, P. 2004. Creativity and urban governance. *Policy Studies*, 25(2), 87–102.

—— . 2007. *Urban Complexity and Spatial Strategies: Towards a Relational Planning for our Times*. London: Routledge.

Hill, D. 2006. Working for Wran, in *The Wran Era*, edited by T. Bramston. Leichhardt, NSW: Federation Press, 59–66.

Hillier, J. 2002. *Shadows of Power: An Allegory of Prudence in Land-Use Planning*. London: Routledge.

Hillier, J. and Searle, G. 1995. Rien ne va plus: fast track development and public participation in Pyrmont-Ultimo, Sydney. *Sydney Vision: UTS Papers in Planning* 3, Planning Program, Faculty of Design, Architecture and Building, University of Technology Sydney.

Hogarth, M. 1996. Peninsula's high-rise rush causing council concern. *Sydney Morning Herald*, 9 July, 2.

Hopkins, P. 2010. All eyes on Barangaroo. *Sydney Morning Herald (Commercialrealestate)*, 20–21 November, 1.

Huxley, M. 2002. 'This suburb is of value to the whole of Melbourne': Save Our Suburbs and the struggle against inappropriate development. *Working Paper* No. 6, Institute of Social Research, Swinburne University of Technology.

King, M. and Cadavini, F. 1994. *Concrete City*, video recording. Sydney: Frontyard Films.

Laffin, M. and Painter, M. 1995. *Reform and Reversal*. South Melbourne: Macmillan Education Australia.

Lenaghan, N. 2010. Barangaroo detailed. *Australian Financial Review*, 12 November, 53.

Madanipour, A. 2010. Connectivity and contingency in planning. *Planning Theory*, 9(4), 351–68.

Maher, K. 1993. Letter for RAIA (NSW Chapter) to Department of Planning, 22 March.

McGuirk, P. 2004. State, strategy, and scale in the competitive city: a neo-Gramscian analysis of the governance of 'global Sydney'. *Environment and Planning A*, 36(6), 1019–43.

Moore, C. 2002. Public consultation complaint to Ombudsman. *Bligh News*, 83, 8 February.

Moore, M. 2011. Architects slam Premier's decision to ignore rethink on Barangaroo. *Sydney Morning Herald*, 3 October, 3.

Munro, K. 2011. New Barangaroo vision pulls hotel back from harbour. *Sydney Morning Herald*, 19–20 February, 6.

Murdoch, J. and Abram, S. 2002. *Rationalities of Planning*. Aldershot: Ashgate.

Needham, B. 1996. Planning doctrine and continuity and change. *Planning Theory*, 16, 68–70.

New South Wales Government. 1995. *Cities for the 21st Century*. Sydney: Department of Planning.

—— . 2005. *City of Cities*. Sydney: Department of Planning.

149

Newman, P. and Kenworthy, J. 1989. *Cities and Automobile Dependence: An International Sourcebook*. Aldershot: Gower.

O'Brien, G. 1992. Save us from the yuppies, say Pyrmont residents. *Sydney Morning Herald*, 7 September, 8.

——. 1993. Pyrmont: blurring the vision splendid. *Sydney Morning Herald*, 18 December, 5A.

Painter, J. 1997. Regulation, regime, and practice in urban politics, in *Reconstructing Urban Regime Theory: Regulating Urban Politics in a Global Economy*, edited by M. Lauria. Thousand Oaks, CA: Sage, 122–43.

——. 2000. Pierre Bourdieu, in *Thinking Space*, edited by M. Crang and N. Thrift. London: Routledge, 239–59.

Punter, J. 2005. Urban design in central Sydney 1945–2002: laissez-faire and discretionary traditions in the accidental city. *Progress in Planning*, 63(1), 11–160.

Pusey, M. 1991. *Economic Rationalism in Canberra: A Nation Building State Changes its Mind*. Cambridge: Cambridge University Press.

Salusinszky, I. 2011. Liberals wobble over hub. *Weekend Australian Inquirer*, 21–22 May, 7.

Sanyal, B. (ed.) 2005. *Comparative Planning Cultures*. London: Routledge.

Sartor, F. 2011. *The Fog on the Hill*. Melbourne: Pan Macmillan Australia.

Schmidt, V. 2008. Discursive institutionalism: the explanatory power of ideas and discourse. *Annual Review of Political Science*, 11, 303–26.

Schön, D.A. and Rein, M. 1994. *Frame Reflection: Toward the Resolution of Intractable Policy Controversies*. New York: Basic Books.

Searle, G. 1996. *Sydney as a Global City*. Sydney: Department of State and Regional Development and Department of Urban Affairs and Planning.

——. 2005. Power and planning consent in Sydney's urban consolidation program, in *Consent and Consensus: Politics, Media and Governance in Twentieth Century Australia*, edited by D. Cryle and J. Hillier. Perth: API Network, 297–317.

——. 2007. Sydney's urban consolidation experience: power, politics and community. *Research Paper* 12, Urban Research Program, Griffith University.

Searle, G. and Bounds, M. 1999. State powers, state land and competition for global entertainment: the case of Sydney. *International Journal of Urban and Regional Research*, 23(1), 165–72.

Searle, G. and Bunker, R. 2010. Metropolitan strategic planning: an Australian paradigm? *Planning Theory*, 9(3), 163–80.

Searle, G. and Byrne, J. 2002. Selective memories, sanitised futures: constructing visions of future place in Sydney. *Urban Policy and Research*, 20(1), 7–25.

Swartz, D. 1997. *Culture and Power: The Sociology of Pierre Bourdieu*. Chicago, IL: University of Chicago Press.

Tovey, J. 2010. City councillors launch anti-Barangaroo group. *Sydney Morning Herald*, 1 June. [Online]. Available at: http://smh.com.au/national/city-councillors-launch-antibarangaroo-group-20100531-wrgr.html [accessed 4 December 2011].

Travis Partners. 1991. *Ultimo-Pyrmont Planning Study*. Sydney: Department of Planning.

PART 3
Culture and Its Dimensions

Preface to Part 3

Deborah Stevenson

According to the influential work of Raymond Williams 'culture' refers variously to 'a general process of intellectual, spiritual and aesthetic development; … the works and practices of intellectual and especially artistic activity' and to 'particular way of life'. It was the third of these definitions – culture as a way of life – that gained currency in cultural studies, and the related field of cultural policy studies. Indeed, this expanded understanding of culture provided a language for talking about sites of struggle and forms of inequality that were cultural rather than economic. It also articulated with a democratic impulse to move away from narrowly conceived ideas of culture as 'art' to embrace and value a range of creative practices and forms from the popular to the multicultural. Significantly, too, this definitional shift also made it possible to conceptualize culture as also encompassing dynamic and pervasive processes.

In thinking about the dimensions of culture and its relationship to urban policy and planning attention also turns to the issue of urban culture which is usually understood as referring to the range of meanings, ways of life and social practices that are associated with the modern city. What emerges as important are the diversity and complexity of cities and the multiplicity of ways of living in and valuing them. Also, in studying urban cultures it is possible to discern regularities and commonalities, as well as the specificities and rhythms that mark a particular city as unique.

It is some of these themes, intersections and assumptions that are the focus of the chapters in this Part of the book, which begins with Deborah Stevenson's consideration of the way in which the idea of citizenship is mobilized or assumed within many of the most influential texts of cultural planning. The chapter argues that there are three key ways in which the idea of citizenship is invoked – the civic, social inclusion and cosmopolitanism. First, there is a view that cultural planning is a tool for reviving the 'lost' traditions, spaces and politics associated with the idea of the civic. The roots of this perspective lie in a shifting social democratic agenda, as do attempts to see cultural planning as a way to foster social inclusion. Citizenship understood in terms of social inclusion and the civic was part of a 'third way' intended to bridge the ideological Left and Right. Sitting slightly to one side of the civic and social inclusion strategies for citizenship is cosmopolitanism understood variously as a disposition and an ability to consume.

Justin O'Connor picks up on several of the themes and tensions discussed in Stevenson's chapter, notably the move within urban cultural policy from the cultural to the creative industries. O'Connor describes the emergence of the cultural industries in the 1970s as an 'opening up' of cultural policy to incorporate the products and experiences provided by the commercial sector. In the late 1990s, however, there was a shift to talking about the creative industries which he says marked a profound change in the nature and rationale of cultural policy. First, he suggests the cultural industries were promoted as part of a social democratic agenda and not as an economic end in itself. In addition, where cultural industries frameworks used economic tools and the language of economics as strategies to 'protect against' the failures that are 'intrinsic to the market', creative industries models are focused on the 'failure to achieve market success'. Finally, O'Connor argues that cultural industries approaches embraced markets in the context of 'socio-cultural practices' and the mixed economy while the creative industries is the embrace of neoliberal rational choice.

An important dimension of culture and the agendas of urban and cultural planning and any reconfigured urban planning is heritage in its built, lived and landscape forms. Indeed, as G.J. Ashworth explains in his chapter, heritage planning as a sub-section of cultural planning is simultaneously a resource, process and outcome. In exploring the different ways in which heritage is used in place management Ashworth ponders whether or not heritage planning should be considered a special type of planning, bracketed off from the planning and management of place. In pointing out that all places are made at the intersection of history and culture, he argues that it is necessary for heritage to be fully integrated into urban planning processes in order to create places that are environmentally and economically sustainable as well as meaningful and valued by local communities.

Concluding this part of the book is a significant case study from Masayuki Sasaki on the use of creative city and cultural cluster policies in the provincial post-industrial Japanese cities of Kanazawa and Yokohama. In Kanazawa the clustering of museum and crafts is intended to create a unique space and the cultural and economic contribute to both local economic and social development and cultural value. Yokohama renovated a cluster of old bank buildings and warehouses on the waterfront that are now spaces for young artists and cultural workers. The need to support systems and processes that promote both individuality and creativity is highlighted in the chapter which concludes with a call for the development of an Asian Region creative city network as the impetus for the emergence of a new form of 'Creative Asia'.

The chapters in this section of the book combine to illustrate the multidimensional nature of culture and the extent to which culture and cultural practices are embedded in, and shaped by, cities and their spaces. In recent years, an increasing number of governments around the world have recognized the potential of culture and the arts to rejuvenate cities, create places and support local economies and have implemented a range of strategies intended to achieve these ends. Strategies may include building museums and art galleries and establishing designated cultural precincts; but they can also aim to foster creative industries and cities models or mobilize an expanded definition of culture to shape local citizenship and nurture built heritage. In probing some of these issues, this section contributes significant insights into the attempts to use culture and cultural practice instrumentally in the building of cities. What is clear is that it is no longer appropriate for urban cultural strategies to be separate not only from urban planning and the broader strategies of local government, but from local communities and significant global and regional policy processes.

Culture, Planning, Citizenship

Deborah Stevenson

Introduction

Cultural planning is a strategic approach to city building and reimaging, and community cultural development that, at its most modest, involves establishing arts precincts and nurturing local creativity. More ambitiously, it often also includes supporting a 'creative cities' agenda, repositioning and expanding the arts as the 'creative' or 'cultural industries', and advocating a range of initiatives to attract and satisfy footloose capital and the so-called 'creative class'. It would be misleading to represent cultural planning as a cohesive body of thought or set of policy interventions and, indeed, the name is not even universally used. Nevertheless, it is the case that worldwide since the late 1970s and early 1980s there have been various attempts by local governments to use the arts and cultural resources strategically in the development of precincts, cities and regions and the term cultural planning is often used generically to refer to such approaches (Evans 2001).

Against a background of varying political configurations and local histories, cultural planning has increasingly also been positioned as capable of achieving an extensive range of aesthetic, social, economic and urban outcomes. Of particular note are claims that such strategies can foster an ethic of mutual responsibility, rejuvenate cities, and rebuild local communities and economies. Woven through this expansive agenda is a recurring view that cultural planning also has the capacity, indeed, some would say, responsibility, to shape and create an urban citizenry. Unclear, however, is just who this citizenry is and what the complex of rights and responsibilities that defines membership might be. Indeed, the citizen of cultural planning is contradictory, multifaceted and numerous, variously forged in the lived and imaginative spaces of the local and the global, the urban and the transnational, the 'included' and the marginal.

Through a consideration of underpinning assumptions, discourses and inter-relationships, this chapter examines some of the ways in which influential cultural planning texts (and the texts that have influenced cultural planning) have directly and indirectly mobilized and engaged with citizenship as both concept and goal – this is citizenship as a category to be constituted, assumed or applied by government through its policy and planning processes. To this end, the chapter is organized around three central themes each of which is important to these conceptualizations of citizenship –

'the civic', 'cosmopolitanism' and 'social inclusion'. It argues that although each theme is discrete they contribute in different ways to a view of the citizen as being active and locally engaged. Finally, and mindful of Franco Bianchini and Jude Bloomfield's (1996) proposition that citizenship developed through urban cultural policy has the potential to bridge the divide between the political and ideological Left and Right (the collective and the individual), a thread running through the chapter is the extent to which the cultural planning agenda of the Left regarding citizenship not only interacts and connects with, but actually supports, that of the Right. In particular, the chapter suggests that the citizen that comes into focus within many of the texts of cultural planning is one that is defined in terms of a series of opposites that are not merged or transcended but involve the subjugation of one to the other.

Recreating the Civic

It is possibly because, as Nick Stevenson (2003: 57) puts it, 'the contours of citizenship are progressively shaped by the social and political fabric of the city' that cultural planning has been couched in the language of citizenship from the outset. Indeed, citizenship was a fundamental preoccupation of the original British exponents of cultural planning and traces of this concern remain, including in many derivative versions developed elsewhere. The centrality of citizenship to British cultural planning would appear in part to be an outcome of the tradition's origins within the British Labour Party (Bianchini et al. 1988, Mulgan and Worpole 1986, Worpole 1992) and notions of citizenship in early cultural planning treatises are imbued with a range of meanings which are grounded quite specifically in the history of the labour movement, an agenda for social justice, support for the welfare state, and a concern with local political configurations including the relationship between the tiers of British government. Also, as Bianchini and Bloomfield (1996) discuss, in the 1980s and 1990s many Labour local governments in the UK attempted deliberately to use cultural planning to strengthen civic identity, in part, in an effort to build a form of shared city identification. Significantly, this was a form of identification and citizenship that was understood as being forged in the city centre – the 'civic heart' of the city.

According to Ken Worpole (1992: 4), the city centre is the 'essence' of place, the 'focus of civic identity'. It is in this context that contemporary placemaking and urban revival strategies, such as the development of cultural precincts or 'quarters' become implicated in discussions of citizenship and democracy. In the balance is the belief that an empowering 'civic culture' (Montgomery 1990: 105) was once instrumental in shaping democratic politics and local citizenship (Bianchini and Bloomfield 1996) and has the potential to do so again. Notwithstanding Colin Mercer's (1991: 9) call for the 'nature and meaning of the … civic realm' to be redefined, a task of cultural planning was (somewhat romantically) to facilitate or create the conditions for its revival.

The centrality of the 'civic' to foundational cultural planning thinking is illustrated by the following passage from Geoff Mulgan and Ken Worpole's (1986: 27) pioneering manifesto *Saturday Night or Sunday Morning? From Arts to Industry: New Forms of Cultural Policy*, which both links citizenship to the civic and points to the political tradition that underpins its original usage within British cultural planning:

The key word in post-war Labour Party vocabulary was 'civic'. It expressed the strong sense of active citizenship which came out of the war; it expressed a sense of there being such a thing as a 'civic culture' – the reciprocal responsibility between state and citizen, and amongst citizens towards each other. 'Civic responsibility' and 'civic pride' were transformed into 'civic halls', 'civic baths', 'civic gardens', 'civic theatres' and so on ... This is where the heart of such cultural policy as there was at a local government level was expressed: through very patrician forms of municipal provision.

This particular understanding of the citizen and the allied concept of the civic established a rationale that justified the scope and objectives of cultural planning as it came to be promoted first in Britain and then elsewhere. It is important also to note that this citizen is regarded as being 'active', which is a conception that, as discussed below, put the onus on the choices and decisions of individuals, and opened the way for subsequent pronouncements focused on social inclusion and access to opportunity. Importantly, active citizenship is also a discourse that was appropriated by the British Conservative Party during the Prime Ministership of Margaret Thatcher in a way that fused citizenship with the idea of the responsible consuming individual. This was the citizen of 'post-society' individualism and 'moral responsibility' (Bianchini and Schwengel 1991: 218).

Within the texts of cultural planning the civic is both a political and a physical space and many of the cultural strategies that were pivotal to reviving it and local citizenship were those devised to protect what was regarded as the integrity of public buildings and public space, in particular those in the inner city. (See also, Watson, Low, and Paddison, Chapters 1, 17 and 18, this volume.) Interventions such as the promotion of cultural activity and the development of cultural precincts and other leisure and recreational spaces, were expected to stimulate urban culture, animate the city and protect its spaces, at the same time as being conceptualized as part of an agenda to revive the core traditions and spaces of the civic, including local participatory democracy and fostering a 'collective morality' (Mulgan 1989: 263). The ideas of the civic and the citizen that were mobilized within cultural planning welded with those associated with city imaging and precinct development from the outset. They also became entangled with the idea of the morally responsible active citizen that was being conjured by the Right in Britain at the time as cultural planning was gaining currency.

Either as an element or a consequence of being integrated into an expanded nexus of interests and associations, the 'devolution of power to community groups' (Bianchini and Schwengel 1991: 233) – the revitalization of local democracy – was also seen within cultural planning as being important to nurturing the civic and building active citizens. Furthermore, the promotion of consultation and participation strategies and practices also became entwined with ideas of citizenship as being fermented in the spaces of the civic and grounded in assumptions about traditional working-class solidarity (Mulgan 1989). Mulgan's (1989: 276) advocacy of cultural planning was in part the result of his concern that the 'cultural roots of democratic, public life [were] withering' as a consequence of the privatization of public space. He warns that: 'Any plans for the creation of convivial, communicating cities inevitably find themselves struggling with a long erosion of the traditional political structure of the city within which people think, argue and organise' (1989: 275). A foundational goal of cultural planning, therefore, was to address this demise by implementing strategies intended to redevelop and animate public space, and foster local community engagement and participation. Charles Landry (2003: np) talks about 'civic creativity' as being the application of, 'imaginative problem solving to objectives of public good. The aim is to generate a continual flow of innovative solutions to problems which have an impact on the public realm.' These opinions connect with the broader view that cities play a critical role as places of public debate and the exchange of ideas with Jurgen Habermas (1991) being one commentator who has documented in detail the link between political debate and the public sphere and changes to both over time (Stevenson 2013a). The public sphere is not coterminous with public space (although it is often treated as such), but it does contain components of spatiality.

Sitting uneasily within initial expositions of cultural planning and their pronouncements regarding the (re)animation of the civic, were a number of more conservative influences and discourses which were also deeply implicated in prescriptions to revive local citizenship and its precincts and spaces, and address what was regarded as the decline of civic life. In this context, neoliberal economics and idealized notions of the Roman *piazza* and the Greek *polis* emerged as important.

The Urban *Civitas*

In his influential book on urban culture, the American social commentator Richard Sennett (1990) explores the emergence of modern urbanism as a process of 'wall building' whereby urban life has become 'trivialized' to such an extent that it separates subjectivity from the experience of the outside world. At the root of this change, he argues, is fear. The desire to construct urban spaces in a way that protects (some) city dwellers from the threat posed – or perceived to be posed – by other users of the city, is an outcome of this fear. The result, says Sennett (1977: 338) in an earlier work, has been an eroding of the 'balance between public and private life, a balance between an impersonal realm in which men [sic] could invest one kind of passion and a personal realm in which they could invest another'. Sennett regards the Greek *polis* as a use of urban space and an acceptance of the 'reality' of urbanism that are superior to the current situation. He does not deny that the contemporary urban experience of difference and confusion is new, but argues that attempts to insulate people from the reality of this urbanism are. He calls for simpler urban spaces to again be designed and built that will 'produce … more social complexity, more social interaction' (Ravlich 1988: 473).

Many early cultural planning commentators shared Sennett's concerns and some considered the Greek *polis* and its particular organization of (civic) space as being the archetype of urban life and politics. Mulgan (1989: 275) also has pointed out that there is an association between politics and city life that can be traced back to the Greek *polis*, and this, in turn, is a model for its revival in the political and physical spaces of the contemporary civic. It was by using Sennett's work as his point of departure that influential cultural planning exponent Colin Mercer (1991: 1) came to suggest that cultural planning was a tool capable of being used to revive the 'outside' as a 'dimension of human experience' and thereby addressing what he says Sennett had identified as a fundamental urban problem – the disequilibrium between, '… *urbs* and *civitas*, stones and rituals, shelter and emotions, commerce and citizenship, outside and inside'.

Cultural planning based on the animation of the city and the revival of the civic came to be positioned as a means by which 'the walls' which separate urban dwellers could be removed and a 'third way' established that either reconciled or built connections between opposites. Mercer (1991: 2–3) suggests that this quest has several pivots including treating the city as a 'stage', utilizing and supporting local 'cultural capital' and, importantly, linking economic, social and cultural objectives. In a later work specifically on cultural policy, planning and citizenship, Mercer (2002: xix) argues that 'citizenship is what cultural policy is – or should be – about' because, as he explains, it is concerned with '… the resources which define, enable, constrain and shape (both positively and negatively) that most fundamental of human capacities: identity'. Evident here is also a subtle shift in focus if not away from the communal and the collective to the individual and identity, then surely their conflation. This repositioning is important because, as Bauman (2011: 15) has observed in a very different context, the idea of identity has become something of a 'surrogate for community' which in turn has led to the privileging of the individual over commonality, of autonomy over interdependence. It is in a society of individuals, Bauman suggests, that people increasingly come to fear (and actively seek to avoid) difference, diversity and the 'other', and herein lies the challenge for cultural planning, which is positioned as capable of achieving two competing outcomes – creating a citizenry through, and in, the spaces of the civic and the communal, whilst also fostering via the same processes and legitimizing discourses often very individualized forms of identity, cityscapes and conceptions of cultural diversity. Both the merging and balancing of opposites are inherently fraught and conflicting endeavours but they are at the core of the mission of cultural planning.

In the foundational texts of cultural planning, local cultural development and nurturing the civic or participatory realm and its spaces are central, as is the reconciliation of opposites, and the active citizen is imagined variously, often simultaneously, as being part of a local community and in terms of a shifting set of identities. The task thus becomes one of using cultural strategies to create new public or civic spaces that will be the crucibles for the emergence of reconfigured forms of civic identity and citizenship. This is a view of citizenship that floats free of idealized notions of working-class solidarity and community and their collective spaces (although, importantly, these themes and legitimizing discourses remain resonant), to foreground instead cultural identity and a reconceptualization of the space of the civic in terms of diversity, consumption and identity. It may be that cultural planning aims to foster community and identity simultaneously and to move beyond fear and wall building to celebration and engagement, but where one approach stresses homogeneity and commonality as the bases of civic identity and solidarity, the other is concerned with heterogeneity and difference, the individual rather than the communal. These are opposites that cannot readily be balanced or reconciled and, as a result, the civic increasingly is mobilized in the service of identity and the individual. In this regard, a theme that emerges as important is cosmopolitanism, which is understood in terms both of the city (its amenity and spaces) and the disposition of the urban citizen.

The Cosmopolitan Citizen

With the privileging of identity over community has come an associated concern to support the social, cultural, racial and ethnic diversity of 'citizens'. The spatial, and in particular the urban, dimensions of diversity are paramount because it is in the city and specifically in its public spaces that difference in all its expressions is most evident (D. Stevenson 2003). The city is where, as Simmel (1995/1903) famously argued at the start of the twentieth century, the freedom to be different is possible. To this end, cultural planning is positioned as being something of an instrument for fostering diversity through the creation and animation of public space and for negotiating and neutralizing the tensions that form in the context of the coexistence of difference. What is at stake is the linking of the 'universalism of individual human rights with the particular rights of minority groups' (Bianchini and Bloomfield 1996: 88) and the establishment and maintenance of spaces within which both commonality and diversity can be expressed and encouraged. The civic, as an idea and a space, again comes to prominence. The ideal places and spaces of cultural planning are increasingly conceptualized as intercultural sites of tolerance and understanding, while the ideal urban citizen is an individual who is cosmopolitan in outlook and taste.

A number of interesting concepts have developed to explain urban diversity with Leonie Sandercock's (2003) metaphor of the 'mongrel city' being perhaps the most evocative in pointing to the complexity of a '… new urban condition in which difference, otherness, fragmentation, splintering, multiplicity, heterogeneity, diversity, plurality prevail' (2003: 1). Cosmopolitanism 'understood as implying a particular stance towards difference in the world, one that involves an openness to, and tolerance of, diversity' (Young, Diep and Drabble 2006: 1687) is a recurring notion in recent attempts to theorize the contemporary city in terms of culture, identity and difference. And public space – its uses, purpose, design and management – is often at the centre of these discussions. Indeed, public space as the location of encounters with difference is often bracketed with cosmopolitanism as a 'foundational element of any city' (Sassen 2010: 490). It is where 'differing diversities' (Bennett 2001) are at their most obvious and, potentially, most volatile.

Cosmopolitanism is commonly understood as an orientation or ethos that transcends geopolitical borders and traditional social categories; it is a disposition (*habitus*) or set of dispositions associated with possessing the cultural capital to appreciate, and engage with, cultural difference and 'the global' (N. Stevenson 2003). It is in many ways the triumph of identity over community. The idea of cosmopolitan citizenship thus assumes a reworking, if not a rupturing, of the established taken-for-granted relationship between citizen and nation/place, because as is claimed repeatedly, the cosmopolitan is a 'citizen of the world'. Beck and Grande (2008: 11–12) suggest that the meanings ascribed to cosmopolitanism are simultaneously 'old' and forward pointing, combining an 'appreciation of difference and alterity with attempts to conceive of new democratic forms of political rule beyond the nation-state'. For them, cosmopolitanism describes a 'specific way of *dealing socially with difference*' (2008: 12) that is de-territorialized in that it transcends national or local contexts. In their survey of the field, Woodward, Skrbis and Bean (2008: 3) point out that conceptualizations of cosmopolitanism are fluid and 'can mean anything from an attitude or value, to a regime of international governance, or a set of epistemological assumptions about the nature of social structures'. It is a 'what' and a 'who' (Woodward, Skrbis and Bean 2008: 2) – a form of politics and an openness of self that is premised on mobility, transnationalism and transience. Cosmopolitanism conceived thus also challenges the fixity and certainty of home, place and community – core ideals of traditional cultural planning.

If cosmopolitanism is beyond borders and geography then it must by implication exceed the city and associated understandings of urbanism and the civic which, as discussed above, reside at the centre of its imagined, active citizen. Like the notion of 'cosmopolitan citizenship', the 'cosmopolitan city' appears to be something of an oxymoron. But this is not so because, frequently, the city is regarded as being pivotal to the development and expression of cosmopolitanism, sometimes only because it is in the city that the cosmopolitan elite is said to reside. The cosmopolitan citizen invoked through cultural planning texts may be globally aware and mobile, but mobility is as much about stopping points and locations as it is about movement, and cities are critical stopping points. Their precincts are the locales where cosmopolitanism is lived, experienced and made meaningful. Everyday cosmopolitanism is regarded as 'multiple', fractured and shaped by local circumstances and a range of socio-cultural factors (Woodward, Skrbis and Bean 2008: 18). Not only are major cities the spaces of lived cosmopolitanism, but they are also seen as the fulcra of significant global networks and associations, including migration. So-called 'world' or 'global' cities are defined in terms of a myriad of interconnections and exchanges that are city-to-city and not nation-state-to-nation-state – cities are 'nodes within a network that is globally interconnected, while being simultaneously locally disconnected' (N. Stevenson 2003: 57). These observations point to important themes in understandings of cosmopolitanism and, in particular, cosmopolitan citizenship, that have resonance in cultural planning discourses, especially those that regard cosmopolitanism as a quality (disposition) of urbanism. There are two entwined dimensions that are relevant – one that emphasizes the need to accept that difference is a critical part of a lively and engaged intercultural city, and another that mobilizes the idea of the urban creative class.

Creative Citizenship

The starting point for a cultural planning approach to fostering cosmopolitanism is the recognition that cities are places of ethnic and racial diversity and so either are, or have the potential to be, sites of (inter)cultural tolerance, dialogue and innovation – the locales of active citizenship. Cultural planners Phil Wood and Charles Landry (2008: 317) claim that while '[d]ynamic cities have always attracted migrants', what matters is not their presence but attitudes towards them and the ways in which these attitudes are expressed in everyday routines and practices of engagement and avoidance. The presence of diversity does not necessarily foster an urban culture that is tolerant, or indeed, 'dynamic', because with cultural diversity comes the 'potential for conflict' (Wood and Landry 2008: 317) and as discussed, above, 'wall building'. Attitudes towards diversity vary from 'active hatred' to 'active interaction … and co-creation' (Wood and Landry 2008: 16). The centrality of activity, elective engagement and initiative to this imagined citizen is again noteworthy.

The task of cultural planning according to its advocates is to find ways of promoting and constructing the spaces (physical and communicative) within which 'intercultural' meetings and exchanges can occur in order to foster understanding and tolerance, and create engaged, active citizens. (See also, Sasaki, and Evans, Chapters 12 and 13, this volume.) These are the spaces of the reimagined civic and the reconfigured public sphere. For Wood and Landry (2008: 319) a successful city is one that implements policies and processes that will promote and support intercultural mixing, exchange and understanding, with the most successful cities and urban cultures supposedly being those that embrace diversity as a resource to be nurtured and an opportunity to be grasped. (See also Bianchini, Chapter 22, this volume.) This success is described as multifaceted and evident across a range of indicators and it is actively played out in the spaces of the civic, which, increasingly, are understood as precincts that are sites of sociality and consumption.

As a place of engagement and interaction, the intercultural creative city framed within the discourses of cultural planning is thus at the forefront of a renegotiation of citizenship and urbanism, but this is a form of citizenship conceptualized not as transcending the local but as melding the local with the global. In other words, opposites are again being balanced/merged with location being understood in the context of the global. Wood and Landry (2008: 273) point out that there are some who argue that the intercultural citizen has (or should have) a status that is shaped in response to a set of local-global priorities. The more cogent point, however, is that the form of citizenship that emerges in relation to the intercultural is one that is cosmopolitan in outlook and disposition at the same time as being locally committed. Although having values that are formed through a global viewpoint, they are said, nevertheless, to identify strongly with the city in which they live, work and recreate. This commitment and identification is deeply political. Indeed, Wood and Landry (2008: 284) say that they 'cannot emphasize enough the importance of restoring the political (both formal and informal) to the heart of civic life' in the intercultural city.

The citizen of the intercultural city is conceived as being actively engaged in the local politics of everyday urban diversity and its consequences. This is diversity as it is encountered at the level of the street and the precinct and understood through a sensibility that is international (cosmopolitan). The intercultural city thus differs from the multicultural city, which Wood and Landry argue can be a place of ghettoes and parallel coexistence. An intercultural conceptualization of citizenship and urban diversity also extends beyond more usual multicultural models in that its principal assumptions are not only that racial and cultural difference must be recognized, accepted and, indeed, encouraged, but that it is necessary to put in place mechanisms to foster mutual understanding, common aspirations and the sharing of space. Cultural planning is one such mechanism. Where multiculturalism emphasizes difference, the intercultural is concerned with where, through what processes, and in which spaces difference intersects. That said, it is difficult to imagine a situation where it might actually be possible to talk about an 'intercultural city' as opposed to 'intercultural spaces' within a city – particular parks, markets and precincts that 'work' as the locales for intermingling and cultural exchange. Even cities that contain enlivened and diverse public spaces will almost certainly also contain enclaves of homogeneity bordered by fences and gates, both real and imagined. The intercultural city is thus in effect about precincts and neighbourhoods and the activities that occur within them.

Discourses of diversity and cosmopolitanism are also present in the work of Richard Florida (2003) which has become influential within cultural planning, being key ingredients of those aspects of city life that supposedly appeal to the (cosmopolitan) creative class. Florida's is a formula for urban renewal and economic prosperity that pivots on adopting local cultural policies to 'attract, retain and even pamper a mobile and finicky class of "creatives", whose aggregate efforts have become the primary drivers of economic development' (Peck 2005: 740). This creative class is made up of a core group of people engaged in such fields as the arts, research and science, and a peripheral group that comprises those working in areas including law, health, business and finance (Florida 2003: 8). The creative class supposedly is defined by its ability to be innovative and flexible in its work and by having a disposition that is outward looking and cosmopolitan. In Florida's (2003: 226) schema the creative class chooses to live, work and consume in cities (and precincts within cities) that are demographically diverse and where variety 'of thought and open-mindedness' is accepted. Its members 'actively seek out places for diversity and look for signs of it when evaluating communities' (Florida 2003: 226). One could readily read this as the intercultural (creative) city, and Landry (2006) has certainly pointed to this connection. For Florida (2003: 227) '[d] iversity also means "excitement" and "energy" and creative-minded people enjoy a mix of influences. They want to hear different kinds of music and try different kinds of food. They want to meet and socialize with people unlike themselves, to trade views and spar over issues.' The extent, however, to which the result is fundamental cultural change and exchange or simply an expanded suite of food and consumption options remains an open question.

Following Florida's blueprint, in order to be economically prosperous cities are counselled to implement strategies and plans that will develop the urban, social, cultural and economic infrastructure that will attract the creative class. (See also, Miller, Chapter 3, this volume.) Cities must be cosmopolitan if they are to appeal to the cosmopolitans. In turn, the presence of the cosmopolitan creative class supposedly animates a city and makes it a place where businesses will want to locate. To this end, cities are encouraged to build art galleries, restaurants, cultural precincts and other forms of 'soft' creative infrastructure, including, public space, pedestrian zones, effective street lighting and 'appropriate' (in design and placement) street furniture, and cultural precincts are key sites in this regard. Jamie Peck (2005: 741) is not alone in suggesting that Florida, '... mixes cosmopolitan elitism and pop universalism, hedonism and responsibility, cultural radicalism and economic conservatism, casual and causal inference, and social libertarianism and business realism'. The creative class blueprint is, however, also implicitly a recipe for gentrification and the displacement of often-lower-income (apparently non-creative, non-cosmopolitan) residents because the spaces that the creative class is attracted to are often those in previously low socio-economic areas located in or near the city centre (Stevenson 2013a). Although a core aim is to foster urban diversity and cosmopolitanism, the result can also be the opposite – wall building, displacement and the reproduction of sameness.

The cultural planning approaches of Florida, and Wood and Landry and others grounded as they are in intriguing mixtures of urban planning, economics, sociology and motivational psychology, have proved to be seductive, accessible and highly marketable. Jim McGuigan (2009) is one of many to note the eagerness of local governments around the world to '[do] a Florida thing' and implement this approach to cultural planning. These consultants and their blueprints have successfully captured the attention of city leaders, artists and urban planners alike and been influential in shaping not only local cultural planning approaches, but also debates about the idea of the city and what is meant by 'city-ness', civic culture and urban citizenship. Culture and creativity supposedly can be measured, developed and then traded in an international marketplace comprised of cities eager to compete with each other on the basis of image, amenity, liveability and cosmopolitanism (Richards and Palmer 2010). Also underpinning such cultural planning approaches is a promise of positive social outcomes and it is here that social inclusion emerges in cultural planning as a dominant theme in the imagining of the active citizen.

Including the Citizen

In order to understand social inclusion and its mobilization within cultural planning texts, it is necessary to return to the connection between British cultural planning and the Labour Party (discussed above) and specifically the 'third way' schema of the former Blair government. The third way was an ideological and pragmatic shift, a strategic response to the challenges seen to be posed by: globalization; individualism; the collapse of the political division of Left and Right; emerging forms of (niche) political affiliation; and ecological awareness (Giddens 1998). It was a hybrid position explicitly intended to transcend (or bridge) the divide between the traditional Left and the neoliberal Right (Rose 2001). According to David Byrne (2005: 151): 'The key themes of the third way are accommodation (seen as inevitable) with the agendas of globalizing corporate capitalism, equality of opportunity rather an outcome, and a concentration on the creation of wealth rather than its redistribution.' To this end, the third way reframed many of the core social and political assumptions and principles of the Left, including those associated with the role of the state, the nature of citizenship and community, the goal of social justice (Everingham 2003) and the civic and the public sphere. It privileged a form of participatory democracy and active citizenship that was highly individualized and framed in terms of reciprocity between citizen and state.

Significantly, the third way shifted away from a concern with social justice that was traditionally at the centre of the mission of the Left, its ideas of the civic and approach to building citizens, to mobilize the discourse of social inclusion, which assumes an active citizenry. Even though social inclusion and social justice are frequently used interchangeably including in cultural planning texts, they are grounded in different ideological assumptions. Where social justice is premised on a commitment to social equity, social inclusion is concerned with social order; social justice is understood in terms of a set of structural relationships that limit the ability of some social groups to access social, economic and cultural resources, while social inclusion refers to the desire of, and relative opportunities available to, individuals to 'participate' in society; social justice requires an interventionist state with a redistributive agenda and a concern with social outcomes, while social inclusion legitimizes mutual obligation, 'small' government and equal opportunity (Everingham 2003). The aim of the social inclusion agenda of the third way, therefore, was to put in place a range of policy initiatives intended to give the marginal(ized) the 'opportunity' to become full members of society. In other words, citizenship within a social inclusion framework is conceived as active, not passive; it is about individuals not communities (Rose 2001).

As a third way strategy, cultural planning advocates fostering the creative industries and attracting the cosmopolitan creative class as a way to facilitate social inclusion and create active citizens by making it possible for people to participate in the cultural economy as producers and/or consumers. Indeed, social inclusion within the texts of cultural planning frequently privileges the economy to such an extent that participation in society (full citizenship) is often understood as being achieved through active participation in the economy. And if it does not explicitly privilege the economy its third way porousness creates the space for this outcome. Culture and creativity, therefore, are regarded as forms of capital, capacities that can be developed to ameliorate social exclusion with the core assumption being that through their active participation in the economy, the marginal have the opportunity to become integrated into society as 'full' citizens.

Woven through cultural planning strategies are prescriptions for addressing urban decline, creating public and quasi-public spaces (precincts and quarters) and supporting the creative industries in order to achieve social inclusion and foster active citizenship (Mercer 2002, Worpole 1992). As discussed above, there was a view within early cultural planning texts that the physical decline of the civic had led to an erosion of the core ingredients of active citizenship – the physical spaces of democracy and the 'free' exchange of ideas (Mercer 2002, Worpole 1992). These themes are also relevant to the related goal of social inclusion and with respect to the third way Giddens (1998: 79) has suggested that 'government can and must take a major part in renewing civic culture', 'civic liberalism', and in 'recapturing … public space'. The reimagined civic is the physical and imaginative space of creativity, cosmopolitanism, consumption and citizenship. Cultural planner Lia Ghilardi (2001: 129) explains some of these connections and their importance to fostering social inclusion:

> The cultural industries support services developed within those agencies focused on issues of access, social inclusion and participation as much as on business generation. Social inclusion here is understood as an incentive to cultural production and as a way of fostering civic pride, and a sense of local identity and ownership. … The philosophy behind the above developments is that of a 'productive' use of diversity to create a sustainable skills base and a culture of innovation capable of yielding economic rewards for everybody. This is an approach that sees cultural diversity not as a problem to be controlled by top-down policies, but as an asset for the development of the local community.

The key themes are all here – inclusion, community, identity, the civic and diversity – mobilized in the service of the cultural economy and the facilitation of active citizenship.

Conclusion

Cultural planning is frequently promoted as a way of achieving a broad range of social, economic, artistic and urban goals. Significantly, too it is often positioned as a strategy for reconciling opposites including, the global and the local, opportunity and outcome, and community and the individual. As an aspect or consequence of this expansive agenda, cultural planning is seen as playing an important role in creating citizens and the spaces within which citizenship forms. Indeed, citizenship within cultural planning is frequently understood as linking the individual and the collective – as capable of retaining the social concerns of the Left at the same time as embracing entrepreneurial approaches to urban, and cultural, development. Pivotal is a view of the citizen as actively participating in the spaces of local democracy, economy and society, with the role of cultural planning being to provide the context for within which this can occur. Often this means animating public space through the development of cultural precincts and the provision of 'soft' creative infrastructure, as well through gentrification and establishing entertainment and consumption zones. Cultural planning in its broadest and narrowest manifestations is framed as being a facilitator of opportunity, an invitation to citizenship. At the same time responsibility for cultural and public provision has shifted from the government to the private sector and understandings of citizenship have become entwined with the ability to consume. The chapter argued that it is possible to identify three separate but intersecting themes within the cultural planning literature and treatises that are important to this positioning of the citizen – the civic, cosmopolitanism and social inclusion.

Initially, a concern to use cultural planning as a tool to build local citizens was an outcome in part of its foundations within in the British Labour Party, which embedded in cultural planning the importance of the civic (as a physical and political space) and the provision of public spaces and facilities in the civic heart of the city; but this agenda also supports entrepreneurial and city-imaging strategies often focused on consumption. The citizen conceived in terms of the civic has become one that is constituted principally in the public and quasi-public precincts of the (gentrified) inner city and, in spite of being couched in the language of democracy and community, is highly individualized, mobile and middle class. The chapter goes on to say that the idea of cosmopolitanism has become particularly relevant in this context, emerging in recent years to be a key dimension of cultural planning including its invocation of the citizen. 'Cosmopolitan citizenship' is both a contradictory notion and shaped through a number of seemingly incongruous but, nevertheless, intersecting discourses including, in particular, those associated with the creative class and the 'intercultural' city. The citizenship of cosmopolitanism also speaks to themes of the global which are in tension with both a notion of citizenship framed in terms of the civic and the city, but also with the third dimension of a cultural planning conception of citizenship – social inclusion.

Citizenship imagined in terms of social inclusion is where cultural strategies and programmes are used to give marginal(ized) groups the opportunity to participate fully in society and by extension become 'active' citizens. Social inclusion is often understood as being something of a synonym for social justice, which was an original aim of British cultural planning; however, as the chapter argues the two have very different objectives and are grounded in very different ideological assumptions, with social inclusion being focused on providing individuals with opportunities to be active citizens while social justice is concerned with outcomes and communities. What is also pivotal is that the achievement of social inclusion is often premised on an engagement in the cultural economy.

At the heart of the way in which cultural planning evokes the citizen is a series of oppositions, including Left–Right, community–identity, global–local and opportunity–outcome, which are played out in relation to themes of the civic, cosmopolitanism and social inclusion. Rather than bridging the divide between opposites, however, all too often the outcome has been the subordination of one to the other with the idea of the active citizen being central. The terrain of citizenship with its overtones of the collective may once have been the province of the Left but this is no longer the case. Now through the language of the 'active citizen', citizenship is also coded individual and linked with the private sector, consumption and identity.

As I have argued elsewhere (Stevenson 2004) cultural planning is premised on a discursive pun. It aims to develop local cultural activity, attract the cosmopolitan creative class to the intercultural city and use the creative industries to fashion a lively economy in ways that will position a town or city as a 'cultural capital'. At the same time, however, these very strategies and objectives are being positioned as fundamental to the development of the 'cultural/creative capital' of the local population in a way that will address social exclusion, foster participation in the economy and create active citizens. The need for an evaluation and reconceptualization of these relationships and goals and of what citizenship means within a cultural planning framework has clearly become urgent and much ground has already been given up. At stake is a fundamental challenge to the theories and the practices of cultural planning and many of its underpinning assumptions, particularly those grounded in the now outmoded third way compromise. Such a challenge is necessary, however, to make possible a reconceptualization that is more than superficial and for cultural planning to become a truly progressive approach to supporting local citizenship and providing the spaces within which this citizenship is expressed and affirmed.

Acknowledgements

A version of this chapter appears in Stevenson (2013b).

References

Bauman, Z. 2011. *Community: Seeking Safety in an Insecure World*. Cambridge: Polity.

Beck, U. and Grande, E. 2008. *Cosmopolitan Europe*. Cambridge, UK and Malden, MA: Polity.

Bennett, T. 2001. *Differing Diversities: Transversal Study on the Theme of Cultural Policy and Cultural Diversity*. Strasbourg: Council of Europe Publishing.

Bianchini, F. and Bloomfield, J. 1996. Urban cultural policies and the development of citizenship: reflections on contemporary European experience. *Culture and Policy*, 7(1): 85–113.

Bianchini, F., Fisher, M., Montgomery, J. and Worpole, K. 1988. *City Centres, City Cultures: The Role of the Arts in the Revitalisation of Towns and Cities*. Manchester: Centre for Local Economic Strategies.

Bianchini, F. and Schwengel, H. 1991. Re-Imagining the city, in *Enterprise and Heritage: Crosscurrents of National Culture*, edited by J. Corner and S. Harvey. London: Routledge, 212–34.

Byrne, D. 2005. *Social Exclusion*. Maidenhead: Open University Press.

Evans, G. 2001. *Cultural Planning: An Urban Renaissance?* London: Routledge.

Everingham, C. 2003. *Social Justice and the Politics of Community*. London: Ashgate.

Florida, R. 2003. *The Rise of the Creative Class: And How it's Transforming Work, Leisure, Community and Everyday Life*. North Melbourne: Pluto Press.

Ghilardi, L. 2001. Cultural planning and cultural diversity, in *Differing Diversities: Transversal Study on the Theme of Cultural Policy and Cultural Diversity*, edited by T. Bennett. Strasbourg: Council of Europe Publishing, 116–27.

Giddens, A. 1998. *The Third Way: The Renewal of Social Democracy*. Cambridge: Polity.

Habermas, J. 1991. *The Structural Transformation of the Public Sphere: An Inquiry into a Category of Bourgeois Society*. Cambridge, MA: MIT Press.

Landry, C. 2003. unpublished contribution to an on-line Opendemocracy debate about aesthetics in planning. [Online]. Available at: www.opendemocracy.net/debates. html [accessed 10 February 2012].

——. 2006. *The Art of City-Making*. London and Sterling, VA: Earthscan.

McGuigan, J. 2009. Doing a Florida thing: the creative class thesis and cultural policy. *International Journal of Cultural Policy*, 15(3), 291–300.

Mercer, C. 1991. *Urbs et Civitas: Cultural Planning as City Planning*. Paper presented to the Urban Environments Seminar Series: The Economics of Place and People, Penrith, August.

——. 2002. *Towards Cultural Citizenship: Tools for Cultural Policy and Development*. Stockholm: Bank of Sweden Tercentenary Foundations and Gidlunds Förlag.

Montgomery, J. 1990. Counter revolution: out-of-town shopping and the future of town centres, in *Radical Planning Initiatives: New Directions for Urban Planning in the 1990s*, edited by J. Montgomery and A. Thornley. Aldershot: Gower, 105–18.

Mulgan, G. 1989. The changing shape of the city, in *New Times: The Changing Face of Politics in the 1990s*, edited by S. Hall and M. Jacques. London: Lawrence & Wishart.

Mulgan, G. and Worpole, K. 1986. *Saturday Night or Sunday Morning? From Arts to Industry: New Forms of Cultural Policy*. London: Comedia.

Peck, J. 2005. Struggling with the creative class. *International Journal of Urban and Regional Research*, 29(4), 740–70.

Ravlich, R. 1988. City limits. *Meanjin*, 47(3), 468–82.

Richards, G. and Palmer, R. 2010. *Eventful Cities: Cultural Management and Urban Revitalisation*. London and New York: Butterworth-Heinemann.

Rose, N. 2001. Community, citizenship and the third way, in *Citizenship and Cultural Policy*, edited by D. Meredyth and J. Minson. London: Sage, 1–17.

Sandercock, L. 2003. *Cosmopolis II: Mongrel Cities of the 21st Century*. London and New York: Continuum.

Sassen, S. 2010. Public space, in *The Endless City*, edited by R. Burdett and D. Sudjic. London: Phaidon, 490.

Sennett, R. 1977. *The Fall of the Public Man*. Cambridge: Cambridge University Press.

——. 1990. *The Conscience of the Eye: The Design and Social Life of Cities*. New York: Alfred A. Knopf.

Simmel, G. 1995. The metropolis and mental life, in *Metropolis: Centre and Symbol of our Times*, edited by P. Kasinitz. London: Macmillan, 30–45.

Stevenson, D. 2003. *Cities and Urban Cultures*. Maidenhead and Philadelphia, PA: Open University Press.

——. 2004. 'Civic gold rush': cultural planning and the politics of the Third Way. *The International Journal of Cultural Policy*, 10(1), 119–31.

——. 2013a. *The City*. Cambridge: Polity.

——. 2013b. *Cities of Culture: A Global Perspective*. London: Routledge.

Stevenson, N. 2003. *Cultural Citizenship: Cosmopolitan Questions*. Maidenhead: Open University Press.

Wood, P. and Landry, C. 2008. *The Intercultural City: Planning for Diversity Advantage*. London and Sterling, VA: Earthscan.

Woodward, I., Skrbis, Z. and Bean, C. 2008. Attitudes towards globalization and cosmopolitanism: cultural diversity, personal consumption and the national economy. *British Journal of Sociology*, 59(2), 207–26.

Worpole, K. 1992. *Towns for People*. Buckingham: Open University Press.

Young, C., Diep, M. and Drabble, S. 2006. Living with difference? The 'cosmopolitan city' and urban reimaging in Manchester, UK. *Urban Studies*, 43(10), 1687–714.

The Cultural and Creative Industries

Justin O'Connor

The concept is not object but territory. It does not have an Object but a territory. For that very reason it has a past form, a present form and, perhaps, a form to come. (Deleuze and Guattari 1994: 101)

Introduction

Approaches to the cultural and/or creative industries tend to take two forms. One identifies a set of institutions and practices (a 'sector' or an 'industry') that demands our attention in some way, often against a background of their previously marginal position. A second takes a more 'constructivist' perspective, highlighting an active process whereby an object is created or assembled by or through policy discourse(s).

Looking back over 40 years of policy and academic (and indeed 'activist') writing on this topic it seems clear that these two approaches are not mutually exclusive; they represent different narratives or rhetorical tropes that have been used (often by the same person) in different situations. The former positions the creative and/or cultural industries as harbingers or catalysts for something new, 'out there', demanding recognition, investigation, promotion; they point to the real, the urgent, the exciting. The latter does not decry the activities to which the concept points (a position taken by cultural conservatives, by a certain kind of Marxism or by mainstream economic sceptics) but the ways in which these have been shaped, co-opted, maybe high-jacked, by different policy agendas. It is then tempting to range these approaches across a scale. At one end we can see purely empiricist attempts to define and statistically map the sector, the more to pin it down and establish its 'value'. At the other the term becomes an 'empty signifier', a stake in the game between competing interests who wish to provide the content that most suits their objectives. As such one can be a 'disinterested' agnostic about the term, or see it as a mere symptom (or mask) of a deeper tendency (such as 'dumbing down', or 'globalization', or 'neo-liberalism' or 'precarity').

However, this would be to position those evoking a catalytic emergent sector as naïve realists and the constructivists as reflexive and critical. There is inevitably something of this involved, in a move from immediate presence to some complex, perhaps contradictory, mediated reality. But rather than a simple passage from 'dupe' to 'worldy-wise', or the unmasking of error or illusion, we could follow Jameson (in the different context of 'realism/modernism') and see the move as a kind of dialectical thickening, or putting the first 'realist' concept 'under erasure', somehow co-present with the 'constructivist' (Jameson 2002). Or, following Latour (2004), we might say our task is less the iconoclastic undermining of immediate 'matters of fact' (exposure of illusory naturalness, revelation of deeper 'invisible' causality) and more the attempt to deepen our understanding of 'matters of concern'. For example, it is glaringly obvious in the constant terminological stumble of 'cultural and/or creative' that our concern here involves something more than simply a new sector 'out there', to be 'nailed down' by yet another definitive definition. Yet that something new *has* demanded our attention – has become a matter of concern – in this last 40 years is unavoidable – the 'digital revolution' providing the most recent such wake-up call (if we still needed it).

In this chapter I want to trace the complex and contested narratives around the 'cultural and/or creative industries', to try to establish if not what they are 'in themselves', then at least why they are a policy stake worth the struggle. That is, how they became 'matters of concern' and what kind of new concern they might now be.

The Cultural Industries

Forty years ago puts us in the early 1970s, when 'the cultural industries' began to emerge as an object of academic and policy concern. (See also Miller, Chapter 3, this volume.) Given the subsequent academic focus on frictions between culture and economics we need to emphasize that, at this time, it was the issue of culture and politics that was primary. Or rather, economics was registered mostly in terms of the social inequality which gave differential access to the media – a problem for liberal pluralist theories and for social democratic notions of the 'public sphere' (Habermas' (1989) influential work on this was published in 1962). This connection between culture and politics was to become ever closer as the notion of 'ideology' began to move beyond its crude reductionist use in the political battles of the Cold War years, acquiring a more complex set of 'cultural' meanings to account for the continued existence of 'capitalism', the 'status quo', the 'establishment' and so on. It was the political consequences of 'the culture industry' that had been most prominent since its inception in a post-war USA by Adorno and Horkheimer (1979). For them the term represented the final subsumption of culture to the logic of monopoly capitalism. No longer just dominated at work the worker was programmed during the leisure hours by 'conditioned response' entertainment that relaxed them in order to get them back on the assembly line in the morning. This thesis was wrongly, though inevitably, lumped in with 'mass society' theory and conservative anti-democratic cultural criticism that by the 1970s had become a well-worn academic and policy trope.

The appearance of the 'cultural industries' as a more positive policy concern at the end of the 1970s was not some 'recognition' of the economic importance of commercial culture. Rather it was an opening up of a new kind of 'cultural political' space within what had previously been viewed by many in the policy establishment as degraded Americanized kitsch. This new cultural political space can be seen clearly in Augustin Girard's influential 1980 paper for UNESCO, written as head of research at the French Ministry of Culture (Girard 1982). Girard points to the huge commercial cultural sector and as a matter of urgency calls on the cultural policy establishment to take note. It was the same call as that made within the Greater London Council's (GLC) new left-wing Labour leadership, elected in May 1981 (Bianchini 1987, Garnham 1990, see also Bianchini, Chapter 22, this volume), and by Mitterand's new Minister for Culture Jack Lange in the same year (Ahearne 2010, Looseley 1995). That is, that the vast majority of cultural consumption now took place outside the subsidized sector; that the consumption of commercial culture was growing at extraordinary rates across all social levels; that traditional, subsidized 'live' cultural forms were (following Baumol and Bowen 1966) economically incapable of satisfying this demand; and therefore a refusal to engage with this commercial sector was elitist and irresponsible. A cultural policy must engage with this sector to be democratic; it needs to engage with it in order to challenge some of the more 'negative tendencies' within it. As a consequence Girard called for more research into the dimensions and dynamics of the sector; but three themes already stood out clearly.

First, a positive charge was now attached to the notion of 'industry' as a collective project; individual artistic practice had to be set within a wide range of professional, managerial and commercial services. Media and communications academics in Europe and North America had already established this within the mass media. In the United States the 'production of culture' school had begun to investigate how both 'popular' and 'high culture' was produced within complex socio-economic 'art worlds' in which the 'artist' was a constructed and contingent position (Hesmondhalgh 2007). Bourdieu's work on cultural production and consumption had begun to open up similar ground in France. In the UK Raymond Williams (1981) had also become interested in the material 'industrial' conditions of cultural production and their historical trajectories. Indeed, a new kind of art history rejected the transcendental notion of the artists and placed the individual genius squarely back within her or his social and historical context (O'Connor 2011). This recognition of the collective social basis of cultural production thus gave a strong democratic valency to the notion of 'industry'.

But, second, this industry was also about markets and profits, which raised difficult issues for cultural policy makers. These issues were outlined clearly in the mid-1980s by Bernard Miege (1989, 1987, 1979) and Nicholas Garnham (1990), both academics who had been close to the policy worlds of Jack Lange and the GLC (see Bianchini 1987). Taking direct issue with Adorno and Horkheimer's account they wanted to give a much more specific account of the cultural industries, not so much as capitalist *ideology* but as capitalist *industries* engaged in the production of cultural commodities at a profit. In contrast to the monolithic 'culture industry' and echoing similar findings by the 'production of culture' school, they identified a much more fragmented and disparate group of cultural *industries*. Their products could cut across the explicit political ideologies of the state in their search for markets; their need to make a profit demanded some degree of innovation not just formulaic repetition; and their very success in reaping profit from the exchange value of cultural commodities related, in part at least, to the ability of such commodities to provide 'use value' to their consumers. This opened up a more contradictory cultural space and it also introduced those 'negative tendencies' of which Girard spoke. These latter included concentration, monopoly, cross-ownership, vertical integration, ever increasing levels of capitalization and so on. Girard had also pointed to 'imbalances' at the international level, anticipating later accounts of globalization. Finally, there was the position of the artists. Artists (or creative workers/professionals as they were being called) had not been absorbed into some Taylorist culture factory, as Adorno had predicted, but remained a largely freelance workforce. For Miege and Garnham the continued independence of the artist was not a hangover from their bohemian past (as Girard suggested) but essential for the profitability of the cultural industries – including free R and D, a 'reserve army of unemployed', flexible staffing and so on.

Third, alongside these negative tendencies we can see in this cultural industries moment a more positive appropriation of new technologies of production, reproduction and distribution. There was a strong sense of seizing hold of a democratic modernity – breaking with Heideggerian anti-technological 'culture critique' as well as the formalist aesthetics of post-war modernist orthodoxy. The 1980s saw a rediscovery of the thematics of inter-war left modernism, which had embraced the future promise held out by the forms and technologies of American and home-grown mass culture. This was a re-appropriation clearly made possible by the energies released by new forms of popular culture that had burgeoned since the 1960s. The embrace of industry and technology was necessarily accompanied by a revalorization of the market. It was clearly not just 'collective' production and technological reproduction/distribution that counted here but its organization outside state subsidy and control, that is, in the market.

Thus the early 1980s saw left-leaning cultural policy makers embrace markets and technology, both of which had previously seemed to mark the boundary between art and commercial culture. Can we see this as a first repudiation of that 'elitist' opposition of arts/industry or culture/capitalism that many such as Hartley (2005) claim for the 'creative industries'? To some extent this is so. The idea of a transcendental art(ist) untainted by commerce and aloof from the world of machines had been systematically undermined. Equally, social democratic politics were now much more open to the idea of markets and much more wary of the state. Garnham (1990), for example, was explicit in his claim that the market was crucial to a modern democratic cultural policy; how else could the production of, and demand for, culture be regulated? After all, how did one embrace commercial culture without somehow embracing the commerce involved?

But there were some key elements that mark it as very different from the 'creative industries' moment of the late 1990s. First, though the economic elements were to be embraced as a crucial dimension of cultural policy, the overall intention was their contribution to a more democratic culture rather than to 'the economy' per se. Girard's call for more economic research was in order to guide intervention. The introduction of economic concepts such as the 'value chain', as well as the serious investigation of employment statistics and industry trends in this period, were to be used primarily to secure cultural ends. Second, these economic concepts and tools were there to correct 'negative tendencies' – issues of monopoly, exploitation, international domination and so on. They were there to protect against market failure – not the failure to achieve market success, as it became, but the failures intrinsic to the market mechanism per se. Third, though markets were embraced they were markets redefined – not the abstract neo-classical rational choice market but embedded socio-cultural practices. They were part of a mixed economy, not so much the Keynesian 'commanding heights' model of the 1950s but one that had emerged from a decade of grassroots democracy and urban social movements, from the rapid decline of the political prestige of the state and the incipient energies of post-Fordism. As we shall see, such an approach worked much better at local level, which is where some of the main strands of cultural industries policy-making developed.

From Arts to Cultural Policy

We also need to situate these developments within the shift, from the 1970s onwards, from 'arts policy' to 'cultural policy'. This is usually presented as a move from a narrow to a broad conception of culture. In the Anglophone tradition Williams' famous statement that 'culture is ordinary' (1958) is invoked here, and of course the shift owes a great deal to the rise of cultural studies since the late 1950s. In France it relates more to Henri Lefebvre's late 1950s work on 'everyday life', and crucially extended by Bourdieu and de Certeau (Ahearne 2010). In general the policy shift is associated with a deepening of democracy – from the post-war social democratic concern to open up 'access to the arts' to more a participatory and interactive 'cultural democracy'. There is no denying the strong democratic content of such a shift but it is useful here to examine a core theme of this narrative, that which links the 'elitism' of art to its separation from 'life'. (See also, Young, Chapter 23, this volume.)

Charges against art's social irresponsibility, elitism, solipsistic individualism, unconcern for the real world and so on are of long standing across the political spectrum, and these had been exacerbated by the formalist tendencies within modernism. The powerful attacks of the community arts movement – and urban new social movements of which they were part (Bianchini and Parkinson 1993, Castells 1983, Mayer 2006) – on the 'arts establishment' revived themes of an older left modernism, but they also coincided with a philosophical and sociological challenge more fundamental than the long familiar critique of 'art for art's sake'. This challenge suggested that the originating claim of Western 'aesthetic' art since the eighteenth century, to a transcendental access to a certain 'truth' was deeply ideological (see O'Connor 2011 for longer discussion). This ideological function was systematically analysed by Bourdieu's 1974 La distinction, a work which more than any other subsequent 'debunking' fixed the equation of art and elitism (Bourdieu 1986). He suggested that the 'disinterest' which Kant saw as the defining characteristic of aesthetic reception, and which via Schiller was to become the basis for the 'autonomy' of art, was an expression of an emerging bourgeois 'habitus' (Bourdieu 1986). That is, it grounded the 'correct' ability to appreciate art in the 'higher' faculties which were freed from direct need or desire. The working class were thus excluded by their subjection to the lower passions and their need to labour. Art's claims to autonomy, and the faculties required to appreciate the 'free play' at its heart, were thus implicated in practices of social distinction and domination.

There are some huge problems with this account which we cannot address here; what is important is how this was used within cultural policy. To a left-leaning cultural politics it suggested that the self-contained, separate 'autonomous' work of art needed now to take its place in a wider social context, in 'everyday life'. At the same time, it needed to recognize its material conditions of production, its relationship to 'economy'. In so doing its Apollonian 'disinterest' would give way to the Dionysian embrace of the 'lower' desires and interests, the business of political demands, the messiness of the market, and the unruliness of contemporary popular culture. Much of cultural studies comes from this. However, I would suggest that it is possible to see the autonomy of art not as secured at the expense of 'life', of the 'fallen' world, but against *culture* – a culture it sees as merely conventional, outworn, reified, debased. Rancière (2009) has persuasively argued that the characterization of 'aesthetic art' as separate from life is incorrect – that to the contrary it shows a constant dissolution of the boundaries between art and life established by pre-modern artistic practices. Jameson (2002) also points to the constant, systematic, often agonistic exchanges between 'art' and 'life' in the modern era. The autonomy claimed by art has a disruptive, transgressive force used not against 'life' but against the fixed, conventional forms of culture that mediate it. In this sense it is art, not 'culture', which asserts the radical heterogeneity of its domain of activity from the conceptual and administrative languages of economics, politics, the law and so on.

This can be clearly seen in two crucial areas of the post-war cultural policy settlement. On the one hand we have an art which had become a privileged representation and exemplification of national cultural identity. On the other we have an 'autonomous art' whose relationship towards such a national identity (or at least the conventional cultural expressions of such) was frequently ambivalent if not antagonistic. These two elements were highly disjointed, as the inter-war years showed, with 'autonomous art' (equated more or less with modernism) suppressed by totalitarian and right-wing authoritarian regimes. Cultural policy after the Second World War was an attempted social-democratization of this national cultural heritage. The decision to subsidize the arts was not (only) to take art out of the grubbiness of the market (although, of course, if you were already in the market then you disqualified yourself from being taken out) but to prevent art becoming the exclusive domain of the wealthy. Hence the crucial links to the expansion of education, public museums, libraries and so on. At the same time the promotion of a new cannon of modern art (increasingly internationalized in the context of the Cold War and the Marshall Plan) involved a set of values much more ambivalent than the conventional humanism of a 'common artistic heritage' evoked by writers and policy makers such as Andre Malraux (see his 1951 *Voices of Silence*). These values included a radical, anarchic experimentation; a concern with the formal demands of the artistic medium; an avoidance of 'uplifting' humanistic content (and indeed content per se); a rejection of social and ethical conventions; and a refusal to position their work in relation to explicit political and economic rationales.

The move from 'art' to 'culture' might thus be seen as a widening of an elitist, autonomous art to embrace the messy, grounded realities of 'ordinary culture'; but it should be clear that this widening was also a migration. It involved the introduction of the themes of 'autonomous art' – what Boltanski and Chiapello (2005) called the 'artistic critique of capitalism' – into 'new left' politics, resulting in new kinds of cultural demands and aspirations. We can see a transfer of many of the themes of modernism away from 'high art' and into 'culture'. In the radical community arts movements retrieving left modernism; in the urban social movements setting up new kinds of spaces and organizations; in the artistic avant-gardes operating on the fringe of popular culture (and vice versa); in the transformation and expansion of higher education and the growth of radical cultural theory within it – in all these we can see not just the culturization of art but also the 'aesthetisization' of culture. The rise of cultural industries was not then a repudiation of autonomous art by 'entertainment' or 'popular culture' but the extension of many of its values into their heart.

Here is the source of that urgency about a new 'out there', a rapid and volatile transformation of the practices of art and 'everyday' culture. Something is happening and we don't know what it is. The signifier 'cultural industries' was a way for policy makers to come to terms with this unknown 'out there'.

This can be seen in those community arts and urban social movements which formed a 'new left' opposition to traditional arts policy, resulting in new kinds of cultural policy-making in mainland Europe (Bianchini and Parkinson 1993), in the new GLC and in radical cultural policy coalitions in North America, Australia and elsewhere. Demands around 'collective consumption' were extended to cultural provision, as well as for more grass-roots control over resources. Such expanded culture became 'aesthetic' – more autonomous, opaque, refractive, abrasive with regards to 'mainstream' culture. This was not simply a replication of the forms of 'difficult' modern art but was part of that transformative promise of the 'artistic critique of capitalism'. That is, critique which stressed not social injustice (though it did not deny this) but the inability of capitalism to satisfy those human demands for a meaningful life that were promised or embodied in the autonomous work of art. Rimbaud's call to *changer la vie* could be seen in Joseph Beuys as well as in the cultural currents of Punk and Post-Punk. Further, this 'artistic critique' was no longer restricted to artists; though subsequently reduced to 'bohemian lifestyle' it initially brought new demands on work, new attitudes to careers, to social conventions, to the life-course and so on (Boltanski and Chiapello 2005). Urban centres especially underwent a sea change, as demands for a more 'meaningful life' produced new cultures of consumption and production. We can see a new *habitus* emerging; new urban cultural milieus in which new cultural aspirations were learned, as were ways of inhabiting these aspirations and of turning these into some kind of income (Raban 1974, Zukin 1982). It was these aspirations to a meaningful, democratic culture in common, coupled with the possibilities of gainful and meaningful work that gave the cultural industries signifier a utopian charge amongst activists, academics and policy-makers as it entered into a very different political era (O'Connor and Wynne 1996).

The Creative Industries

From where then did the creative industries agenda come?

Though the cultural industries was associated with left-leaning governments this has not been the case with the creative industries. The UK's 'New Labour' government coined the term in 1998 (DCMS), borrowing heavily from the forward vision of the Australian Prime Minister Paul Keating's 1994 *Creative Nation* (Oakley 2013). But in Australia, for example, the creative industries agenda has been associated with the neo-liberal Howard government (Turner 2011). Despite suspicions regarding its opposition party provenance it has nevertheless been adopted by the Conservatives in the UK and the Labor party in Australia. Across Europe it has been picked up by a range of political parties; and its rapid adoption across the very different contexts of China and East Asia, South East Asia, parts of Latin America and Africa (Kong et al. 2006, see also, Sasaki, Chapter 12, this volume) suggests the 'empty signifier' again. What does it signify?

One clear referent is 'modernity' or 'the future'; but we might say these are empty signifiers in turn, and that these have been under political dispute since 'conservative' parties became neo-liberal modernisers. However, we can see in the polyvalency of the creative industries the emergence of a right-wing or at least neo-liberal claim to a cultural modernity traditionally associated with the left. We have noted how the creative industries are presented as a reduction of culture to the economy; this is an over simplification if we do not recognize that this is a new kind of culture and a new kind of economy. Claims for a new cultural economy were part of the cultural industries agenda in the 1980s and '90s; indeed, the 1998 launch of the new creative industries agenda in the UK by a minister with newly conferred cabinet status (and whose title included the word 'culture' for the first time, rather than 'arts' or 'heritage') suggested its political apotheosis. The sense of a new post-1960s cultural renewal along with the recognition of a new cultural economic sector bringing local and national benefits was palpable (DCMS 1998). This embrace of the new 'out there' against the older establishments of 'real industry/proper jobs' and publicly subsidized art elites gave this agenda a powerful charge of youthful, generational change. Taking a genealogical approach, however, we might see how elements of the cultural industries agenda were hollowed out and charged differently, or repositioned in a different signifying system changing their meaning.

The use of 'creativity' is a case in point. The change from 'cultural' to 'creative' has been widely discussed. For some it was a recognition of the centrality of culture, simply written under the sign of 'creativity'; the terminological change was pragmatic and not central to the real 'out there' which it designated. For others it was nonsensical: did it describe an input or an output; what was not creative; how was science, technology or business creativity different to that of 'culture'; was there a difference between creative and cultural industries – and art? I do not want to rehearse these here (see O'Connor 2011); I would suggest that 'creative' is quite clearly being used as pertaining to culture – but to an aestheticized culture exemplified by (a now 'democratized') art. Through the term creativity, the autonomy claimed by art against established culture – its challenge to conventions; its avoidance or deliberate flouting of rules; its concern to follow its own aesthetic logic, its specific non-logical methodologies – now becomes part of the symbolic meaning-making capacity of all individuals. 'Creativity' takes a specific kind of aesthetic, autonomous art and turns it into a universal human attribute – now no longer the exclusive property of the artist and one that can be made available for a wider social and economic development.

This can be set against the shift within 'information society' discourse from a concern with an abstract individual cultural capacity to 'process knowledge and manipulate symbols' (Castells 1989) acquired through formal education (and used as a standard measure of the quality of a local workforce) towards a more embedded notion of culture. This wider cultural capacity had complex historical roots which could not be (easily) replicated – and indeed, such 'tacit' or 'embedded' knowledge was part of their competitiveness and resilience *vis-à-vis* mobile global capital. As policy makers became more concerned with the demands of post-industrial innovation this cultural capacity – culture now in the sense of Williams' 'whole way of life' – was now to be mobilized as a key economic resource or identified as dysfunctional drag. Either way what often became important was the capacity to re-invent and mobilize local 'structures of feeling', or to transcend the past, to slough off constrictive social and cultural traditions. In this way Saxenian's (1994) well-known comparative study of Boston and Silicon Valley was crucially instructive. The reason the former became the innovative powerhouse despite the latter's high levels of investment was that it escaped the traditional social, cultural and institutional structures that gave the East Coast city a comparative stiffness. Similarly Granovetter's 'The strength of weak ties' (1973) came to overturn the worries of people like Robert Putnam (2000) about social solidarity and suggested that the lack of strong social bonds allowed for great fluidity of interaction and exchange and thus economic innovation (Currid 2007, Florida 2002).

The cultural capacity for innovation thus went beyond the ability to routinely 'process knowledge and manipulate symbols' toward the ability to operate along the edges of established rules (Castells 1989). Management and business literature began to promote working 'outside the box', deliberately courting failure, chaos and disorder, using para-rational or intuitive knowledge, operating as a maverick and so on (Kelly 1998). These new values or ways of working explicitly drew on the unorthodox and unpredictable practices of artists and visionary scientists. In fact the newly emergent notion of 'creativity' within business language was parasitic on these exemplary figures. This was so not just in the realm of 'blue skies thinking' and the breaking of established paradigms and ways of doing – the new figure of the entrepreneur also picked up the cultural capital associated with the artist as social rebel.

In the 1980s the Schumpeterian entrepreneur made a comeback against the Fordist 'organization man' of the 1950s and '60s. It was part of a re-vamped neo-liberal attack on state corporatism in favour of the small business enterprise. The New Right positioned both itself and the entrepreneur as outsiders and rebels. Entrepreneurs worked at the edges of the system, pushed its boundaries, explored new territories, confronting ossified ways of thinking and doing. Schumpeter's 'creative destruction' therefore had clear links with the dominant account of cultural modernism: its iconoclastic, shock-of-the-new obsession with innovation (Anderson 1984). During the 1980s entrepreneurs and artists often occupied the same place in new management literature – as society's outriders, productive rebels who might glimpse the outline of the future. In these ways – mobilizing a local cultural capacity, using aesthetic art as exemplar for innovation, and transforming the bohemian counter-cultural producer into creative entrepreneur – art and culture, no longer recalcitrant to economic development, become 'resource' (Yudice 2003).

The consequences of these kinds of shifts can be seen in the extensive debates around cultural work (see Oakley 2009). The promise of meaningful, autonomous cultural work has frequently resulted in (self-)exploitation; 'creativity' has been a way of shifting job market responsibility from governments to individuals; the creation of a culture in common easily becomes narcissistic self-promotion and the instrumental exploitation of social networks. I won't add to these extensive critiques here. What is crucial is that urgent 'out there' which the creative industries discourse could mobilize. What transformed the artist-creative from exemplary role-model (avant-garde artistic practice as a model for innovative and entrepreneurial thinking in business) to real economic resource was the growth of the 'cultural economy' itself. Not just the expansion of cultural commodity markets per se – music, television, radio, publishing, film, visual arts, fashion, computer games and so on – but the increased cultural or symbolic content of functional goods and services. Product and interior design, 'experience value' in services, 'attention value' in marketing and public relations, cultural tourism, the growing role of web 2.0 based social networking within all of these – they were all part of that 'culturalization of the economy' which Scott Lash and John Urry had announced in 1994. (See also, Young, Chapter 23, this volume.) Therefore, though 'creativity' in general is deemed a core social value, because cultural or symbolic inputs were now a major source of value right across the economy then the particular skills, mind-sets and working practices of those operating in this risky, volatile and maverick cultural/ creative industries sector would be at a premium.

It might be noted in passing that these kinds of transformations cannot simply be reduced to 'neo-liberalism'. There is a polyvalency around these themes which makes them unstable. For the neo-liberal agenda is not simply the prevalence of the 'free-market'; such an agenda marks the cultural policy struggles under the Reagan and Thatcher era – of de-regulation, cuts in subsidy and the insistence on economic justifications for art. The price of everything and the value of nothing and so on. Neo-liberalism was introduced by conservatives – who saw the 1960s counter-culture as antithetical to their project. This culture was 'anti-business' of course, but it also promoted social and cultural values which were detrimental to the traditional symbols of nationhood under which these early neo-liberal reforms were conducted. Hence the 'culture-wars' and the increasingly 'conservative' image of these inveterate modernizers. It was Clinton and Blair who saw the political availability of '60s counterculture to present a forward-looking agenda in which many of the themes of neo-liberalism could be extended through and within the realm of 'aesthetisized' culture.

The new cultural economy involved new kinds and scales of commodification. But this was not the reduction of cultural 'use value' to the universal equivalence of exchange. This new economy was built on recognition of cultural use value and the skills and processes necessary to organize this. Hence the catalytic role with respect to the wider economy – generally demanding more 'experience', 'attention' and other service-industry qualities – claimed for the creative industries (Cunningham 2004, 2002, Miller, Chapter 3, this volume). The cultural use value is now linked to exchange value by extremely rapid, multiple and sophisticated circuits. Indeed, the integration of web 2.0 technologies into this process over the last decade has radically destabilized any remaining direct ownership of use value by the creative worker. Co-creation involved not just the direct input of consumers into the creative process; the very act of consumption and the technological ability of machines and organizations to track and analyse such consumption allows the generation of new value. The cultural product now gains an almost Adorno-esque objective existence apart from its creator; but it is its very distinctiveness, its evasion of equivalence and disruption of established rules which is the source of its economic value (Lash and Lury 2007).

The opposition of cultural use value to exchange value no longer works as it did. In the creative industries creators don't fully create and rejecting exchange value can only be elitist; or to claim the role of the 'expert' which amounts to the same thing (Hartley 2005). In the notion – developed from Hartley's work – of 'social network markets' there is no source of meaning other than that instantaneously manifested in particular conjunction of personal preferences creating 'value' (O'Connor 2009, Potts et al. 2008). Its combination of methodological individualism and the reduction of cultural value to exchange value represent the arrival of neo-liberal thought at the very heart of cultural theory.

Conclusion

In this chapter I have tried to approach the cultural and creative industries policy not from the perspective of an economic sector to which various technical support policies can be applied. Rather I have tried to outline the ways in which they have emerged as 'matters of concern' for cultural policy. The creative industries 'moment' which began in 1997 combined many different and contradictory cultural agendas around an urgent call to recognize a new 'out there' – one that represented the future, change, renewal. Though often received with some cynicism in the UK around its explicit party political elements (see Harris 2003) it contained an energizing imperative. I have frequently witnessed, in Russia, in East Asia and elsewhere, assemblies of the most econometric policy-makers, calculating the value-added of culture, alongside young, energetic 'creatives' kicking against ignorant and corrupt politicians, global corporations and smug arts establishments which they see as standing in their way – of making a living and making a new culture in common. These moments are not to be denied their power; just as the earlier moments of the cultural industries coalition cannot be dismissed because their value was recouped by property developers and city marketing departments.

In developed countries at least the 'artistic critique of capitalism' has now become domesticated, a resource for economists, developers and high-minded idealists alike. The ability of the established corporate structures of the cultural industries to absorb the new social media and digital challenges, and the rapidity with which the new players became integrated; the association of creative work with new forms of exploitation; the absorption of the creative industries agenda into property development and the paucity of the intellectual and financial resources city governments (with some exceptions) put into their development – all these have very much undermined the transformative energies with which the creative industries agenda was first welcomed.

The ubiquity of cultural commodities and the easy access to the technologies of production and distribution is now taken for granted. Globalization is no longer the sole province of the de-regulators and off-shore outsourcing but also belongs to the post-national 'multitude' which – rather than being assembled *right now* under the banner of McDonalds or Benetton (Hartley 1999) – demands work to give it form (Hardt and Negri 2005). New kinds of cultural practice across the globe, concerned to create new spaces of possibilities and collaboration, can be seen as part of work to invent new kinds of social collectivities. They suggest a movement beyond autonomous aesthetic culture to a recognition of the social and ethical bonds within which this culture is produced. That is building cultural connections in a context – after neo-liberalism – in which 'traditional' (including Fordist industrial) cultures have been strip-mined and de-stabilized; in which 'conventional culture' now includes the urge to self-expression, creativity and innovation (often at the expense of any other consideration). This is clearly what is now at stake with the debates about self-control (Brooks 2011), 'bigger-than-self' thinking, and even 'big society', to put together a social order rocked by four decades of economic modernization. As opposed to a conservative re-assertion of traditional values – deeply compromised by its neo-liberal turn – we might see the issue as creating a society of 'weak ties' conceived not as fragmented individualism but as an open, democratic social bond.

We can see this in the 'ethical turn' in design, where its association with the 'aesthetic' allure of the commodity is giving way to its application to social structures and processes (O'Connor 2012). So too is the general shift across what was once called the creative industries towards sustainability – not just to 'green' production but to local markets and livelihoods and to the cultures which intersect with these. In these processes older artistic values – of craft, time, patience, the determination by the object rather than self-expression, ethical-aesthetic communities – emerge (Gibson 2011). No longer just in North America, Europe and Australia but globally, the social and ethical dimensions of culture have been asserted against the purely economic, and the uncoupling of cultural workers from the agenda of the creative industries has gone on apace, if unevenly. This is the big 'out there' that should be a matter of concern. So far we don't have a signifier for it.

References

Adorno, T. and Horkheimer, M. 1979. *The Dialectic of Enlightenment*. London: Verso.

Ahearne, J. 2010. *Intellectuals, Culture and Public Policy in France: Approaches from the Left*. Liverpool: Liverpool University Press.

Anderson, P. 1984. Modernity and revolution. *New Left Review*, 144 (March–April), 96–113.

Baumol, W. and Bowen, W. 1966. *Performing Arts: The Economic Dilemma*. New York: Twentieth Century Fund.

Bianchini, F. 1987. GLC R.I.P. Cultural policies in London 1981–1986. *New Formations*, 1, 103–17.

Bianchini, F. and Parkinson, M. 1993. *Cultural Policy and Urban Regeneration: The West European Experience*. Manchester: Manchester University Press.

Boltanski, L. and Chiapello, E. 2005. *The New Spirit of Capitalism*. London: Verso.

Bourdieu, P. 1986. *Distinction: A Social Critique of the Judgement of Taste*. London: Routledge.

Brooks, D. 2011. *The Social Animal: The Hidden Sources of Love, Character and Achievement*. London: Random House.

Castells, M. 1983. *The City and the Grassroots: A Cross-cultural Theory of Urban Social Movements*. London: Edward Arnold.

——. 1989. *The Informational City: Information, Technology, Economic Restructuring, and the Urban-Regional Process*. Oxford: Blackwell.

Cunningham, S. 2002. From cultural to creative industries: theory, industry, and policy implications. *Media International Australia, Incorporating Culture and Policy*, 102 (February), 54–65.

——. 2004. Creative industries after cultural policy. *International Journal of Cultural Studies*, 7(1), 105–15.

Currid, E. 2007. *The Warhol Economy: How Fashion, Art, and Music Drive New York City*. New York: Princeton University Press.

DCMS. 1998. *Creative Industries Mapping Document*. London: DCMS.

Deleuze, G. and Guattari, F. 1994. *What is Philosophy?* (translated by G. Burchell and H. Tomlinson). London: Verso.

Florida, R. 2002. *The Rise of the Creative Class*. New York: Basic Books.

Garnham, N. 1990. Concepts of culture: public policy in the cultural industries, in *Capitalism and Communication: Global Culture and the Economics of Information*. London: Sage, 153–67.

Gibson, C. 2011. Cultural economy: achievements, divergences, future prospects. *Geographical Research*. 50:1 doi: 10.1111/j.1745-5871.2011.00738.x.

Girard, A. 1982. Cultural industries: a handicap or a new opportunity for cultural development?, in *Cultural Industries: a Challenge for the Future of Culture*, Paris: UNESCO, 24–40.

Granovetter, M. 1973. The strength of weak ties. *American Journal of Sociology*, 78(6), 1360–80.

Habermas, J. 1989. *The Structural Transformation of the Public Sphere: An Inquiry into a Category of Bourgeois Society*. Cambridge: Polity Press.

Hardt, M. and Negri, T. 2005. *Multitude*. London: Penguin.

Harris, J. 2003. *The Last Party: Britpop, Blair and the Demise of English Rock*. London and New York: Fourth Estate.

Hartley, J. 1999. *Uses of Television*. London: Routledge.

—— . 2005. Creative industries, in *Creative Industries*, edited by J. Hartley. Oxford: Blackwell, 1–39.

Hesmondhalgh, D. 2007. *The Cultural Industries*. 2nd Edition. London: Sage.

Jameson, F. 2002. *A Singular Modernity: Essay on the Ontology of the Present*. London: Verso.

Kelly, K. 1998. *New Rules for the New Economy*. London: Fourth Estate.

Kong, L., Gibson, C., Khoo, L.M. and Semple, A.L. 2006. Knowledges of the creative economy: towards a relational geography of diffusion and adaptation in Asia. *Asia Pacific Viewpoint*, 47(2), 173–94.

Lash, S. and Lury, C. 2007. *Global Culture Industry*. Cambridge: Polity.

Lash, S. and Urry, J. 1994. *Economies of Signs and Space*. London: Sage.

Latour, B. 2004. Why has critique run out of steam? From matters of fact to matters of concern. *Critical Inquiry*, 30(2), 225–48.

Looseley, D. 1995. *The Politics of Fun: Cultural Policy and Debate in Contemporary France*. Oxford and Washington DC: Berg.

Mayer, M. 2006. Manuel Castells' The City and the Grassroots. *International Journal of Urban and Regional Research*, 30(1), 202–6.

Miege, B. 1979. The cultural commodity (translated by N. Garnham). *Media, Culture and Society*, 1(3), 297–311.

—— . 1987. The logics at work in the new cultural industries. *Media, Culture and Society*, 9(3), 273–89.

—— . 1989. *The Capitalisation of Cultural Production*. New York: International General.

Oakley, K. 2009. *Art Works: Cultural Labour Markets. A Review of the Literature*. London: Creativity, Culture and Education.

—— . 2013. A different class: politics and culture in London, in *The Politics of Urban Cultural Policy: Global Perspectives*, edited by C. Grodach and Daniel Silver. Abingdon: Routledge, 21–41.

O'Connor, J. 2009. Creative industries: a new direction?. *International Journal of Cultural Policy*, Special Issue: *Creative Industries Ten Years After*, edited by M. Banks and J. O'Connor, 15(4), 387–404.

——. 2011. *The Arts and Creative Industries*. Sydney: Australian Council for the Arts.

——. 2012. From allure to ethics: design as a 'creative industry', in *Design and Ethics: Reflections on Practice*, edited by E. Felton. London: Routledge, 33–42.

O'Connor, J. and Wynne, D. 1996. Left loafing: cultural consumption and production in the postmodern city, in *From the Margins to the Centre: Cultural Production and Consumption in the Post-Industrial City*, edited by J. O'Connor and D. Wynne. Aldershot: Ashgate.

Potts, J., Cunningham, S., Hartley, J. and Ormerod, P. 2008. Social network markets: a new definition of the creative industries. *Journal of Cultural Economics*, 32(3), 167–85.

Putnam, R. 2000. *Bowling Alone: The Collapse and Revival of American Community*. New York: Simon and Schuster.

Raban, J. 1974. *Soft City*. London: Hamish Hamilton.

Rancière, J. 2009. *Aesthetics and Its Discontents*. London: Polity.

Saxenian, A. 1994. *Regional Advantage: Culture and Competition in Silicon Valley and Route 128*. Cambridge, MA: Harvard University Press.

Turner, G. 2011. *What's Become of Cultural Studies?*. London: Sage.

Williams, R. 1958. Culture is ordinary, in *Studies in Culture: An Introductory Reader*, edited by A. Gray and J. McGuigan. London: Arnold, 5–14.

——. 1981. *Culture*. London: Fontana.

Yúdice, G. 2003. *The Expediency of Culture: Uses of Culture in the Global Era*. Durham, NC: Duke University Press.

Zukin, S. 1982. *Loft-living: Culture and Capital in Urban Change*. London: The Johns Hopkins Press Ltd.

Heritage in Planning:
Using Pasts in Shaping Futures

G.J. Ashworth

The Past in the Future

As heritage is by definition a cultural construction, it is an aspect of a wider culture and thus heritage planning is a sub-division of cultural planning. Heritage can be seen as a resource, a process and an outcome and all three approaches, as they are used in spatial planning, will be considered here. The answer to the straightforward question, 'can the past be used in planning for the present and future?' is at one level self-evident. It is being done globally in the pursuit of many different planning objectives. The more interesting question is, 'what is actually happening and is it what we intend to be happening'?

What is often not realized is that these uses of the past that are now commonplace are relatively recent, stemming from only the last century and a half and are subsequent in that they are uses of heritage that had already been brought into existence for other purposes. Neither the past nor the future exists now. Therefore the way in which the past can be used in the present and future needs explanation. In spatial planning three different paradigms have evolved as approaches of the present towards the past. Although these three paradigms were evolved in an historical sequence, the new did not replace the old but was adopted by some of those involved in the process of heritage creation and use. This incomplete paradigm shift means that at least three quite different ways of treating the past in the present now coexist, often uncomfortably (Graham et al. 2000).

The first, and oldest of these is preservation, which is easy to define as the protection from harm, whether from natural elements, human intervention or neglect, of those attributes of the past that have survived into the present. These may be relict objects, buildings, landscapes and locations as well as recollected narratives, traditions and reminiscences. In the course of the nineteenth century in the industrialized and urbanized parts of the world the ostensibly curious notion was developed of attempting to preserve what could be preserved from what was seen as a fortuitous legacy from a rapidly vanishing pre-industrial society. The passionate pleas of a prophetic, noisy and influential minority were eventually rewarded in most Western countries during the nineteenth and early twentieth centuries by the establishment of government funded agencies charged with listing, creating collections and preserving what they had selected. The aspect of this of most concern here is that physical artefacts and structures were offered legal protection, thus creating the concept of historic monument and site. The purposes were defensive and the timing no doubt had much to do with the unprecedented rapid pace of economic, social and environmental change being experienced. The intention was to save what could be saved before it was too late, the value was intrinsic to the object and the justification was self-evident and indisputable.

The impacts of this official protection on spatial planning are limited and essentially negative. Preservation, the prevention of change, is the antithesis of development, the encouragement of desirable change. The preserved artefacts and sites become at best irrelevant and at worst an obstacle to planning. There are three possible legal reactions. First, the planner avoids the spaces and buildings that have been removed from 'normal' planning and development occurs elsewhere. A spatial fossilization occurs as change is prevented at designated places. Second, if the legislation permits, the preserved structures can be incorporated into the development, which occurs around, above and below them. Although the object itself is untouched, its context and function is radically changed. Third, and common in North America, the protected structure can be moved, which preserves the object but divorces it from its location. Concentrations of such relocated structures are known as 'skansens' (from the location in Sweden of the first such collection by Hazelius in 1891). The authenticity of the object is preserved at the cost of the authenticity of the ensemble, the site and the function.

The term conservation acquired a different meaning from preservation in planning but not in the museum world where conservation has remained a synonym for preservation. The shift in meaning occurred, in the middle years of the last century when the conservation paradigm was defined as 'preserving purposefully' (Burke 1976) which added contemporary functions to past forms and incorporated the present or future purposes of the building or area into the decision to conserve. Also, conservation planning enlarged the focus of attention from individual buildings or sites to ensembles. Historic areas became the focus of planning attention. Conservation legislation in the 1960s and '70s, initially in Western European jurisdictions did not replace the preservation legislation but was incorporated alongside it (Larkham 1996). Development is not prohibited but managed. Consequently the focus was upon designated areas and the addition of function to form as justification and financial support. The idea of adaptive reuse became central to planning intervention in conservation buildings and areas. Thus planners and urban managers superseded architects, and historians, as decisive actors in this process.

Finally, the heritage paradigm is a logical progression from existing paradigms and also perhaps results from their success as preserved monuments and conserved areas multiplied dramatically towards the end of the twentieth century. (See also, Evans, Chapter 13, this volume.) These became so numerous as a proportion of the building stock and so extensive a land-use that they could no longer be treated as exceptions, removed from the normal functioning and management of places.

The simple definition of heritage as 'the contemporary uses of the past' (Ashworth 1991) has quite radical implications for two reasons. First, it focuses on the present, when, and for whose benefit, decisions are made. Both the past and the future are imagined entities created in the present. Preservation assumes an objectively determined past can be preserved for the benefit of an objectively determined future. Heritage accepts that only the present is real. Contemporary needs determine both the inputs into the process, namely selection, packaging and interpretation, and the outputs in terms of the uses of the past in the present. Heritage is thus driven by current needs and tastes, untrammelled by the universal, eternal and inalienable values of the preservationist. The view of the future in heritage is driven by current needs and values, which are projected into a future as imagined from the present.

Second, an important implication is that as a creation of the human imagination, there is no fixed stock of heritage that can be depleted or exhausted. If supply is driven by demand, then in theory heritage is ubiquitous and infinite. The common-sense objection that many heritage resources are overused is countered by the observation that an excess of demand over supply at particular times or places is a consequence of management failure to increase supply. Equally there is no fixed legacy that determines the historic character of places: nowhere is locked into a predetermined past as heritage is a discretionary development option.

How is Heritage Used in Planning?

This question receives a multitude of answers as heritage has become overburdened with overlapping and sometimes contradictory expectations. Answers must first acknowledge that most heritage was not brought into existence to satisfy the needs of spatial planners nor to fulfil the functions that they now expect from it. (See also, Hillier, Chapter 24, this volume.) The original overriding purpose and continuing main motives behind a concern for the past is first, political legitimation of authority and second, the socialization of individuals into contemporary society, fostering the cohesion or inclusiveness of the group and exclusiveness in relation to other groups. These are the reasons why museums, national histories, monuments and statues were originally created and remain the continuing primary justification for their management. The other uses, such as place management are subsequent. Some of these will now be reviewed.

Heritage and Place Identity

There is the explicit idea that history endows places with a distinctive identity. It transforms spaces into places through its inherent quality of uniqueness. Simply, places are different from each other because their history, expressed through heritage, is necessarily different. The implications of this for local planning and the possibilities it offers in support of local planning goals account for the prevalence of this idea as an objective in numerous planning policies.

However there are a number of flaws in the assertion that history, as transformed into heritage, creates identity, which then enhances the value of places and that this quality can be created or managed by planning. There are the obvious difficulties of choice within history itself. Fundamentally, identity is ascribed to places not given by them. Place identity is not a latent quality passively waiting for identification and appreciation. People identify with places and are the active creators. (See also Dovey, and Nyseth, Chapters 15 and 19, this volume.) Therefore which events, personalities, movements and ideas are to be selected from the maelstrom of available possibilities and who is to make such a selection? Identity and place identity are not the same. Particular social or political ideas may be associated with specific places, which are used to represent or manifest them but much of the identification of individuals with each other is not place-bound. It may be that identification with local places is not a universal basic human need but is a preoccupation of a minority and of little significance to a majority. Individuals identify with, or against, places and the diversity of individuals results in diverse identities. Therefore policies for shaping and promoting place identities operate on the assumption that individual place identities can be aggregated into a collective identity. This idea, like collective memory, is a metaphor because individuals not societies identify, and remembered places do not have a single, agreed, immutable, collective place identity. Thus place managers often attempt to create and promote different brand identities for different purposes and markets, most obviously locals and outsiders. This raises the specific questions, 'do people need to identify with places?' and 'is there an identity dividend, an added benefit conferred on users of places by the existence of a strong place identity?'

The Dutch *Belvedere* programme (1999–2009) was an attempt of the national government to discover, strengthen and promote specifically local identities (described in Ashworth and Graham 2005, Ashworth and Kuipers 2002). It identified 60 landscapes and 105 towns that possessed local character, much of which was derived from the past and transmitted through environments, local traditions and even nomenclature. The justification for such national funding of local initiatives was to provide local counter-balance to economic and social globalization. Local place identity becomes a desirable contemporary attribute of places and the fostering of this becomes a task of official local planning agencies.

Creating place identity is implicit in many public policies and raises misgivings. First, do these assumed benefits of a strong positive identity accrue to collectivities through social cohesion or political allegiance or to individuals who receive some automatic benefit from using such places? Second, if some places and their users have more of such a beneficial quality than others, then equable planning policy should compensate the deprived not reward the already fortunate.

Heritage used in the creation and promotion of local identities contains an inherent contradiction. The commodification processes whereby aspects of the past are transformed into heritage products and experiences are themselves frequently global rather than local. The attempt of governments to support local identity may thus lead to the opposite, namely global replication and homogeneity, which will be exemplified later. There is also the danger that shaping a heritage-based identity may stress already obsolete and largely irrelevant social and cultural elements that may inhibit change. Communities and their identities are constantly changing and cannot be frozen in time. Townscapes can be physically and legally preserved but the communities that inhabit them and their local identities cannot.

Heritage as Environmental Amenity

Heritage is expected to contribute to positive environmental amenity (Ashworth 2008a). Although as many ugly buildings as beautiful ones were built, the latter were preserved while the former were demolished. Similarly heritage landscapes exist because they are seen to be aesthetically satisfying and culturally enriching. Is there therefore a 'heritage environment dividend' similar to the 'identity dividend' discussed earlier, namely some added benefit or well-being emanating from such environments and, if so, what are its effects and who receives them? Such environmental amenity might be dismissed as a vague, immeasurable and minor consideration but it does seem to be reflected in real estate values. Although logically the restrictions imposed on owners and users of historic buildings and districts by protective legislation would reduce utility and thereby property values, in practice the reverse has proved often to occur, which may reflect people's spatial preferences. Eyles and Williams (2008) argue for a link between the environmental quality including its perceived historicity, and a sense of well-being, and even good health, amongst its inhabitants. The association of high quality heritage environments with measures of economic or socio-psychological well-being, and the converse association of low quality environments with economic and social dysfunction, may reverse cause and effect. It is not the collectively enriching and personally satisfying environments that create well-being: it is those with economic means that are able to live in such environments.

The many 'league tables' of places are summary reflections of how people as residents, investors or tourists score places as the most 'livable', 'workable', 'visitable' or 'investable' and are both competitive indicators and judgements on management. The role played by heritage environments in such rankings is difficult to disentangle from many other place-attributes, but, notably, the consistent top scorers in such leagues are cities with established heritage reputations (such as Seattle, Boston, San Francisco, Edinburgh or Bath), while those at the bottom are places lacking such reputation. Whether or not a heritage induced, high quality environmental amenity actually confers the anticipated benefits, people behave as if it does, which invites planning intervention.

Heritage as Location Factor

Heritage as environmental amenity can be extended by treating it as a factor in the location decisions of enterprises. Studies pursuing the idea that firms 'will not locate in cultural deserts' (Whitt 1987) stress a heritage-rich environment, meaning places with monuments, heritage areas, museums and historic place associations (see for example Lewis 1990, Miles 1997, Myerscough 1988) are factors influencing decisions to locate. More specifically heritage designated buildings and districts confer both benefits and costs upon their occupiers, the importance of which depends upon the nature of the economic activity. The resulting positive or negative balance encourages location into, or relocation out of, such buildings or areas. On the positive side are usually central locations, high pedestrian footfall and, less easy to quantify, high address recognition, status and even transferable associations from the historicity of the location to the current occupier, such as continuity, reliability and artistic patronage. On the negative side, heritage designated buildings incur higher costs for maintenance and retrofitting modern facilities, and have restrictions on adapting spaces to changing requirements. Heritage areas are frequently also pedestrianized, restricting vehicular accessibility and parking. The historical associations valued by some enterprises may be disadvantageous to others whose image is modernity and progress rather than longevity and tradition. It depends upon how a particular activity values the costs and benefits and thus location in heritage buildings and districts which may be anywhere from highly attractive or completely repellent. The result will be the familiar concentration of certain enterprises in historic areas and, conversely, the absence of others (Ashworth and Tunbridge 2000).

A related aspect is place promotion and branding using heritage elements within wider place marketing strategies (Ashworth and Kavaratzis 2010) intended to attract enterprises, investment, residents or visitors. Using heritage in place marketing campaigns is ostensibly strange: places attempt to improve their presents and futures by referring to their pasts. However the past is increasingly featuring in the marketing even of places where heritage is not a major economic activity or defining characteristic of the place. It is the uniqueness of local pasts and the flexibility of heritage as an imagined construction that renders it ideal for place-product differentiation, stressing unique selling points of the place-product or place-brand. Historic events, personalities, mythologies, folk-lore and customs are widely used in the construction of place-images for place-product marketing and branding to differentiate one place from another, which has advantages even for non-heritage related economic activities. This heritage branding is part of wider cultural branding where places designate themselves as 'cities of culture', even when their economic success and reputation depends upon other activities. Cities such as Glasgow, Rotterdam or Essen in search of some balancing or refocusing seek the appellation because it sharply contrasts with their existing image.

The three main techniques for heritage place-branding, which are generally used in various combination, are: 'personality association' (Ashworth 2010), linking the attributes of an historic person to the place; 'hallmark events' (Hall 1989, Quinn 2005), conferring cultural patronage; and 'signature building' (Ashworth 2009b, Warnaby and Medway 2010), using modern or historic structures to further instant place recognition.

Heritage as Economic Resource

Heritage can be an economic enterprise, part of wider cultural industries, activating resources, producing saleable goods and services, and generating returns in profits and jobs. Thus it has a distinctive geography of production with identifiable spatial location factors, distinctive industrial production structures, patterns of employment, inter-firm linkages and networks (Evans 2001, Hughes and Gratton 1992, Wynne 1992). Heritage enterprises can thus be planned and managed as other cultural enterprises in the public interest by using its distinctive industrial characteristics, vertical disintegration, spatial clustering and specialized labour pool dependence, all of which have geographical dimensions.

The ubiquity and flexibility of the heritage resource renders it ideal for the creation of place-bound tourism products. Heritage tourism is one special case of a heritage service industry but its economic importance has generated considerable management, scientific and political attention. The main problem with using tourism within local development strategies is the asymmetrical relationship of resource providers and product creators. To the tourism industry, heritage is a zero-cost, freely accessible, flexible and inexhaustible resource. To the resource providers, however, often in the public sector, heritage is costly, multi-used and vulnerable to damage or depletion. The challenge for local planners is managing this asymmetry, minimising local external costs and maximising local external benefits of a heritage tourism product that is consumed selectively, rapidly and capriciously (Ashworth 1995, 2009a, Timothy and Boyd 2003).

Heritage in Local Area Regeneration

Finally heritage is used in local area physical renovation, social regeneration and economic revitalization bringing together many of the functions of heritage described above. There are very few cases where heritage has been used as the dominant instrument in repositioning local economies and reshaping local places. Usually heritage is used indirectly in such policies, by contributing to environmental quality, favourable place image and popular identification. Heritage can be an effective way of communicating public intent. Some renovation or cleaning of public buildings, installation of 'historic' street furniture and promotion of a named historic district, cost little but are clear and visible signals that government is aware of an area's circumstances and is acting and thereby encouraging private investment. If impressive enough a single building may convey the intended message. Adaptively reusing a London power station (Tate Modern) or a Paris railway station (Musée d'Orsay) are examples. More modestly, the restoration and re-functioning of Halifax's Piece Hall, is making a statement through form and function of the abandonment of one economic sector and embracing of another, typical of so many former manufacturing towns in north-western Europe.

Conserved and adaptively reused heritage buildings, promoted historical associations and major museums and galleries (which may themselves be new but which house heritage art and artefacts) may all contribute to area regeneration even when the economic development has little directly to do with heritage (Bianchini and Parkinson 1993, Stabler 1996). Such heritage facilities are rarely economically viable in their own right but are, nevertheless, often included in multi-functional projects (Bianchini 1993, Lim 1993, Whitt 1987) because of the perceived positive externalities they contribute to local areas. These externalities include 'animation', encouraging the existence of a vivacious street scene, increasing the numbers of users, their length of stay and perhaps their disposition to spend (Burgers 1995). This 'forum function' of cities (Oosterman 1993) where local planners encourage socio-spatial behaviour through the shaping of outdoor terraces, piazzas and colonnades, are often self-conscious replications of historical urban designs (Ennen 1997). Cultural engineering by creating spaces in cities which are endeavouring to project themselves as culture centres is substantially dependent upon the structure and quality of their physical forms (Tiesdell et al. 1996). In such a scenario, exemplified by London's Covent Garden, Krakow's Rynek Glowny, Toronto's Dundas Square or Copenhagen's Rådhuspladsen, space becomes a theatre where the roles of producer and consumer, player and spectator, are interchanged. Second, there is *cachet*, an aura of social propriety, acceptable taste and continuity stemming from artistic patronage that extends to other coexistent prosaic commercial functions, which may be providing most of the economic returns.

However, three caveats dispel any illusion that heritage offers an automatic, universally applicable route to successful area regeneration. First, linking a particular place with a particular heritage element is an inherently hazardous strategy because heritage is fashion driven (Ashworth 2008b). Today's feted historical personality, event or period is tomorrow's forgotten, outdated or even embarrassing leftover. When, Gaudi or Hundertwasser become a footnote in art history, Barcelona or Vienna may find their former successful association with the buildings and their architects is an obstacle to re-branding. The Nazi era *Kunsthäuser* or Soviet 'Palaces of Culture' now embarrassingly litter German and East European cities.

Second, success in using heritage associations and structures as catalysts in local regeneration depends upon the 'Guggenheim effect' in which a cultural attraction, in this case Frank Gehry's museum in Bilbao, stimulates not only place-recognition but also local positive externalities, thereby reversing long-term structural economic decline that was its original purpose. Third, the heritage development option may be selected in a climate of despair and implemented as a last resort when other measures have been unsuccessful. In these circumstances the ubiquity and flexibility of potential heritage resources, mentioned earlier as an asset, can also be a drawback. Such a universally available instrument for regeneration may be selected regardless of relative strengths and realistic assessments of the competition. There are just too many nineteenth-century industrial cities attempting to use their heritage potential as part of a new *raison d'etre* for them all to reap even the limited success of Lowell, Liège, Manchester, Lille, Halifax, Glasgow, Bradford or Bilbao. Simply, supply outstrips demand.

Where is Heritage Used in Planning?

The creation of heritage places is a high-risk enterprise in which public and private venture capital is invested in the expectation of a return at least as high as alternative investment possibilities. The risk is increased by three characteristics of investment in heritage places. It is usually an all or nothing rather than incremental investment, partial restoration is rarely possible; second, it tends to be front-loaded, that is investment is made immediately but returns may only be apparent later and then may accrue over many centuries; third, there is a mutual dependence of different agencies and investors in the public and private sectors that need to act together to be effective.

The parties involved will try to minimize the risks by replicating what has previously been successful elsewhere. Ironically, places attempting to promote unique local identities by using their necessarily distinctive pasts increasingly are dependent on investors and developers who are global in their operations and visions. Even architects and designers (including the 'A-list' architects) who are recruited to convey global artistic status and recognition to a local project, have an interest in minimizing the chances of failure in high-profile commissions. This tendency towards replication is also evident among local public-sector policy makers who, faced with large and risky investments, are driven by international trends deliberately to seek out replicable global 'best practice'. Even the details of materials, paving, signage and street furniture produce a street-scene dominated by 'catalogue heritage'. The danger of these understandably cautious reactions is that initially original strategies become only local variations of globally evident projects, tending to a generic heritage of off-the-peg planning clichés which no longer express the originally intended locally distinctive identity (Ashworth and Tunbridge 2003). Some of the more common types of heritage development (used singly or in combination) are considered below. A common feature of each of these is the temporal shift from experimental uniqueness to global replication.

The Small 'Monofunctional' Heritage Town.

First, there is the 'gem-city' (Ashworth and Tunbridge 2000, 1990), a rare category of place, where the heritage resource is so remarkable, complete and valued that such places are locked into the perfection of their specific heritage, having little choice about local planning strategies and development options. The heritage resource is not the individual building or space but the totality of the town and its contents and this characteristic makes change very difficult. Any optimism that the costs, especially opportunity costs, of such heritage is offset by earnings from commercial heritage activities, especially tourism, is often dashed by the restrictions that preservation imposes on development in such towns. The heritage resource may be abundant but exploiting its economic potential is almost impossible. World heritage site designation compounds the problem by focusing the attention of preservation agencies (usually ICOMOS) on any perceived diminution of the heritage resource. (See also Duxbury and Jeannotte, Chapter 21, this volume.) The local authority of the Elbe Valley world heritage site (delisted 2009) sacrificed its WHS designation for a much needed transport improvement (the Waldschlösschen bridge). For instance, Regensburg, with a similar problem built a new and needed bridge over the Danube but at an inconvenient distance from the old town.

A different but similarly monopolistic dependence on heritage occurs when a small town deliberately selects some aspect of heritage as its leading economic sector. There are a handful of such successful cases but these have encouraged imitation, often unsuccessfully, because the heritage selected is so specialized that success depends upon being the first. The process is usually externally induced drastic economic change, resulting in local economic collapse, which fuels a creative initiative in which some aspect of local heritage, however tenuous and unlikely, becomes the leading economic sector. Laying claim to an already famous historic personality can be a successful heritage development strategy: Niagara-on-the-Lake (Canada) selected G.B. Shaw, Deventer (Netherlands) chose Charles Dickens and the most spectacular and ostensibly unlikely success is Stratford, Ontario, which shifted from railway engineering in the 1950s to the second largest Shakespeare festival in the world. The lesson of such cases is not the universal applicability of such heritage strategies but their quite specific local conditions for success. The key variables are the determination of local communities to succeed, the presence of creative individuals, the good fortune of external circumstances of time and place, and being the first.

Regenerated Docklands and Waterfronts

It is no surprise that one of the most globally prevalent heritage districts are waterfronts, the historic forms of which have been renovated and functions revitalized. In few other parts of the city has fundamental technological change impacted so decisively upon the functioning of an area. New developments in shipping and cargo handling moved commercial docks away from the central sites they had occupied often since the founding of the city. In the tourist-historic waterfront model (Tunbridge and Ashworth 1992) the crucial time period determining the future of waterfronts is the gap between the departure of the dockside commercial functions and the arrival of a leisure-based heritagization of the resulting 'zone of discard'.

Almost uniquely the origins of this type of heritage-driven revitalization are attributed to a particular person and place. James Rouse (1914–96) and associates were instrumental in the redevelopment of the Boston waterfront (completed in 1976). Centred upon the historic Faneuil Hall and including Quincy Market this development became an archetype for similar developments combining building conservation, and waterfront historic associations with leisure retailing (Hoyle et al. 1988). In the United States, Fisherman's Quay/Ghiradeli Square, San Francisco (1980), Harborplace, Baltimore (1980) and Riverwalk, New Orleans (1984) followed, and this 'Rousification' of the waterfront was exported worldwide.

The success of such developments conceals somewhat the difficulties. Typically waterfronts have accessibility problems, often becoming separated from the rest of the city by major transport arteries built along the available waterfront so that cities had to be re-orientated towards an abandoned, inaccessible and often unattractive waterfront. The past has been used in different ways in these developments. In some cases new leisure catering and retailing have been accommodated in the renovated piers and adapted quayside buildings (as in Historic Properties, Halifax, Queen's Quay, Toronto, or Waterfront, Seattle); the association of the waterfront with historic events through nomenclature, commemorative route marking, interpretive plaques and signage is very common, as such places have usually been central to a city's history and in North America or Australia often the place of original settlement. However, there are also examples of demolition and redevelopment. Portland's Pioneer Place (1990) is historic only in its name, as is Cape Town's Victoria & Alfred (1988). Both, however, use commemorative marking of historical events as the only remaining heritage component in an otherwise modern development.

Restored historic ships are often incorporated into developments, such as in Halifax, Boston, Portsmouth, London, Bristol and Leith. New Orleans' Riverwalk features functioning replicas of stern-wheel Mississippi paddle-boats. Again there is no assumption of success and there are cases where historic ships have not become the centre of heritage development (Hamilton and Buffalo). Although heritage related waterfronts are now commonplace, it was not always so and may be deliberately eschewed. London's 1951 Festival of Britain South Bank was a deliberate break with an unwanted and for many discredited past. Ironically this future-oriented district has become over time an eclectic mix of post-war modernism (Festival Hall), adaptive reuse (Tate Modern) and the postmodern (Millennium Bridge). In Europe the stimulus for waterfront revitalization was from 1985 often the 'European City (after 1999 Capital) of Culture' designation, as in Antwerp (1993), Salonika (1997), Bergen (2000), Porto

(2001) and Liverpool (2008), which all seized the opportunity to renovate, revitalize and reincorporate their waterfronts. However, others ignored this opportunity (Amsterdam, 1987; Dublin, 1991; Rotterdam, 2001; Tallinn, 2011).

The Mediterranean 'Fishing Harbour'

This is a variant of waterfront development where the historic function is fishing not trade. It is typified by a quayside promenade, lined with restaurants (often specializing in fish), leisure shopping and hotels. Boats are a décor accessory and possible tourism facility. The historicity is implicit rather than stated although sometimes enhanced by some heritage harbour structure such as fort, customs house, mole or lighthouse. Kyrenia, North Cyprus, is an archetype but the development has been replicated not only around the Mediterranean but also elsewhere in coastal locations attempting to become self-consciously 'Mediterranean' such as Cornwall, UK (Looe/Polperro), Western Australia (Albany) and South Africa's Cape Peninsula (False Bay, Fishhoek, Muizenberg).

Festival Markets

The term festival market, popularized by Rouse, is frequently an element in waterfront redevelopment but has a longer history as the mundane marketing, especially of perishable food, became a leisure attraction. Pike Place, Seattle was saved from redevelopment by popular protest as early as 1971. However, it was in 1976 that the renovated Quincy Market, Boston became the centrepiece of waterfront regeneration mutating from a daily necessities provision market to a recreational market experience. This use of historic buildings was not inevitable. For instance, both London and Paris relocated their traditional central food wholesaling market to larger more accessible peripheral sites. The problem of reusing the sites was similar but the solution was different. In Paris demolition and rebuilding in modernist style resulted in the cultural centre of Forum Les Halles (1971–9). In London's Covent Garden, nineteenth-century buildings were retained and reoccupied for speciality retailing (1980), which quickly became a major tourism attraction, whilst Forum Les Halles never achieved such popularity and is scheduled for redevelopment.

The Heritagized High Street

The pedestrianized heritage High Street, flanked by conserved buildings, is epitomized by 'period' cobbled surfacing, street furniture, signage and shop-front design. The function is fun-shopping and catering especially for merchandise associated in some way, however oblique, with heritage. The archetype is Elm Hill, Norwich, saved from demolition by local campaigners as early as 1927 and renovated and reoccupied by heritage tourism shops and services. Among numerous imitators are Böttcherstrasse, Bremen (1954), Shambles, York, Stokstraat, Maastricht (1973) and Gastown/Water Street, Vancouver (1971). The heritage is usually non-specific and just a generic atmosphere of unspecified antiquity. Despite their different origins they all have similar locations and retailing functions, serving a recreation market whether local or tourist and employing a similar catalogue of heritage design clichés (Ashworth and Tunbridge 2000, 1990).

High Amenity Town House Residential Areas

As monument lists became inflated in the last decades of the twentieth century, a necessarily growing percentage of designated monuments were smaller domestic residential buildings rather than palaces and churches. The term adaptive reuse had previously meant inserting, often creatively, a new use into a form created and used for some quite different function, but now it meant continuing the residential use for people from a different time. These heritage districts are strongly associated with gentrification, which is a process whereby typically low-quality, low-cost but accessible working-class housing is renovated and upgraded by middle-class private investor-occupiers on a house by house basis leading to a rise in structural quality, property values and ultimately change in the neighbourhood social composition.

The Jordaan district of central Amsterdam was a poor working-class neighbourhood prone to anti-government riots (1835, 1886, 1917 and 1934). Gentrification from the 1970s turned it into a highly desirable (and expensive) downtown residential neighbourhood with little reminiscent of a probably unwelcome past. Similarly in France, where the slogan *'renovation est deportation'* was coined, the Marais in Paris became the first (1964) national *secteur sauvegardé*. Many provincial towns followed, such as Strasbourg's Petite France or Colmar's Petite Venise. The Bergkwartier, Deventer, was an early example (1967–77) and subsequent archetype of such district renovation in other medium-sized Dutch towns (Goudappel 1978). Although many neighbourhoods have such a history of gentrification it is by means universal. Louisburg Square, Boston for example was almost from the city's origin an elite area (Beacon Hill) whose status just continued. Similarly the Georgian additions to Bath (Circus 1754–68, Royal Crescent 1767–74) were built to be leased to relatively wealthy residents and have largely continued to be so.

The 'Plaza Mayor'

First identified as a feature of Spanish cities (Ford 1985), this is a combination of architecturally imposing façades which predominantly house restaurants and cafes, enclosing an open-air public courtyard that is used as a dining, meeting or just 'happening' space. It has spread from Spanish and Italian cities and has been self-consciously imported into a climatically less sympathetic northern Europe making use of what were previously central produce market spaces (for example Grote Markt, Groningen; Brink, Tilburg; Marktplatz, Bremen). An elongated version of this, similarly with a Spanish origin, would be the *Ramblas*, which is a linear pedestrian boulevard that is used not only for access but for recreation and diversion. It combines some aspects of the park with public 'happening space' flanked by commercial leisure and entertainment activities.

Museum Districts

Individual museums are often used as signature buildings either as heritage in themselves (Louvre, Hermitage, Rijksmuseum) or as deliberately arresting anti-historicist buildings housing heritage artefacts (Paris Beaubourg 1977; Bilbao Guggenheim 1997). Such 'signatures of the city' are intended to restore local civic pride and promote a new city-wide brand in pursuit of economic or social repositioning (Hamnett and Shoval 2003). This 'Guggenheim gambit' can be extended by deliberately creating a museum district, endowing a locality with a signature function for the entire city. Although there are few functional associations between museums to encourage spatial clustering and visitors rarely engage in 'comparative shopping', the benefit seems to be in city-wide district branding. This is not a recent innovation. A number of European cities in the second half of the nineteenth century accepted public responsibility for accommodating what was seen as worthy culture-building major museums, galleries, libraries, exhibition halls, theatres, concert halls and opera houses. Capital cities in particular assumed responsibility for 'showcasing' the nation to itself and to the world. These were often clustered to form a museum district, such as London's South Kensington, the Rijksmuseum complex in Amsterdam, Berlin's Museum Island and the Brussels' Kunstberg. More recently some cities have tried to create such districts in order to extend their image or, as in Frankfurt-Main's Museum Riverbank, to use culture to balance a place-image dominated by financial services.

The Artist Colony

This is a phenomenon that is not created by heritage or usually by planners. Indeed such colonies of creative individuals may be antipathetic to heritage (which is seen as the constraints of the past) and may spontaneously emerge through the absence of both planning and conventional social restraints. However, once in existence such spaces may attract the attention of urban planners and can evolve into heritage.

The archetype is perhaps the Christiania district of Copenhagen where an abandoned military area was occupied by squatters in 1971. A self-declared 'Freetown' developed, attempting to set its own rules and pursue a self-conscious freedom of experimental artistic expression in an often uneasy relationship with the authorities. Inherently, many similar instances of the use of abandoned factory buildings (tolerated or ignored by planning authorities) are temporary and can develop in one of two ways. Such 'gritty neighbourhoods' may be conventionally redeveloped, with the displacement of the existing occupiers and their previous artistic function becomes no more than a possibly embarrassing memory. However, they may evolve quite differently. The practising artists attract other artists and those who associate themselves with art. Life-styles evolve that favour residential locations with access to urban cultural and social facilities and flexible and cheap studio and workshop space (the 'artscape' of Evans 2001). As the district becomes better known, and even associated with a place-bound 'school', it becomes a general arts and crafts district that produces saleable products as well as a distinctive artistic atmosphere that attracts residents, businesses and ultimately tourists. It has thus becomes a valued city heritage district. (See also, Montgomery, Chapter 20, this volume.) At some point in this progression the *avant-garde* artists depart to restart the process elsewhere. A well-established case is Montmartre, Paris, but examples are found as widely as Cabbage Town, Toronto, Taos, New Mexico (Wilson 1996), Usedom, Germany (van Hoven, Meijering and Huigen 2004) and St Ives, Cornwall.

The role of planning in such creative districts is ambiguous. Many local authorities attempt to respond to Landry's (2000) plea for 'creative cities'. However, the history of such districts suggests that they emerge in the absence of planning or even despite it rather than because of it. Usually planners only become aware of such districts once they have evolved into more conventional residential and commercial places retaining only a lingering atmosphere of artistic and social freedom, a *quatier latin* for an imagined *la vie Bohème*.

Is Heritage Planning a Special Type of Planning?

The discussion above has outlined the wide range of possible contributions of heritage to the solution of diverse place management issues. Notable is the diversity of aspects of the past that are used, the roles they play and the objectives they are expected to achieve. These can be reduced to a single central question, 'is heritage planning a special type of planning that undertakes the tasks involving heritage, as inventorized above, in the pursuit of collective goals?'

It is evident from most of the above discussion that heritage elements are rarely the sole or even most important components in planning strategies but are most often and effectively used in combination with other elements. Heritage often acts as a catalyst, aiding or accelerating processes of change in quite different attributes of the place. This being so, the answer would appear to be that heritage in planning is not a separate, exotic addition to the existing toolbox available to place planners but an integral part of managing the past in the present. However logical this may be, practice is often different, contradicting the above argument for three reasons. First, heritage planning needs specialized knowledge and expertise. Apart from a working knowledge of the past as history, chronology and quarry of personalities, events and cultural productivity, the contemporary use of these requires a knowledge of the legal basis to monuments, areas and more recently below-ground, preservation, and the specialist technical knowledge of material conservation that often determines present possibilities. These skills require specialist knowledge that is different from that found among most planners.

Second, many working with historical events and heritage artefacts and structures value them for more than their contemporary instrumental use. Indeed many who are concerned with the creation and maintenance of the resources from which heritage is created, may have little interest in their subsequent use and may even disapprove of it. Third, if heritage planning was just one type of spatial planning alongside other facets of planning, then those who perform these tasks would function within planning agencies and departments in coordinated policies. This is frequently not the case. Many of the attributes of the past, discussed above, are the responsibilities of various national ministries or agencies or are scattered among local departments of public works, heritage, culture, museums or even education or marketing rather than in planning departments. This reflects an underlying and uncomfortable reality of using heritage in contemporary local planning, namely that heritage is multi-used. It was not usually brought into being or maintained to serve contemporary planning needs and other purposes, whether political, social or cultural, compete for its use.

There is still a lingering belief among some planners that heritage planning is a special sort of planning, applicable only in special sorts of heritage places, namely a few specially dedicated heritage districts. These special places are removed from 'real' planning to be treated by a special heritage planning, leaving 'ordinary' places to be treated by 'ordinary' planning. The instances discussed above contradict this view. All places have history and all people cultural experiences: thus heritage is a ubiquitous discretionary development option. No place is locked into or out of the past by a spurious separation of places into exclusive categories of heritage-rich or heritage-poor. It is not a unique strategy applicable in unique places for narrow preservational purposes: it is an integral part of wider place management strategies, potentially applicable everywhere and anywhere.

References

Ashworth, G.J. 1991. *Heritage Planning: Conservation as the Management of Urban Change*. Groningen: Geopers.

——. 1995. Managing the cultural tourist, in *Tourism and Spatial Transformations: Implications for Policy and Planning*, edited by G.J Ashworth and A.G.J. Dietvorst. Wallingford: CAB International, 265–84.

——. 2008a. In search of the place identity dividend: using heritage landscapes to create place identity, in *Sense of Place, Health and Quality of Life*, edited by J. Eyles and A. Williams. Burlington, VT: Ashgate, 185–96.

——. 2008b. Grote Markt Groningen: re-heritagisation of the public realm, in *City Spaces – Tourist Places: Urban Tourism Precincts*, edited by B. Hayllar, T.Griffin and D. Edwards. London: Butterworth-Heinemann, 261–74.

——. 2009a. Do tourists destroy the heritage they have come to experience? *Tourism Recreation Research*, 34(1), 79–83.

——. 2009b. The instruments of place branding: how it is done?. *European Spatial Research and Policy*, 16(1), 9–22.

——. 2010. Personality association as an instrument of place branding, in *Towards Effective Place Branding*, edited by G.J. Ashworth and M. Kavaratzis. London: Elgar, 222–33.

Ashworth, G.J. and Graham, B.J. 2005. *Senses of Place: Senses of Time*. Aldershot: Ashgate.

Ashworth, G.J. and Kavaratzis, M. 2010. *Towards Effective Place Branding*. London: Elgar.

Ashworth, G.J. and Kuipers, M.J. 2002. Conservation and identity: a new vision of pasts and futures in the Netherlands. *European Spatial Research and Policy*, 8(2), 55–65.

Ashworth, G.J. and Tunbridge, J.E. 1990. *The Tourist-Historic City*. London: Belhaven.

——. 2000. *The Tourist-Historic City: Retrospect and Prospect on Managing the Heritage City*. London: Elsevier.

——. 2003. Whose tourist-historic city? Localising the global and globalising the local, in *Globalisation and Contestation*, edited by M. Hall. London: Routledge, 12–34.

Bianchini, F. 1993. Remaking European cities: the role of cultural policies, in *Cultural Policy and Urban Regeneration: The West European Experience*, edited by F. Bianchini and M. Parkinson. Manchester: Manchester University Press, 1–20.

Bianchini, F. and Parkinson, M. (eds) 1993. *Cultural Policy and Urban Regeneration: The West European Experience*. Manchester: Manchester University Press.

Burgers, J. 1995. Public space in the post-industrial city, in *Tourism and Spatial Transformation: Implications for Policy and Planning*, edited by G.J Ashworth and A.G.J. Dietvorst. Wallingford: CAB International, 147–58.

Burke, G. 1976. *Townscapes*. Harmondsworth: Penguin.

Ennen, E. 1997. The Groningen museum: urban heritage in fragments. *International Journal of Heritage Studies*, 3(3), 144–56.

Evans, G. 2001. *Cultural Planning: An Urban Renaissance?* London: Routledge.

Eyles, J. and Williams, A. 2008. *Sense of Place, Health and Quality of Life*. Burlington, VT: Ashgate.

Ford, L.R. 1985. Urban morphology and preservation in Spain. *Geographical Review*, 75(3), 265–99.

Goudappel, H.M. 1978. *Bergkwartier Deventer 1967–77*. Deventer: Ankh-Hermes.

Graham, B., Ashworth, G.J. and Tunbridge, J.E. 2000. *A Geography of Heritage: Power, Culture and Economy*. London: Arnold.

Hall, C.M. 1989. The definition and analysis of hallmark tourist events. *Geojournal*, 19(3), 263–8.

Hamnett, C. and Shoval, N. 2003. Museums as flagships of urban development, in *Cities and Visitors: Regulating People, Markets, and City Space*, edited by L.M. Hoffman, S.S. Fainstein and D.J. Judd. Oxford: Blackwell, 219–36.

Hoyle, B., Pinder, D. and Hussain M. (eds) 1988 *Revitalizing the Waterfront: International Dimensions of Dockland Redevelopment*. London: Belhaven.

Hughes, H. and Gratton, C. 1992. The economics of the culture industry, in *The Culture Industry: The Arts in Urban Regeneration*, edited by D. Wynne. Aldershot: Avebury, 96–107.

Landry, C. 2000. *The Creative City: A Toolkit for Urban Innovators*. London: Earthscan.

Larkham, P.J. 1996. *Conservation and the City*. London: Routledge.

Lewis, J. 1990. *Art, Culture and Enterprise*. London: Routledge.

Lim, H. 1993. Cultural strategies for revitalising the city: review and evaluation. *Regional Studies*, 27(6), 589–95.

Miles, M. 1997. *Art, Space and the City: Public Art and Urban Futures*. London: Routledge.

Myerscough, J. 1988. *The Economic Importance of the Arts in Britain*. London: Policy Studies Institute.

Oosterman, J. 1993. Welcome to the pleasure dome: play and entertainment in urban public space: the example of the sidewalk cafe. *Built Environment*, 18(2), 155–65.

Quinn, B. 2005. Arts festivals and the city. *Urban Studies*, 42(5–6), 927–43.

Stabler, M. 1996. Are heritage and tourism compatible? An economic evaluation of their role in urban regeneration, in *Tourism and Culture Towards the 21st Century*, edited by M. Robinson, N. Evans and P. Callaghan. Newcastle: University of Northumbria, 417–38.

Tiesdell, S., Oc, T. and Heath, T. 1996. *Revitalising Historic Urban Quarters*. Oxford: Architectural Press.

Timothy, D.J. and Boyd, S.W. 2003. *Heritage Tourism*. London: Pearson.

Tunbridge, J.E. and Ashworth, G.J. 1992. Leisure resource development in cityport revitalisation: the tourist-historic dimension, in *European Port Cities in Transition*, edited by D.S. Hoyle and D.R. Pinder. London: Belhaven, 76–200.

van Hoven, B., Meijering, L. and Huigen, P.P.P. 2004. Escaping time and place: an artist community in Germany, in *Senses of Place: Senses of Time*, edited by G.J. Ashworth and B.J. Graham. Aldershot: Ashgate, 155– 64.

Warnaby, G. and Medway, D. 2010. Semiotics and place branding: the influence of the built and natural environment in city logos, in *Towards Effective Place Branding*, edited by G.J. Ashworth and M. Kavaratzis. London: Elgar, 205–21.

Whitt, J.A. 1987. Mozart in the metropolis. *Urban Affairs Quarterly*, 23(1), 15–36.

Wilson, C. 1996. *The Myth of Santa Fe: Creating a Modern Regional Tradition*. Albuquerque, NM: University of New Mexico Press.

Wynne, D. 1992. *The Culture Industry: The Arts in Urban Regeneration*. Aldershot: Avebury.

Case Study Window – Cultural Cluster, Capital and Cityscape: The Cultural Economy of Japanese Creative Cities

Masayuki Sasaki

The Era of the Creative City

With a major shift toward globalization and the knowledge-based economy, the industrial city is already declining. A great deal of attention is being given to the development of a new type of city, 'the creative city'. These cities are characterized by the formation of clusters of creative industries, such as film, video, music and arts. (See also Miller, and O'Connor, Chapters 3 and 10, this volume.) These are also cities where 'the creative class' made up of high-tech experts, artists and creators prefer to live.

The concept of 'the creative city', both in theory and in practice, is at the heart of this chapter. This concept refers to a mobilization of the 'creativity' inherent in art and culture to create new industries and employment opportunities. In addition to addressing the problems of homelessness and the urban environment, it is believed that such an approach can foster a comprehensive urban regeneration.

In academia this concept first attracted attention through the works of Peter Hall, an internationally renowned authority on urban theory, and Charles Landry, an international consultant (Hall 1998, Landry 2000). In Japan and Asia, the author has played a leading role in promoting this concept in both theory and practice through his research and policy work (Sasaki 2001, 1997).

Part of the broader diffusion of the creative cities ideal has come through the launch of UNESCO's 'Global Network of Creative Cities' in 2004, and interest has quick spread beyond the confines of Europe and America to Asia and developing countries throughout the world. Prior to this, UNESCO adopted the *Universal Declaration on Cultural Diversity* (2001) for the purpose of reducing cultural standardization caused by the impacts of globalization. (See also, Duxbury and Jeannotte, Chapter 21, this volume). Now 34 cities globally and three in Japan, Kanazawa, Kobe and Nagoya, are registered with the global network.

Cities in Asia, especially Japanese cities, with their long history of bureaucratically led developmentalism at the centre of urban and regional politics, have suffered as neoliberal globalization has transformed industries and threatened social welfare systems. Environmental, employment and housing crises have also become more acute in this era of neoliberalism. At the same time, the businesses and families that have been central to coping with social crises in the past are no longer functional. In these times of crisis and recession, it seems that the need for fundamental social reconstruction from the grass roots has arrived.

While promoting global research on urban problems from the perspective of creative cities, we must be careful not to force a Western conception of the creative city ideal on our study of Japanese cities. Instead we must rethink the concept of creative cities in light of the myriad problems facing Japanese cities with the hope of creating a new urban society and a new urban theory based on 'cultural cluster, cultural capital and cultural cityscape' appropriate to the Japanese context.

Rethinking Creative City Theory

The creative cities idea emerged as a new urban model with the European Union's 'European City of Culture' or 'European Capital of Culture' projects. In these cases the creativity inherent in art and culture were utilized to create new industries and employment opportunities while also tackling environmental problems and homelessness. In short, this was a multifaceted attempt at urban regeneration. And the work of Charles Landry and of this author has put the issues of social sustainability at the centre of their respective visions of the creative city. In addition, Richard Florida has suggested that US cities should implement policies to attract the type of people he defines as a 'creative class', who he sees as needed to sustain the new creative industries (Florida 2002). (See also, Stevenson, Chapter 9, this volume.)

Florida has also advocated his own creativity index consisting of eight indices in three fields: talent, technology and tolerance. This index has created a stir among urban theorists and policymakers throughout the world. Among these three categories, Florida himself has stressed tolerance. Especially noticeable has been his gay index, in which the regional proportion of gays and lesbians to the entire nation is measured by location quotient (Florida 2005). His gay index has become a symbol strongly suggestive of the creativity of social groups like the open-minded, avant-garde young artists called Bohemians. Florida contends that this group displays the American counter-culture's fundamental opposition to highbrow European society, as in American musicals compared to European operas and American jazz and rock versus European classical music. The impact of Florida's unconventional theory has led to the common misperception that cities prosper as people of the creative class, such as artists and gays, gather (Long 2009, Zimmerman 2008).

Creative Cities and Culture-Based Production Systems

Other theorists, however, have noted that attracting people of the creative class does not automatically make a creative city. As Allen Scott maintains, for the development of creative industries that serve as economic engines for a creative city, it is imperative to have a large workforce with specific skills and the necessary industries to support that workforce (Scott 2006). And if the city's economy does not have a marketing capability that enables it to develop on the world market, sustainable development will prove elusive. In common with Scott, University of Minnesota professor, Ann Markusen, attaches importance to the role of the cultural and economic sectors of the city in the context of the knowledge/information-based economy. At the same time she criticizes Florida, saying that his argument lacks a development theory applicable to particular local economies. She contends that although export-oriented economic theories have long been in the mainstream as development theories for local economies, in this era of knowledge and information-based economies, economic development in import-substitution industries is more desirable (Markusen and Schrock 2006a, 2006b).

Markusen credits Jane Jacobs as the pioneer of this theory, and contends that cities pursuing export-oriented economic development through mass-production are liable to have insufficient consumption within the region and limited fields of industries. On the other hand, she advocates an import-substitution model that is centred on cultural industries to enhance consumption in the region and help bring about a diversified workforce and more sophisticated human capital to develop new knowledge/information-based industries. Markusen insists therefore that it is important to analyse the role artists play in creative cities on multiple levels – socially, culturally and economically (Markusen and King 2003).

Jane Jacobs' analysis of Bologna, Italy provides a good illustration of these principles in practice (Jacobs 1984). Bologna is a city with a flexible network system of small-scale production facilities that has repeatedly demonstrated a faculty for innovation and improvisation. With these principles in mind, we could define the creative city as a city that cultivates new trends in arts and culture, promotes innovative and creative industries through the energetic creative activities of artists, creators and ordinary citizens, contains many diverse 'creative milieus' and 'innovative milieus', and has a regional, grass-roots capability to find solutions to social exclusion problems such as homeless people (Sasaki 2001). The conditions needed for the realization of a creative city can be further clarified into the list of six conditions stated below.

First, it is a city equipped with an urban economic system not only in which artists and scientists can freely develop their creativity, but where workers and craftspeople can also engage in creative, flexible production, and in the process withstand the threats of global restructuring.

Second, it is a city equipped with universities, vocational colleges and research institutes which support scientific and artistic creativity in the city, as well as cultural facilities like theatres and libraries. It also has a very active non-profit sector featuring cooperative associations and establishments through which the rights of medium–small craftsperson's businesses are protected. Such a city would also have an environment where new businesses can be set up easily and creative work is well supported. Above all a creative city will have the necessary social infrastructure to support creative individuals and activities.

Third, it is a city in which industrial growth improves the 'quality of life' of the citizens and provides substantial social services. Therefore it stimulates the development of new industries in the fields of the environment, welfare, medical services and art. In other words, it is a city with a well-balanced development of industrial dynamism and cultural life, where production and consumption are also in harmony.

Fourth, it is a city that has a right to stipulate the spaces where production and consumption develop, and where the urban environment is preserved. It is a city with beautiful urban spaces to enhance the creativity and sensitivity of its citizens.

Fifth, it is a city that has a mechanism of citizen participation in city administration that guarantees the versatility and creativity of its citizens. In other words it is a city with a system of small-area autonomy supported by large-area administration that can take charge of large-range management of the region's environment.

Sixth, it is a city equipped with its own financial administration that sustains creative, autonomous administration along with personnel who excel in policy formation.

In addition to these general terms based on the empirical analyses of Bologna, Italy and Kanazawa, Japan, I define a 'cultural mode of production model' (see Figure 12.1).

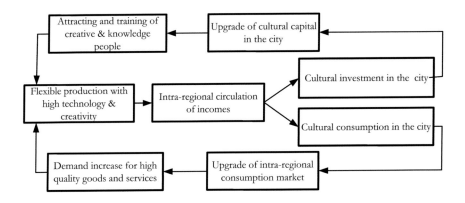

Figure 12.1   Cultural mode of production model

Source: Masayuki Sasaki.

The 'cultural mode of production' is the well-balanced system of cultural production and cultural consumption that takes advantage of accumulated cultural capital to produce products and services high in economic as well as cultural value in a system where consumption stimulates production (Sasaki 2007). This definition, however, requires further elaboration in light of the research of Ann Markusen and Andy Pratt. This method to create new industry for the development of the city economy through high-quality cultural capital may be called the 'cultural mode of production utilizing cultural capital'.

The 'cultural mode of production' that is Kanazawa's goal consists of the following elements.

- production of goods and services with high cultural value added through the integration of the skills and sensibilities of the artisans with high-tech devices in the production process;
- creation of a tightly knit, organic, industry-related structure of companies developing endogenously in the region, ranging from the cultural-goods industries to the high-tech, software and design industries; in order to
- circulate income obtained outside the region within the region, aiming for new cultural investment and consumption;
- cultural investments to go to the construction of museums and the support of private design research centres and orchestras, and so on, with the increased cultural concentration in the city resulting in the development and establishment in the region of high-tech/high-touch creative human resources as the players in the cultural mode of production;
- cultural consumption to upgrade the quality of local consumer markets and stimulate demand for the cultural mode of production through consumers who have the ability to enjoy goods and services that have abundant cultural and artistic qualities.

In Japan and other developed societies, where a mass production/consumption system of cars and hi-tech electronics has declined in the current global economic crisis, it seems that a shift towards the creative economy based on the 'cultural mode of production utilizing cultural capital' is desirable.

Cultural Cluster, Cultural Capital and Cultural Cityscape

Andy Pratt, an academic specialist on cluster policies for the cultural and creative industries, notes that family-operated and small-sized businesses are in the absolute majority in such cultural industries. Furthermore, in order to survive on world markets, it is imperative for these industries to have a network of horizontal cooperation with each other. He points to three characteristics in comparison with ordinary industrial clusters. The first is the importance of the qualitative content of the networks of the entities constituting the cluster, especially the process of 'tacit knowledge' exchange and its spill over. The second is that among corporate transactions that are part of the cluster, the importance of non-monetary transactions based on relations of mutual trust increases. Third, for the formation of the creative cluster, it is important to analyse not only its economic and social contributions, but also how such industries fit in the broader cultural context of the city or region (Pratt 2008, 2004). (See also Montgomery, Chapter 20, this volume.)

In other words, for creative industries – whose lifeblood is the creativity, skill and talent of individuals – to form a cluster, it is imperative to have a 'milieu' in place where creativity can be nurtured and flourish. In creative city theory it is the 'creative milieu' and 'social structure of creativity', and above all the social, cultural and geographical context that are truly vital for the effective integration of industrial, urban and cultural policy. Florida also points out the importance of the 'creative milieu', but he does not analyse deeply the economic aspect of a creative and cultural cluster.

At the same time, the cultural economist David Throsby argues that arts and culture may have a more pervasive role in urban regeneration through the fostering of community identity, creativity, cohesion and vitality via the cultural characteristics and practices that define the city and its citizens. And he points out the importance of cultural capital that embodies and gives rise to both cultural value and economic value in the city. In this respect, the consideration of heritage or cityscape as cultural capital can provide a means of integrating the interests of conservationists, who are concerned with the protection of cultural value, and economists, who look at heritage projects as the problem of the allocation of scarce resources between competing ends. He emphasizes the positive relationship between cultural capital and sustainable development of the city. In other words he argues that the coexistence of preservation and sustainable development is enabled by grasping the concept of the cityscape as cultural capital (Throsby 2001).

Furthermore, Landry suggests that cultural heritage and cultural cityscape are the sum of our past creativities and the results of creativity, and it is this which helps maintains urban society and propels it forward. Culture is the panoply of resources that show that a place is unique and distinctive. The resources of the past can help to inspire and give confidence for the future. And cultural heritage is reinvented daily whether this is through the refurbishment of a building or an adaptation of an old skill for modern times: today's classic is yesterday's innovation. Creativity is not only about a continuous invention of the new, but also how to deal appropriately with the old (Landry 2000). (See also, Ashworth, and Hillier, Chapters 11 and 24, this volume.)

On the other hand, cultural capital and cultural cityscapes form the creative milieu that attracts Florida's creative class to the city, and fosters the formation of a cultural cluster, the engine of sustainable development in a creative city.

I will now evaluate the creative city in Japan from the viewpoint of cultural cluster, cultural capital and cultural cityscape.

Kanazawa as a UNESCO Creative City: Creative City Challenges in Japan

New, experimental policies have emerged in Kanazawa (Sasaki 2003) and Yokohama in Japan as representative models at the same time as the creative cities trend has gained currency in the West.

In terms of population, surroundings, and defining characteristics, the city of Kanazawa has much in common with Bologna, Italy. Kanazawa (see Table 12.1) is a human-scale city of 450,000 that is surrounded by mountains that are the source of two rivers that run through the city. Kanazawa has also preserved its traditional beautiful cityscape and traditional arts and crafts. As a mid-sized city Kanazawa has maintained an independent economic base while also maintaining a healthy balance in terms of development and cultural and environmental preservation. Shortly after the end of the Second World War Kanazawa established the Kanazawa Arts and Crafts University. In addition to nurturing traditional arts and crafts, the city has also produced leaders in industrial design, and examples of local talent have become innovators in the field of

Table 12.1   Characteristics of Kanazawa and Yokohama

	Population	Economic Aspects	Cultural Aspects Budget (yen per capita)	Creative City Initiative
Kanazawa (UNESCO Creative City)	450,000 Human-scale city	Small artisan and medium-sized companies. Traditional crafts	Traditional and contemporary art. 4,000 yen	Business circle. Citizen group. Mayor office
Yokohama 2004	3,600,000 Modern, large city	Large companies. Port, car, hi-tech industries	Contemporary art. Art NPO. 2,500 yen	Mayor office. Art NPO

traditional crafts. Kanazawa has also become a national leader in historical preservation, as is evident in the meticulous preservation of the Tokugawa-era castle town district.

In addition to preserving the historical cityscape and traditional arts and crafts, Kanazawa has produced leading orchestra conductors and chamber music ensembles. Other civic achievements in the area of cultural creativity include the nurturing of local artists through the establishment of the citizens' art village and the Twenty-First Century Art Museum.

At the same time that the trend toward globalization quickly intensified in the latter half of the 1980s, the textile industry that sustained Kanazawa's high growth rates through the years went into decline. In September 1996, however, the Kanazawa Citizens' Art Village opened in a vacated spinning factory and adjacent warehouses. The Mayor of Kanazawa opened this 24-hour facility in response to citizen requests for a public arts facility that they could use in the evening to midnight hours after they had finished their daytime responsibilities. The facility itself is composed of a drama studio, music studio, 'eco-life' studio and art studio that occupy four separate blocks of the old spinning compound. Two directors elected by citizens oversee management of each studio. The active use and independent management of the facility is a remarkable example of a participatory, citizens' cultural institution in contemporary Japan. In sum, through the active participation of the citizenry, abandoned industrial facilities were used to construct a new cultural infrastructure, a new place for cultural creativity.

Another example of reimagining existing facilities and utilizing them in creative ways in Kanazawa would be the Twenty-First Century Art Museum that opened in October 2004 (Figure 12.2). This contemporary art museum is located in an area of the central city that many feared would lose its vitality when the prefectural offices moved from this area to the suburbs. In addition to collecting and exhibiting contemporary art from throughout the world, the new museum began to solicit and feature locally produced traditional arts and crafts. Further to this fusion of the global and the local along with the modern and traditional, the new museum pursued a policy of stimulating local interest and talent in the arts. To this end the first museum director, Mino Yutaka, solicited local schools and the general citizenry to participate in educational tours he dubbed 'museum cruises'. In the first year, the museum attracted around 1.5 million visitors – three times the population of the city. Furthermore, the revenue generated from these tours exceeded 10 billion yen. From 2008 the museum also sponsored open-air exhibits, which livened up a relatively quiet part of town and allowed people to view the work of local artists and studios that produced both contemporary and traditional works. Such policies are a shining example of creatively fusing the traditional and the modern through culture as part of urban regeneration.

Figure 12.2   The Twenty-First Century Contemporary Art Museum of Kanazawa
Source: Masayuki Sasaki.

Located around this contemporary art museum are more than 30 museums, public or private, large and small, as in Figure 12.3. In addition, arts and crafts shops and studios have accumulated around the museum cluster.

There are 22 types of traditional arts and crafts industries in Kanazawa, approximately 900 establishments and some 3,000 employees. This occupies approximately 20 per cent of the establishments in the city with some 6 per cent of the employees. The cluster of craft studios and shops is comprised of many extremely small establishments. In addition, a studio and 74 shops are located within a 5km radius of the old Kanazawa castle located in the inner city.

With the museum at the centre of industrial promotion efforts in the area of fashion and digital design, the city of Kanazawa has been promoting development in the creative industries. Thus we can see how the promotion of art and culture has led to the development of new local industries in contemporary Japan.

The city of Kanazawa is an excellent illustration of how the accumulated creativity in a city with a high level of cultural capital can be used to promote economic development. With a history as a centre of craft production in the Edo era, Kanazawa also clearly illustrates the historical stages of economic development from craft production, to

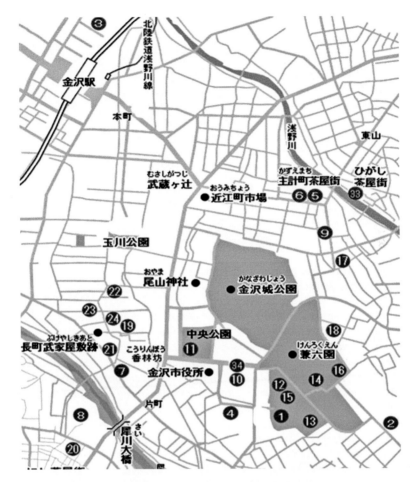

Figure 12.3 Museum cluster in the downtown area of Kanazawa

Source: Masayuki Sasak.

Fordism (mass production), and finally to a new era of culture-based production in the contemporary creative cultural industries (see Figure 12.4).

The creative city strategies of Kanazawa also demonstrate the importance of citizen and government collaboration in forums such as the creative cities council that brings together experts from various fields, and people from inside and outside government to deliberate and decide on matters of public policy. Such a forum and mode of deliberation and decision making is clearly congruous with the ideal of urban creativity. The experiences of Kanazawa outlined above are befitting of a UNESCO Creative City in the craft category. In October 2008 the city applied to UNESCO and was registered in June 2009.

In 2009, facing the challenges posed by the current global financial crisis, the city of Kanazawa implemented the 'Monozukuri (craftsmanship or art of manufacturing)' Ordinance for the protection and promotion of the traditional arts and crafts, and other new industries. The Mayor of Kanazawa describes its aims as follows:

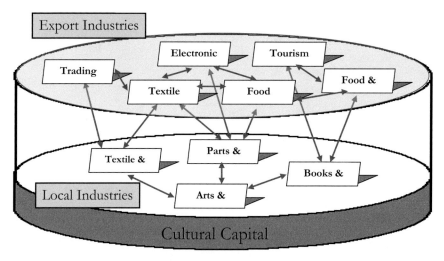

Figure 12.4   Culture and economy of Kanazawa city
Source: Masayuki Sasaki.

> *I think that the present society has lost sight of the meaningfulness of work and the basic way of life. In such an age, we should re-evaluate and cherish the spirit of 'craftsman' which leads to the creation of values. Without such efforts, we might lose our solid foundation of societies. Fortunately, the city of Kanazawa has a broad base of 'the milieu of craftsmanship' handed down from the Edo Period. The arts of Kanazawa's traditional craftworks include, among other things, ceramic ware, Yuzen dyeing, inlaying and gold leafing. We aim to protect and nurture the traditional local industries while working to introduce new technologies and innovative ideas. We also applied to UNESCO's Creative Cities Network for Crafts and Folk Art category. The Ordinance is intended to recognize anew 'the importance of craftsmanship' and 'the pride in craftsman' so that the region as a whole can support 'craftsmanship' industries in order to realize 'the lively city, Kanazawa'. The Ordinance applies to the fields of agriculture and forestry as well. Therefore, we are planning to develop an authorization system for Kanazawa brand agricultural products and to open the Kanazawa Forestry Academy. We are also aspiring to build cooperation between businesses and universities through the opening of institutes for research and promotion of Kaga-yuzen silk dyeing and Kanazawa gold leafing craftwork. I assume that diversified 'craftsmanship' will pave the way for diversified urban development. (Author's interview with the then Mayor of Kanazawa, October 2010)*

As described above, in the city of Kanazawa, both the administration led by the Mayor and private efforts are ongoing and are termed the 'two wheels of one cart'.

The Creative City Yokohama Experiment: Creative City Challenges in Japan

In stark contrast to the image of Kanazawa as an Edo-era castle town with a long and rich history, is the image of Yokohama, a port city that is 150 years old and has become one of Japan's largest urban centres (see Table 12.1). At the height of the bubble economy the city of Yokohama pursued a large-scale waterfront development project to create a new central business district with the aim of shedding its image as a city of heavy industry. However, with the collapse of the bubble economy and subsequent construction boom in central Tokyo, Yokohama suffered a double blow. From the beginning of 2004, however, Yokohama embraced a new urban vision and embarked on a project to reinvent itself as a 'creative city of art and culture'.

The contents of this new urban vision were fourfold: 1) to construct a creative environment where artistic and creative individuals would want to live; 2) to build a creative industrial cluster to spark economic activity; 3) to utilize the city's natural assets to these ends; and 4) to utilize citizen initiative to achieve this vision of a creative city of art and culture. By 2008 the city aimed to attract close to 2,000 artists and nearly 15,000 workers to its creative industrial cluster.

In 2004 former Mayor Nakada opened a special 'Creative City Yokohama' office. At the centre of the new offices activities has been the establishment of several 'creative core' districts in the general vicinity of the port. These creative cores utilize numerous historic buildings such as old bank buildings built in 1929 and once vacant offices to house new 'creative spaces' for citizen artists and other creative individuals. The 'Bank ART 1929' project was the start of this ambitious undertaking (see Figure 12.5). This project runs under the guidance of two non-profit art organizations that were selected via a competitive process and are in charge of organizing an array of exhibits, performances, workshops, symposiums and various other events that have attracted participants from Tokyo as well as Yokohama.

Since its inception the creative corridors have expanded as they have incorporated numerous vacant buildings and warehouses in the vicinity. Around this symbolic Bank ART 1929 building are located over 150 small offices in the genre of fine art, film and picture, design, town planning, photography, music, drama so on. As a result, many young artists in other genres and creators gathered and formed a creative cultural cluster. As of March 2007 the economic ripple effect of the creative corridors for the local economy is estimated to be in the range of 12 billion yen. In July 2007 an arts commission composed of public and private individuals and institutions was established to support and attract artists and other creative individuals to the region.

Of the numerous activities initiated in Yokohama, the experimental 'Kogane Cho Bazaar' of Yokohama is a good example. This event is held in an area formerly known in the chaotic period of the immediate post-war years for gangs and prostitution to what became a shopping district with over 250 shops. In recent years, however, many shops had closed down and the area was in decline. Many young students and artists collaborated with local businesses in the bazaar's projects. The diversity on display during the planning sessions for this event created a clear illustration of how cultural projects can lead to social inclusion. Indeed, these planning events featured the

Figure 12.5   'Bank ART 1929'

Source: Masayuki Sasaki.

participation of local residents, university students, artists and all manner of specialists to create an art event to enliven an area blighted by a plethora of vacant shops.

Finally, as 2009 marks the 150th anniversary of the opening of the port of Yokohama, an international creative cities conference has been opened with a purpose of building a creative cities network in Asia.

The case of Yokohama is remarkable in the sense that the policy aim of utilizing the creativity inherent in art and culture for the purpose of urban regeneration also led to a restructuring of the politics related to cultural policy, industrial policy and community development. In other words the new organizations that emerged to revitalize Yokohama as a city of art and culture transcended the bureaucratic sectionalism that typically plagues policy formation and administration in the fields listed above while also constructively engaging non-profit organizations and citizens in the formation and administration of policy. Throughout Japan it seems that urban policies and projects based on art and culture have given rise to a socially inclusive politics.

Conclusion

Comparing the examples of the above two cities, the medium-scale historic city of Kanazawa is making steady progress towards a creative city based on Bologna-type social capital with the initiative of the local businesses and citizens, involving the municipal government, while Yokohama is succeeding in forming an attractive and creative neighbourhood to attract the Florida-type creative class, and has also attained a positive outcome in the administrative efforts with mobility and cross-sectional cooperation led by the Creative City Headquarters. However, Yokohama has yet to establish a close partnership with local businesses. Briefly, theoretically, Yokohama is a case in which a city voluntarily chose the 'cognitive-cultural' system of production (Scott 2008) after the collapse of the Fordist and neoliberal paradigms. On the other hand, Kanazawa has experienced a more continuous and smooth evolution of historical local traditions.

In this way, Kanazawa and Yokohama became leaders and initiated many practical initiatives for the creative city project in Japan. Following the creative city network of UNESCO, the Agency for Cultural Affairs started an award system from 2007 with four cities selected each year.

On 4 February 2012, mayors and policymakers of 32 municipalities gathered in Tokyo and adopted an agenda to establish a creative cities network of Japan within a year.

In general, however, the urban cultural policy related to Japanese creative city projects seems to lack the strength and coherence of similar policies in the West. This in turn suggests that Japanese cooperatives, social enterprises, art related non-profit organizations, and other such organizations do not have the same level of social prestige and influence as their Western counterparts. However, as we have seen, there are definite signs that a grassroots movement in the area of creative urban policies is definitely gaining steam throughout Japan.

In summary, the policy implications exhibited in the above case study of Japanese creative cities are as follows.

First, it is necessary to conduct an intensive analysis of the embedded culture of the city, increase the shared awareness of fusing contemporary arts with traditional culture, clarify the need to become a 'creative city', and elaborate a creative city concept for the future, with an understanding of the historical context of the city.

Second, in developing concepts, 'artistic and cultural creativity' must be recognized as factors that have an impact on many other areas, including industry, employment, social welfare, education, medical care and the environment. In order to link cultural policy to industrial policy, urban planning and welfare policy, the vertical administrative structure must be made horizontal, ordinary bureaucratic thinking must be eliminated and organizational culture must be changed.

Third, cultural capital must be recognized as part of the basic social infrastructure in the knowledge and information society, and strategic planning must be carried out to inspire the creativity of citizens. Specifically, diverse creative milieus for the cultural cluster must be established in the city and creative producers must be fostered to take charge of this task.

Fourth, for the sustained development of the creative city, the promotion of the culture cluster is indispensable. It is essential to obtain the cooperation of a broad selection of citizens, including business leaders and non-profit organizations, perhaps in the form of a Creative City Promotion Council. Most important for the promotion of creative cities is the establishment of research and educational programmes for developing the necessary human resources.

Developments in the creative cities field in Japan in the midst of worldwide crises and drastic social and economic restructuring suggest some new issues to consider in the field of creative cities theory.

One issue to consider is the movement away from a mass production industrial society toward a creative society of culturally based production where cultural value and economic value are united. A related issue is the high level of cultural diversity required for this social transformation.

Furthermore, with regard to cities in Asia with their shared history of large-scale heavy industries at the heart of economic development policies, we must consider the necessary transition toward more compact cities. At the same time, we must also come to understand, appreciate and preserve the tangible and intangible cultural capital inherent in the traditional urban culture of each individual city.

The second issue to consider is the need to face the problem of social exclusion directly, and provide the social infrastructure, including a real and diverse 'creative milieu' to foster and ensure the active participation of the citizen in urban policy (Sasaki 2010).

The need to create a social system that respects and promotes both individuality and creativity to the utmost degree is vital to the success of tackling both of the issues outlined above. Building an educational and industrial system that fosters and promotes creativity will be central to the new creative economy with equal regard to cultural, social and economic value.

In order to realize and to develop creative cities, not only do we need the global level inter-city network promoted by UNESCO, but we also need to learn from partnerships seen at the Asia Pacific regional level and at the national level as well.

When a creative city network in the Asian region is established to support these activities, a new form of 'Creative Asia' will emerge.

References

Florida, R. 2002. *The Rise of the Creative Class: And How it's Transforming Work, Leisure, Community and Everyday Life*. New York: Basic Books.

—— . 2005. *Cities and the Creative Class*. Routledge: New York.

Hall, P. 1998. *Cities in Civilization: Culture, Innovation, and Urban Order*. London: Weidenfeld & Nicolson.

Jacobs, J. 1984. *Cities and the Wealth of Nations: Principles of Economic Life*. New York: Random House.

Landry, C. 2000. *The Creative City: A Toolkit for Urban Innovators*. London: Comedia.

Long, J. 2009. Sustaining creativity in the creative archetype: the case of Austin, Texas. *Cities*, 26(4), 210–19.

Markusen, A. and King, D. 2003. *The Artistic Dividend: The Hidden Contributions of the Arts to the Regional Economy*. Minneapolis, MN: Project on Regional and Industrial Economics, University of Minnesota.

Markusen, A. and Schrock, G. 2006a. The artistic dividend: urban artistic specialization and economic development implications. *Urban Studies*, 43(9), 1661–86.

—— . 2006b. The distinctive city: divergent patterns in American urban growth, hierarchy and specialization. *Urban Studies*, 43(8), 1301–23.

Pratt, A.C. 2004. Creative clusters: towards the governance of the creative industries production system? *Media International Australia*, 112, 50–66.

——. 2008. Creative cities: the cultural industries and the creative class. *Geografiska Annaler: Series B, Human Geography*, 90(2), 107–17.

Sasaki, M. 1997. *The Economics of Creative Cities* (in Japanese). Keisou Shobou.

——. 2001. *The Challenges for Creative Cities* (in Japanese). Iwanami Shoten.

——. 2003. Kanazawa: a creative and sustainable city. *Policy Science* (Ritsumeikan University), 10(2), 17–30.

——. 2007. Towards an urban cultural mode of production: a case study of Kanazawa, Japan, in *Urban Crisis: Culture and the Sustainability of Cities*, edited by M. Nadarajah and A.T. Yamamoto. Tokyo: United Nations University Press, 156–74.

——. 2010. Urban regeneration through culture creativity and social inclusion: rethinking creative city theory. *Cities*, 27(Supplement 1), S3–S9.

Scott, A.J. 2006. Creative cities: conceptual issues and policy questions. *Journal of Urban Affairs*, 28(1), 1–17.

——. 2008. *Social Economy of the Metropolis*. Oxford: Oxford University Press.

Throsby, D. 2001. *Economics and Culture*. Cambridge: Cambridge University Press.

Zimmerman, J. 2008. From brew town to cool town: neoliberalism and creative city development strategy in Milwaukee. *Cities*, 25(4), 230–42.

PART 4
Planning Practices

Preface to Part 4

Greg Young

In this Part the authors range across contemporary planning practices as they interpret them in response to key, critical issues such as place identity, development, sustainability, creativity and cultural resilience. In spite of the great diversity of themes a common strand running between the chapters is the recognition of the multiple conundrums faced by planning. For example, Evans explores the need to rise above planning's homogenous standards, its development control functions and reductive valorization of land and exchange values, Pieterse asks how ideas about sustainability may be recast in more radically democratic terms in order to create a viable political project to increase the relevance of planning, Dovey reflects on how places may evolve from a closed, purified and static identity towards identities that are open, multiple and dynamic and Baycan and Girard consider a manifesto and programme to promote local grass roots sustainability in cities internationally at the same time as accommodating common factors such as globalized demands from corporations and pressures from tourism.

Graeme Evans in his chapter calls for the embedding of cultural planning in mainstream planning systems and in the education and training of planners and related professionals. He also recommends a planning that considers the whole of a population and the full diversity of culture's manifestations. These recommendations follow from his recognition of the limitations associated with the many 'toolkit models' for cultural planning of the recent past that have so often proved to be 'time-limited' in nature and from the growth of a micro-level approach to place-making and strategic policy-making which in the UK, Europe, Canada and Australia includes the concomitant use of external consultants.

Edgar Pieterse describes the crisis of urban life in cities of the global South where less than one half of the labour force is absorbed through the formal economic system and national governments struggle to protect jobs and induce employment growth while this is deemed to be anti-competitive in a world of increasing economic integration. Beyond this, however, Pieterse poses the burning question of how it is possible to sustainably reconnect young adults between the ages of 15 and 29 with society, nature and their surroundings through social and environmental activities.

Kim Dovey unravels the contradictions of planning and place identity manifest in a number of suburban case studies in Melbourne, Australia and in terms of his theoretical exposition of the contrast between open and closed ontologies used to define place. He notes that for good reasons the concept of place has played a key role in planning and other discourses as a term with multi-disciplinary and multi-scalar application and relevance but also not without the inherent contradictions and ambiguities he outlines.

In this respect, Dovey subtly redefines the terms of a common social conundrum whereby a perceived community environmental loss is felt by residents in the face of developmental intensification.

Tüzin Baycan and Luigi Fusco Girard describe the utilization of culture in sustainability practices through an analysis of the 'slow city movement' which originated in Italy in 1999 as *cittaslow* and consists of member towns with populations of under 50,000 inhabitants as benchmarking exemplars of good sustainability practices and good governance. The slow city movement has objectives ensuring sustainable development at the local level and promoting the movement and manifesto globally. While doing this it seeks to avoid the problematic aspects of the 'fast world' of globalization that promotes standardization and homogeneity in living and practices and to promote local identity and distinctiveness in all aspects of culture including local products and gastronomy. The goal of the *cittaslow* movement is to inculcate cultural planning and cultural resilience in communities world-wide not only through its role as a social movement but also as a model for governance.

Each chapter in this Part is therefore to a greater or lesser degree a manifesto of some kind. In the case study window on the international slow city movement it is a clear-cut strategic manifesto but this has parallels in a sense in the chapters from Evans, Pieterse and Dovey where considered approaches are proposed to create improved and more systematic and reflective planning responses. For example, while Evans sees cultural planning in his chapter as a policy and practitioner's 'field' rather than a single policy or planning method he argues that its ability to contribute to sustainable development depends on it being embedded in mainstream development and resource allocation. He believes a cultural governmentality based on community-led practices should be engaged through cultural planning approaches given the often brittle and ephemeral nature of cultural toolkit paradigms. Pieterse on the other hand sees the delineation of individual urban systems and the use of taxonomies, typologies and categorizations in understanding them as unavoidable if rather foolhardy in a global context of contingency and indeterminacy. In an anti-heroic manifesto he argues that planning can become a sly and subversive player if recast through a radical democratic lens imbued with an understanding of the cultural questions surrounding the core of urban dialectics.

Dovey acknowledges the cultural constructedness of place and provides a closely reasoned argument for environmental and social change viewed as part of a continuum that may embrace a more sustainable city not dependent on a static and closed concept of place and neighbourhood character and identity. This is also a preferred context where change may occur without triggering in residents an ontological crisis of identity theft.

Tüzin and Baycan argue that although the wholesale strategy of slowness was developed for small cities it can be transferred under certain circumstances to medium-sized cities and has smart techniques and original perspectives relevant to cities of all sizes. A common theme running through all chapters is the need for greater creativity and more considered and philosophical approaches to planning. This chimes with the fact that planning as a discipline draws on considerably more than a basis in technical expertise and rational knowledge.

Cultural Planning and Sustainable Development

Graeme Evans

... if we were to ask Patrick Abercrombie today about his feelings on culture and creativity, I think he would raise an eyebrow and smile about the idea that planners can have any direct impact on culture in the city. Then, after a minute of reflection, he would suggest that without creativity planners cannot fulfil their ambitions to create cultural environments for the cities of tomorrow, and he would agree that only creative planning provides a survival strategy for the cultural identity of the European city and its multi-cultural citizens in a period of globalisation. We have to work on it. (Kunzmann 2004: 383)

Introduction

A decade or more has elapsed since the notion of cultural planning was distinguished from the more functional planning for arts and culture within amenity, heritage, land use and economic resource planning systems. The 'cultural turn' in town and spatial planning which emerged from the 1970s and 1980s has taken longer to establish itself than say in geography (Barnes 2001, Crang 1997), political economy (Soja 1999) and sociology (Chaney 1994). Planning as a hybrid practice has adopted culture – at least partially – to better inform more sustainable approaches to scarce land use and amenity needs and to respond to diversity and planning cultures. On the other hand it has also rediscovered culture in order to address aspirations and visions of city planning itself as a 'work of art' (Boyer 1988, Munro 1967, Talen and Ellis 2004). Different approaches and rationales for cultural planning have therefore emerged, rather than a single grand theory or conceptual model (see also Bianchini, Chapter 22, this volume). This is in part due to the epistemological and definitional differences evident in these approaches towards 'culture' and 'planning' and their combination, and in part to cultural politics that views the very idea of planning for culture as inherently instrumental in maintaining cultural and growth hegemonies – both market and state-led. Kunzmann's aspiration (above) towards creativity in planning also conflates the practice of town planners with complex notions of city culture, environment and identity. The notion of the creative city has also extended the cultural planning sphere to the creative industries ('economy') and issues of place-making, design quality and sustainability. Place-making – like New Urbanism (CEU 2004) – whilst usefully focusing on the street, urban design, mixed-use, local accessibility, 'sense of place' and so on, can also be seen as a rejection or failure of planning in terms of the wider scale and the whole, distributed population (Evans 2010). Everywhere is a 'place', not just the valorized spaces which tend to feature in place-making agendas and guidance (APA 2011, Markusen and Gadwa 2010), and which are preoccupied with town centres and regeneration sites (*physical*, *economy*) at the expense of the everyday, residential areas and cultural activity (*people*).

This ground – the creative city and creative places – has very much been occupied by proponents such as Landry (*Art of City Making*, 2006) and in the last decade, Florida (*Cities and the Creative Class*, 2005), and a host of intermediaries working between the urban environment, urban culture and urban society (Evans 2009). This includes 'cultural planners' – whether emanating from traditional arts and cultural development, or from town planning and the new hybrid with architecture – urban design and its scalar practice, masterplanning. A criticism of this movement has been its detachment from the practice and provision of arts and culture itself – particularly its organization, consumption and participation. So that whilst these 'activists' and more engaged academics (for example Markusen and Schrock 2006) argue for an approach that builds on local cultural assets and activity, with concepts such as 'natural cultural districts' (Stern and Seifert 2007), in Grodach's view they: 'tend not to deal with the potential inherent in a comprehensive approach that coordinates and builds on existing arts infrastructure and programs across geographical and institutional divides' (Grodach 2008: 2).

Key approaches to cultural planning have however developed, which attempt to engage local communities and artists/arts groups in the often politicized grand planning and *projects* (Evans 2001, Freestone and Gibson 2006), which exist within the abstract context of city and regional level planning and national policy – both urban and cultural. The more inclusive approach to cultural planning, as now widely promoted, can be traced to advocates, notably in Australia, Western Europe, Canada and the United States; in cultural city and regional planning, including post-industrial cities; and in areas undergoing housing and population growth and also decline, such as in 'shrinking cities' (for example Detroit, Liverpool, East Germany). Cultural planning has had, therefore, to accommodate urban densification and growth in new town and city extensions, as well as in depopulating cities in some cases in the same region. Handbooks and toolkits have emerged as a consequence of these challenges in order to support practitioners – planners and cultural intermediaries (Evans 2008). International agencies such as UNESCO, the EU and the Council of Europe (for example Intercultural City) have followed suit with an increasing range of guidance and networks. The emphasis here is generally on process, and ostensibly a more inclusive and territorial approach to culture and cultural assessment. These models therefore represent an empirically-based heuristic approach to what is a complex and contested arena. Cultural planning can thus be seen as a policy and practitioner 'field' rather than a single policy or planning method.

This chapter provides an exposition of the evolution and modes that have emerged in cultural planning from a sustainable development perspective. This is applied in terms of community cultural planning and responses in the form of cultural planning guidance and toolkits. The extent to which cultural planning might contribute to sustainable development and 'communities' through a whole population approach is discussed, addressing aspects such as spatial planning, culture and governmentality. The limitations of these initiatives in terms of planning practice and sustainability are then highlighted, concluding that whilst progress has been made towards the 'spatial turn' in culture and planning, this has not been sufficiently embedded in mainstream development planning or resource allocation processes.

Culture and Sustainable Development

The paradigm of sustainable development has both a supranational origin and reach, as well as a global and local dimension (WCED 1987). The latter has been promoted in policy terms through Local Agenda 21 (LA21) and a closer alignment of planning with sustainable development objectives. The emergence of the sustainability concept has developed from a political perspective which looks for solutions to the most powerful societal needs: for economic development; for environmental protection; and the imperatives of social justice, cultural diversity and access to public and cultural space (Mitchell 2003). Not surprisingly, sustainable development necessitates policy changes in many sectors and greater coherence between them, requiring: 'integration of objectives where possible; and making trade-offs between objectives where integration is not possible' (Dalal-Clayton and Bass 2002: 7). Sustainable development objectives therefore act in different ways and scales – at global, national and local levels, but should ideally be consistent between these – also a goal of 'planning' – local, spatial and regional. There are a wide range of sustainable development approaches which reveal different challenges faced by individual countries and their response to these. Hence, although sustainable development is a global challenge, it can only be handled and operationalized by national and local practice. This is also true of cultural planning, not least where sustainability is also a consideration and objective.

Culture has been largely absent from these discourses, which have been dominated by economic and social impacts on the environment (and vice versa) and managed 'sustainable' growth. However, initiatives such as Agenda 21 for Culture (UCLG 2009) have more recently promoted the centrality of culture to sustainable development:

> ... 'Culture in sustainable development' is not only about 'using artists to raise concern on climate change' or about 'building cultural venues that are efficient in the use of energy and natural resources'. It is not only about the income that cultural industries can bring to the economy. It is not about 'asking more' to the cultural circles. These are very important questions that need to be addressed, but they do not articulate the core question. The role of culture in sustainable development is mainly about including a cultural perspective in all public policies. It is about guaranteeing that any sustainable development process has a soul. This is the core question. (UNESCO 2009: 6)

This sentiment draws on the notion that culture is a 'fourth pillar of sustainability', as well as a component of local amenity and quality of life (Hawkes 2001). Where culture *has* been placed more centrally in this dialectic is in developing countries and regions. Here culture (and cultural development) is more closely associated with community, society and everyday life (Williams 1983). In so-called developed countries in contrast: 'the tacit acceptance of the arts and heritage version of culture has marginalized the concept of culture and denied theorists and practitioners an extremely effective tool' (Hawkes 2001: 1, Nurse 2006). It is perhaps in the area of human behaviour (and by extension, planning) that sustainable development and culture coincide. It is at the individual, institutional and corporate levels that decisions over sustainability are made both day to day and in resource allocation and planning for the future. The context within which behaviour and action are taken also involves the systems of governance that apply at various scales. The fourth 'cultural' pillar therefore extends the economic, social and environmental dimensions, placing governance as central and emphasizing inclusion (LA21 above) as the key social element in sustainable development (Figure 13.1).

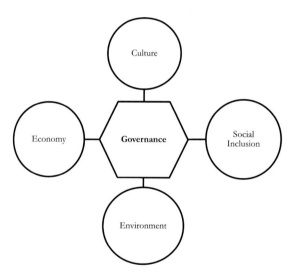

Figure 13.1   Culture, governance and sustainable development

These twin meta-themes of governance and sustainability provide a key challenge, but also fundamental underpinning of the planning and culture dialectic – and their coalescence. They are arguably essential goals and processes that validate the cultural planning paradigm. Bennett goes as far to say that culture itself should be thought of as 'inherently governmental' with 'culture used to refer to a set of practices for social management deployed to constitute autonomous populations as self-governing' (1995: 884). This dispersed form of power (Foucault 1991) seeks to 'shape, guide or affect (our) conduct ... to act on the conduct of conduct' (Gordon 1991: 2), thereby influencing the actions of others through processes which encourage self-regulation.

Versions of the following definition are therefore commonly adopted for the holistic 'cultural planning' and 'planning culture' approach: 'cultural planning is the strategic and integrated planning and use of cultural resources in urban and community development' (Mercer 2006: 6, and see Ghilardi 2004). This in turn implies:

> *an approach based on broad definitions of 'culture' and 'cultural resources', which encompass heritage, local traditions, arts, media, crafts, topography, architecture, urban design, recreation, sports, entertainment, tourism and the cultural representations of places; and a culturally sensitive approach to urban and regional planning and to environmental, social and economic policy making. (Baeker 2010: 16)*

This has some resonance with the unifying idea of 'Cultural Commons' which identifies culture(s) as shared resources for both cultural expression (Bertacchini et al. 2011) and social progress. However, these approaches have tended not to engage with the spatial planning, land use/environment or distributive aspects of cultural activity and facilities, particularly in relation to a wider community, as opposed to communities of interest – for example neighbourhoods, cultural groups, heritage, regeneration and growth areas and sites (Gray 2006). They have also not addressed sustainable development principles or imperatives. This is reflected in the approach to cultural planning undertaken as part of community development, local heritage and neighbourhood regeneration projects (Bianchini and Ghilardi 2004, Guppy 1997) and where: 'cultural landscapes become subject to this manipulation through regeneration programmes and the incentives they selectively employ' (Evans and Foord 2003: 17).

The tools to undertake a wide-ranging inventory and the scope of assessments are thus restricted in practice, due to the time-limited nature of such cultural audits and planning exercises, the imperatives of local agencies (funding/funder-led), and a poor information and knowledge base (and knowledge management and exchange) of history/trends, previous studies and data access and sharing. The risk of such exercises is that they can reflect expressed need and a bias towards those active (and vocal) beneficiaries, rather than the community as a whole (see for example, Bristol waterfront – Bassett, Griffiths and Smith 2002, and Tower Hamlets, East London – Evans and Foord 2003, Evans, Foord and White 1999, Landry 1997). Kunzmann goes as far to say that: 'integrated cultural planning may not be the right approach as it tends to raise expectations beyond reality. Tiresome culture-related shopping lists are not helpful' (2004: 399). Furthermore, the dynamics of population and social change, including migration and mobility ('churn'), increasingly undermines the snapshot audit and perennial profiling and consultation exercises – for example borough plan, arts and cultural strategies, amenity/facility standards, 'satisfaction' surveys – on which public planning has traditionally relied.

Community Cultural Planning

How we define communities, neighbourhoods and culture also requires an historic and geographic sensitivity, particularly when notions of cultural heritage, legacy ('ownership'), access and users/non-users of cultural facilities also reflect social distinctions and divides (Chan and Goldthorpe 2007). The notion of 'community' itself presents a shifting array of definitions no less so than 'heritage'. Bell and Newby (1976) developed four uses for this concept: (1) the idea of community as belonging to a specific topographical location; (2) as defining a particular local social system; (3) in terms of feeling of communitas or togetherness; and (4) as an ideology, often hiding the power relations which inevitably underlie communities. Drawing on these earlier perspectives, these communities can be typified in terms of place, and in terms of interest (including 'communion' – that is, attachment to a place, group or belief), with interest groups often identified in terms of particular locations, meeting places, facilities or venues, including faith centres, whether or not they are frequented by local, or by visiting groups.

Improved profiling of communities in social and geographic terms, as discussed below, in the context of cultural activity and facilities and their relationships, goes some way to capturing the cultural assets and potential of an area – at various scales. Thus offering some comparative evidence and basis from which to plan, identify gaps, clusters and flows (that is, of activity, people and economy). Cultural mapping of areas with no discernible cultural provision can also give the impression of a 'wasteland', and that culture is therefore 'absent', whether rural or post-industrial landscapes. Mapping approaches through oral history, literature, poetry, environmental art (Lacy 1995) can capture the cultural legacy and heritage of areas that may be 'lost', displaced, reflected in artefacts, museum collections, or held in memory. This can include the former residents, workers and visitors, as well as artists. As Mercer quotes (2006: 5): 'We must excavate the layers of our city downwards, into its earliest past … and thence we must read them upwards, visualising as we go … We need to be able to fold and integrate the complex, histories, textures and memories of our urban environments and their populations into the planning process. We need to do some *cultural mapping* – tracing people's memories and visions and values – before we start the planning' (my emphasis Geddes, quoted in Hall 1988, see also Bianchini, Chapter 22, this volume). This approach, it is acknowledged: 'needs to be a consultative and participatory process involving all interested groups within the local and artistic community' (Guppy 1997: 4). This latter point is important, since formal land-use planning and regional spatial strategies do not directly engage with the arts, voluntary or cultural constituency, but who may nonetheless be key to the programming and organization of both facilities and activities. This therefore requires 'a committed municipal leadership that recognizes the contribution of artists and arts organizations to local economic development and community well-being' (Grodach 2008: 5). With this also in mind: 'cultural spatial planning needs to avoid a rigid top-down approach. Some of the most exciting cultural projects and partnerships develop from unanticipated areas and grow in unexpected directions. Spatial planning of cultural facilities needs to be flexible to accommodate creative spontaneity' (Culture South West 2005).

The emphasis placed on 'strategic', 'integration' and 'cultural resources' also requires engagement at several levels of government/policy and with sectoral interests – and their underlying classification and valuation systems. At a regional and city-wide level, the realities and priorities of private and institutional land-ownership, infrastructure and national interests dominate. Today, regional cultural and planning agencies continue to struggle with how to define regional and strategic level cultural facilities and distribution, against local authority parochial interests, legacies and in the absence of a national cultural map or cultural planning standards. In the UK, this is despite earlier attempts to develop model planning policies for arts, culture and entertainment (ACE) in borough unitary development plans (UDPs) which directly addressed these spatial and strategic distinctions. Here, the regional London Planning Advisory Committee (LPAC 1990a and 1990b) had joined forces with the regional arts board for London to develop strategic policies for culture that could be interpreted in the planning process and by professional planners. The adoption of ACE also reflected the more eclectic range of activities – community, public, commercial – as well as their local and strategic role and potential.

This planning initiative argued that one way cultural policies can be incorporated with the environmental planning process is through the development of mechanisms and themes by which cultural planning measures may be included in periodic borough and city plans. In London the requirement of each of the 33 local planning authorities to produce ten-year Unitary Development Plans (UDPs) for their borough, provided the opportunity for the interpretation of culture in town planning proper. The term 'unitary' referred to the fact that these represented a single tier of statutory planning, both strategic and local in effect, in the absence of a regional tier of government and therefore city plan. These borough plans were subject to local inquiry and consultation before their adoption. This made the Unitary Development Plan the primary consideration in development control, effectively making it an enforceable blueprint for each borough: 'the approach shall leave no doubt about the importance of the planled system' (Cullingworth and Nadin 1994: 58).

More recently, in the short-lived English regional spatial strategies, 'Culture' has, however, struggled to find a place, whilst in Montreal, Duxbury observes:

> there is a disturbing absence of culture in the new visions for Montreal as a 'City of Innovation' and 'Knowledge City'. Cultural activities, and the innovative energies they embody and develop, are incorporated into these visions and plans. In general, it appears that cultural and heritage activities and resources are recognized and valued insofar as they attract the scientists, and other knowledge workers the city is recruiting. However, cultural activities are not seen as part of the knowledge and innovation milieu itself. (2004: 1)

Where culture is mentioned in city and regional plans, this is generally in the context of the regional economy (in the guise of the creative or digital industries), social/ regeneration or health impacts, with cultural assets normally referring to 'heritage' and in the context of tourism. In practice, culture has been placed in a peripheral and reactive role in regeneration and planning generally[1] (Evans 2005). As Young also observed: 'while culture is embedded in geographies, societies and histories, its voice is weak in planning. In fact culture rarely seems to speak meaningfully in planning at all' (2006: 43).

---

[1] This absence still persists, with the UK government's Draft National Planning Policy Framework (2011) noting heritage, sport, leisure and recreation, but not culture: 'despite such widespread recognition of the role of culture in the planning system. the planning framework does not reference the arts or culture in its definition of sustainable development, the land use principles underpinning plan making, or the strategic priorities for local plans …' (Arts Council 2011: 2).

Whilst model planning policies can be useful in borough and city plan formulation, and help identify the way that art and culture can contribute to planning objectives, more detailed analysis and guidance is needed in order to inform the more challenging resource allocation process. It is also needed in order to turn plans into practice and deliver effective and sustainable resource allocation. Where this is not the case the classic 'white elephants' are more likely to occur, giving cultural investment and projects a bad and 'unsustainable' press, or even worse, damage the viability of existing cultural provision (Evans 2005). This is particularly the case where property-led regeneration and investment dominates the development agenda – manifested in site-based, area-based, city centre, waterfront, cultural icons and major mixed use schemes, and which tend to skew the whole population planning process and public spending options. In practice this tends to produce a reactive planning and cultural planning response, itself resulting in large-scale cultural and regeneration schemes which are otherwise non-planned and fail to reflect community aspirations or need, and to divert public funding from other areas (Borja and Castells 1997, Evans 2001). As Grodach observes: 'local arts planning initiatives tend to be economically-motivated, project-based, and geographically uneven. In most cities, planning around the arts exhibits a bifurcated pattern with one set of strategies focusing on the downtown and another more diffuse approach that thinly spreads resources throughout the rest of the city' (2008: 1).

Identifying the drivers and agents of change – policy, programmes, agencies, investors/ developers – and growth, is therefore increasingly important in planning for the future and in order to have some influence over the development and resource allocation process. Change of course does not always mean 'growth' – some inner-city as well as rural areas are also 'shrinking', as residential areas change land use, and occupants (especially families and young people/students) move out. By incorporating the mapping of physical and human (social, economic) activity – as discussed in more detail below – to inform a needs assessment as a basis for planning, option appraisal

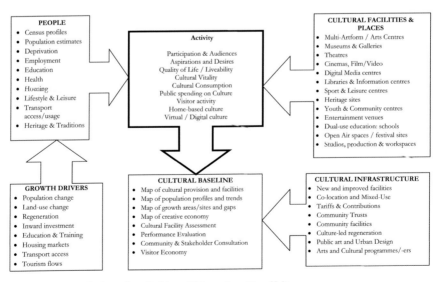

Figure 13.2   Populating the Cultural Planning Toolkit

and scenario-building (including drivers of change and growth), this three-stage process can be under-pinned by consultation with communities at various levels, not just limited to agencies, intermediaries and interest groups (Figure 13.2).

For example, cultural mapping can employ consultative methods such as Geographic Information Systems Participation (GIS-P) with small (focus) groups working with large-scale maps that can be annotated with perceptual as well community information. This local knowledge and opinion can be digitized back into inter-active maps containing geo-demographic, facility, transport and other data. This technique, which draws on the earlier Planning for Real exercise using simple board games, models and maps, is used successfully from primary schoolchildren to pensioners, and around urban design, transport and heritage interpretation (Cinderby, Forrester and Owen 2006) as well as in conflict sites and resolution. Visualizing and animating land and townscapes, as well as human activity and behaviour in the environment, and in terms of cultural activity and aspirations, can also benefit from the direct involvement of artists, crafts and designer-makers, whether as interpreters, catalysts or visionaries (Visionary Thurrock 2004). Community and public arts practice, long established, would appear to have a renewed importance in helping to bridge the current development and planning process and pressures for new and high-density housing, through involvement in cultural mapping. This includes the adaptation of the Design Charrette – originally the preserve of artists and designers/architects – into extensive visioning and community planning and iterative design events (Condon 2008).

Sustainable Communities and Cultural Planning

In the UK, sustainable development has been operationalized in two ways. Firstly through the proxy of 'quality of life' where an extensive set of indicators – social, economic and environmental – has been created to measure and monitor performance over time (and in Canada – see Duxbury 2001 and FCM 2001). These indicators are applied at varying spatial scales – local ('quality of life counts'), regional and national. Culture (including sports, parks, heritage) tends to feature in these indicators in terms of access to services and satisfaction with provision, that is, benchmarks against which cultural provision and satisfaction can be measured and compared. Secondly in strategic and planning terms, the Sustainable Communities Plan sought for the first time to integrate the everyday environment, particularly housing, with the economy and accessibility, guided by notions of mixed-used, compact city living (Foord 2010). This was also in the context of population growth requiring both densification as well as new housing and urban extensions. Thus sustainable communities were to be: 'well designed and built – including appropriate size, scale, density, design and layout, including mixed-use development, high quality mixed-use durable, flexible and adaptable buildings, with accessibility of jobs, key services and facilities by public transport, walking and cycling, and strong business community with links to the wider economy' (ODPM 2005: 3).

The significance of this approach was that culture, whilst literally absent – as usual – was at least an implicit consideration in both quality of life measures and in the planning of sustainable communities. It came to be an explicit one as culture featured in housing growth areas and the planning of cultural amenities engaged with the development process. This responsive position provided a catalyst for cultural planning that on the one hand challenges the master-planning and mega-event imperatives, and on the other, seeks to embed culture in the planning and resourcing processes. A particular manifestation of this was 'Creating Cultural Opportunities for Sustainable Communities', an initiative jointly funded by the government's Communities and Local Government (CLG) and the Investing in Communities (HM Treasury) programme. The stakeholders involved were a collective of cultural partners (arts, heritage, museums and libraries, sport, tourism) whose main aim was to create a national Cultural Planning Toolkit – a set of guidelines, good practice and principles – to inform the assessment and development of cultural needs within the context of new or growing communities. This comprehensive Toolkit took a whole population approach to the iterative mapping, needs assessment and planning process noted above (Figure 13.2), combining people and places with change, underpinned by a wide range of quantitative and qualitative data, spatially visualized where possible (Evans 2008). Complementary local toolkits have been developed in the Thames Gateway region in England, against the background of the largest growth in house-building and population growth and movement in post-war Britain (TGNK 2006). A key objective of these planning tools was to enhance the sustainability and well-being of both new and existing communities by facilitating the efficient development of, and enabling access to, a diverse, vibrant and creative local culture. By providing the planning system with guidelines for cultural and leisure planning and related social infrastructure (for example health, education, community amenities), the toolkit sought to ensure that facilities necessary to support a sustainable community are provided and fit for purpose – thus enhancing quality of life. A key strategic objective of the Cultural Planning Toolkit was therefore to support the work of the local planning authorities and delivery organizations tasked with managing areas undergoing population growth and change, including priority areas defined in the national Sustainable Communities Plan.

A similar approach is also evident in Vancouver where, stimulated by the planning for the 2010 Winter Olympics, the city developed comprehensive cultural mapping and planning toolkits (CCNC 2006a, 2006b). In Australia and New Zealand, cultural planning resource sites have gone further in terms of community input and inclusion, allowing local areas and communities to write their own cultural histories and profiles, linked to facility maps and images. A GIS-based cultural atlas in Western Sydney created a web resource allowing the user to zoom in to images, video, audio, stories and links to documents and producing trails and tours. In Queensland a locally generated web resource provided maps and links to culture in terms of places, people, events, tours and the history of an area. Each of the region's 18 Local Councils received a copy of the Cultural Map application system, a range of media software, a training programme and support from the project initiator. In London (UK) the CultureMap web-based resource has likewise produced online maps of cultural facilities, population profiles and audience penetration from participating arts venues in London. This specialist planning tool was developed by Audience London with the support of Arts Council England, responding to the need both to map cultural provision and to link this to actual usage and population typologies and catchments. CultureMap created a series of web-accessible maps of arts and community cultural provision and audiences for a range of participating cultural facilities at city-wide down to ward levels, alongside demographic and other population data. Primary data is also generated by collaborating arts venues capturing audience profiles in a common format. This online tool can also reveal gaps in provision and participation, drawing on secondary and primary surveys revealing interesting correlations between audience and venue types and locations (Brook 2007), presenting valuable evidence and information on the relationship between certain arts provision and attendance in a demographic and spatial context. Other regional arts marketing projects have also developed similar profiles of cultural activity, linked to regional development and cultural consortia sites. A particular application of this approach is in library provision and usage, where the library card acts as a vital source of user information, including children (that is, books, videos/CDs borrowed and other facility usage) linked to the home address. The latter can locate users in socio-economic, housing and other lifestyle groups. This is in contrast to say, museum or theatre 'education visits', which are recorded as a single homogenous group (for example school/year/class) not as individuals or their home context.

In order to better inform cultural plans and improve cultural mapping and audit exercises, city and provincial authorities in Canada – Toronto and Vancouver in particular – have also developed online inventories of cultural facilities using a cultural and ownership typology, as well as online databases of performing and public art installations that provide location, capacity and operational information. Most recently in the UK, the Culture Ministry commissioned Culture and Sport Physical Asset Mapping guidance and toolkit resources for local areas looking to develop better knowledge about their local supply of culture and sport (DCMS 2010). The toolkit identifies a range of readily available sources of data, allowing areas to obtain a good picture of their supply without commissioning expensive work. It also provides data definitions and frameworks for allowing local areas to generate comparable definitions of asset types, as well as for recording new data resulting from focused data collection. This ensures data comparability between areas and allows a richer picture of culture and sport supply to emerge over time, reducing duplication and increasing data use and re-use. A particular objective of this exercise was to mainstream and make cultural data compatible with national datasets on social, environmental and other planning (for example land use) data. For an assessment of these various guidance, toolkits and inventory resources see Evans (2008) and Evans, Foord and Shaw (2007).

Conclusion

What these cultural planning models and tools – which were designed to develop comprehensive cultural baseline, mapping and consultation processes – have in common is a response to change, whether regeneration (event-based, major sites), new housing (urban villages, brownfield, mixed-use), cultural development and more effective resource planning in the case of local and city authorities. They frequently arose through specific initiatives – policy, funding, efficiency – rather than a systemic change to the planning system or culture, although most cultural planning approaches have explicitly sought to engage the planning system and profession in their guidance and methods. Certainly we have observed a spatial turn in cultural policy and planning over the past five years, in part facilitated by GIS and spatial visualization techniques and take-up (but less so spatial data analysis, with a few exceptions, for example CultureMap, above).

However, their initiative-led and special event status has rendered them time-limited and therefore not sustained – victims of funding expiry, political and regime change, or just obsolescence. This is evident by the fact that web links to these resources are no longer active, host organizations and agencies no longer exist and event roadshows move on (and regeneration and legacy promises fade, for example Vancouver 2010). What this signifies is that there has been a failure to embed cultural planning into the mainstream planning system, including the education and training of planners and related professionals (for example architects, public administrators). Here the adoption of an increasingly micro-level approach to place-making, or strategic policy-making is preferred to more comprehensive planning and a cumulative knowledge/evidence base. This is also reflected in the reliance on external consultants (in the UK, Europe, Canada and beyond) to undertake periodic or special project cultural plans and strategies, with the lack of knowledge and skills transfer that this practice infers. This also creates an inconsistent range of approaches, classifications and data in contrast to say, standard land-use classification, economic and employment other social indicators. Efforts at integrating culture within sustainable development principles and practice have therefore had only a limited effect. Or rather, the level of knowledge and point in the learning curve has been advanced, but this is not universally transferable or well distributed across localities, practice and policy realms. It has already proved to be fragile in the face of shocks such as economic recession, political uncertainty and unfettered and unsustainable (and unplanned) growth, as witnessed in countries such as China.[2]

If culture and governance – and a more community-led practice of 'cultural governmentality' – can be seen as mediating forces in reaching some equilibrium between the three pillars of sustainable development, planning practice and principles, they should arguably engage with these through cultural planning approaches. This entails planning that is both consultative, informed and democratic in considering both the whole population and culture in all of its diverse and collective manifestations and desires. This would appear a necessity given the difficulties that initiative-led and toolkit paradigms have had in influencing planning and development imperatives and therefore practice and outcomes. Returning to some basic principles – bringing sustainable development and community aspirations down to the everyday uses and experience of space, 'habitus' (Lee 1997) social exchange, cultural expression and 'ways of life' – we can present planning as both the facilitating and mediating process rather than through its reductive valorization (land/exchange values), homogenous standards (amenity, space, design) and control (of development, conservation) functions. In sustainable cultural planning, cultural activity, programmes, traditions and engagement together drive facility access, provision, heritage protection and spatial equity – not the other way around. As Lefebvre observed, we do not 'use' a sculpture or work of art, we live and experience it (1974).

---

2   China has invested in hundreds of designated 'culture creative clusters' and buildings on former industrial sites, for example Cyber Recreation District (Beijing), Creative 100 (Qingdao) and Creative Island (Dalian) – primarily property/regeneration driven rather than creative content and needs based (Keane 2007).

References

American Planning Association (APA). 2011. *Community Character: How Arts and Cultural Strategies Create, Reinforce, and Enhance Sense of Place*. Chicago, IL: American Planning Association.

Arts Council of England. 2011. *Response to the CLG Consultation on the Draft National Planning Policy Framework*. London: ACE.

Baeker, G. 2010. Rediscovering the wealth of places: cultural mapping and cultural planning in Canadian municipalities. *Plan Canada*, Summer, 16–18.

Barnes, T. 2001. Retheorizing economic geography: from the quantitative revolution to the cultural turn. *Annals of the Association of American Geographers*, 91(3), 546–65.

Bassett, K., Griffiths, R. and Smith, I. 2002. Testing governance: partnerships, planning and conflict in waterfront regeneration. *Urban Studies*, 39(10), 1757–75.

Bell, C. and Newby, H. 1976. Communion, communalism, class and community action: the sources of new urban politics, in *Social Areas in Cities*, Volume 2, edited by D. Herbert and R. Johnson. Chichester: Wiley, 189–207.

Bennett, T. 1995. The multiplication of culture's utility. *Critical Enquiry*, 21(4), 861–89.

Bertacchini, E., Bravo, G., Marelli, M. and Sangata, Q.W. 2011. *Cultural Commons: A New Perspective on the Production and Evolution of Cultures*. Cheltenham: Edward Elgar.

Bianchini, F. and Ghilardi, L. 2004. The culture of neighbourhoods: a European perspective, in *City of Quarters: Urban Villages in the Contemporary City*, edited by D. Bell and M. Jayne. Aldershot: Ashgate, 237–48.

Borja, J. and Castells, M. 1997. *Local and Global: Management of Cities in the Information Age*. London: Earthscan.

Boyer, C. 1988. The return of aesthetics to city planning. *Society*, 25(4), 4–56.

Brook, O. 2007. Response 1. *Cultural Trends*, 16(4), 385–8.

Chan, T.W. and Goldthorpe, J.H. 2007. The social stratification of cultural consumption: some policy implications of a research project. *Cultural Trends*, 16(4), 373–84.

Chaney, D. 1994. *The Cultural Turn: Scene-Setting Essays on Contemporary Cultural History*. London: Routledge.

Cinderby, S., Forrester, J. and Owen A. 2006. A personal history of participatory geographic information systems in the UK context: successes and failures and their implications for good practice. *Royal Geographical Society Annual Conference*. London, September.

Condon, P. 2008. *Design Charrettes for Sustainable Communities*. Washington, DC: Island Press.

Council for European Urbanism (CEU). 2004. *General Agreement on Mixed-Use for Cities*. Council for European Urbanism. [Online]. Available at: www.ceunet.org/mixeduse.htm [accessed 30 September 2012].

Crang, P. 1997. Cultural turns and the (re)constitution of economic geography: introduction to section one, in *Geographies of Economies*, edited by R. Lee and J. Wills. London: Arnold, 3–15.

Creative City Network of Canada (CCNC). 2006a. *Cultural Mapping Toolkit*. Vancouver: 2010 Legacies Now.

—— . 2006b. *Cultural Planning toolkit*. Vancouver: 2010 Legacies Now.

Cullingworth, J.B. and Nadin, V. 1994. *Town and Country Planning in Britain*. 11th Edition. London: Routledge.

Culture South West. 2005. *All Together Now: Culture and the Planning System in the South West*. Conference Report. Winter Gardens, Weston-super-Mare, UK.

Dalal-Clayton, B. and Bass, S. 2002. *Sustainable Development Strategies: A Resource Book*. London: Earthscan.

DCMS. 2010. *Culture and Sport Asset Guidance and Physical Asset Mapping Toolkit*. London: Department for Culture Media and Sport. [Online]. Available at: http://www.culture.gov.uk/what_we_do/research_and_statistics/7290.aspx#Culture [accessed 30 September 2012].

Duxbury, N. 2001. *Exploring the Role of Arts and Culture in Urban Sustainable Development: Journey in Progress*. Working Paper. Hull, Quebec: Department of Canadian Heritage.

——. 2004. *Creative Cities: Principles and Practices*. Background Paper F47, Canadian Policy Research Networks Inc. Ottawa.

Evans, G.L. 2001. *Cultural Planning: An Urban Renaissance?*. London: Routledge.

——. 2005. Measure for measure: evaluating the evidence of culture's contribution to regeneration. *Urban Studies*, 42(5/6), 959–84.

——. 2008. Cultural mapping and sustainable communities: planning for the arts revisited. *Cultural Trends*, 17(2), 65–96.

——. 2009. Creative cities, creative spaces and urban policy. *Urban Studies*, 46(5/6), 1003–40.

——. 2010. Accessibility and urban design: knowledge matters, in *Urban Design Research: Method and Application*, edited by M. Aboutorabi and A. Wesener. Birmingham: Birmingham City University, 26–37.

Evans, G.L. and Foord, J. 2003. Culture planning in East London, in *Cultures and Settlement: Vol. 3. Art and Urban Futures*, edited by N. Kirkham and M. Miles. Bristol: Intellect Books, 15–30.

Evans, G.L., Foord, J. and Shaw, P. 2007. *Cultural Planning Toolkit: Review of Resources: Guidance, Toolkits and Data*. London: Cities Institute. [Online]. Available at: www.citiesinstitute.org/library/v58000_3.pdf [accessed 30 September 2012].

Evans, G.L., Foord, J. and White, J. 1999. *Putting Stepney Back on the Cultural Map: An Investigation into the Potential for Local Cultural Activity*. London: London Borough of Tower Hamlets.

Federation of Canadian Municipalities. 2001. *The FCM Quality of Life Reporting System: Second Report Quality of Life in Canadian Cities*. March, Ottawa: FCM.

Florida, R. 2005. *Cities and the Creative Class*. New York: Routledge.

Foord, J. 2010. Mixed use trade-offs: how to live and work in a 'compact city' neighbourhood. *Built Environment*, 35(5), 47–62.

Foucault, M. 1991. Governmentality, in *The Foucault Effect: Studies in Governmentality*, edited by G. Burchell, C. Gordon and P. Miller. London: Harvester Wheatsheaf, 87–104.

Freestone, R. and Gibson, C. 2006. The cultural dimension of urban planning strategies: an historical perspective, in *Culture, Urbanism and Planning*, edited by J. Monclus and M. Guardia. Aldershot: Ashgate, 21–41.

Ghilardi, L. 2004. *Only Connect: Cultural Planning for Urban and Community Development*. [Online]. Available at: www.artscouncil-ni.org/conference/cultplan.rtf [accessed 30 September 2012].

Gordon, C. 1991. Governmental rationality: an introduction, in *The Foucault Effect: Studies in Governmentality*, edited by G. Burchill, C. Gordon and P. Miller. London: Harvester Wheatsheaf, 1–52.

Gray, C. 2006. Managing the unmanageable: the politics of cultural planning. *Public Policy and Administration*, 21(2), 101–13.

Grodach, C. 2008. *The Local Arts Planning System: Current and Alternative Directions.* [Online]. Available at: http://commons.newvillagepress.net/commons/new-village-online/the-local-arts-planning-system-current-and-alternative-directions/ [accessed 30 September 2012].

Guppy, M. (ed.) 1997. *Better Places, Richer Communities: Cultural Planning and Local Development – A Practical Guide.* Sydney: Australia Council for the Arts.

Hall, P. 1998. *Cities and Civilisation: Culture, Innovation, and Urban Order.* London: Weidenefeld & Nicholson.

Hawkes, J. 2001. *The Fourth Pillar of Sustainability: Culture's Essential Role in Public Planning.* Melbourne: Common Ground.

Keane, M. 2007. *Created in China: The Great New Leap Forward.* London: Routledge.

Kunzmann, K. 2004. Culture, creativity and spatial planning. *Town Planning Review*, 75(4), 383–404.

Lacy, S. 1995. *Mapping the Terrain: New Genre Public Art.* Seattle, WA: Bay Press.

Landry, C. 1997. *Cultural Industries Strategy for Tower Hamlets.* Stroud: Comedia.

—— . 2006. *The Art of City Making.* London: Earthscan.

Lee, M. 1997. Relocating location: cultural geography, the specificity of place and the city habitus, in *Cultural Methodologies*, edited by J. McGuigan. London: Sage, 126–41.

Lefebvre, H. 1974. *The Production of Space.* Oxford: Blackwell.

London Planning Advisory Committee (LPAC). 1990a. *Strategic Planning Policies for the Arts, Culture and Entertainment*, Report No.18/90. London: LPAC.

—— . 1990b. *Model UDP Policies for the Arts, Culture and Entertainment Activities.* London: LPAC.

Markusen, A. and Gadwa, A. 2010. *Creative Placemaking.* Washington, DC: National Endowment for the Arts.

Markusen, A. and Schrock, G. 2006. The artistic dividend: urban artistic specialisation and economic development implications. *Urban Studies*, 43(10), 1661–86.

Mercer, C. 2006. *Cultural Planning for Urban Development and Creative Cities.* [Online]. Available at: www.kulturplan-oresund.dk/pdf/aktivitetsplan/230506_Shanghai_cultural_planning_paper.pdf [accessed 30 September 2012].

Mitchell, D. 2003. *The Right to the City: Social Justice and the Fight for Public Space.* New York: Guilford Press.

Munro, T. 1967. *The Arts and Their Interrelations.* 2nd Edition. Cleveland, OH: Western Reserve University Press.

Nurse, K. 2006. *Culture as the Fourth Pillar of Sustainable Development.* London: Commonwealth Secretariat.

Office of the Deputy Prime Minister (ODPM). 2005. *Sustainable Communities: People, Places and Prosperity.* London: ODPM.

Soja, E. 1999. In different spaces: the cultural turn in urban and regional political economy. *European Planning Studies*, 7(1), 65–75.

Stern, M.J. and Seifert, C. 2007. *Cultivating 'Natural Cultural Districts': Social Impact of the Arts Project.* Philadelphia, PA: University of Pennsylvania.

Talen, E. and Ellis, C. 2004. Cities as art: exploring the possibility of an aesthetic dimension in planning. *Planning, Theory and Practice*, 5(1), 11–32.

Thames Gateway North Kent (TGNK). 2006. *Sustainable Culture, Sustainable Communities: The Cultural Framework and Toolkit for Thames Gateway North Kent*. Medway: TGNK.

UCLG. 2009. *Culture and Sustainable Development: Examples of Institutional Innovation and Proposal of a New Cultural Policy Profile*. Barcelona: United Cities and Local Governments.

UNESCO. 2009. *Culture and Sustainable Development: Examples of Institutional Innovation and Proposal of a New Cultural Policy Profile, Culture 21*. UNESCO. [Online]. Available at: www.agenda21culture.net [accessed 30 September 2012].

Visionary Thurrock. 2004. *New Models of Cultural Facilities* (Charrette No. 3). Thurrock, Kent.

Williams, R. 1983. *Towards 2000*. London: Pelican.

World Commission on Environment and Development (WCED). 1987. *Report of the World Commission on Environment and Development: Our Common Future*. Oxford: Oxford University Press.

Young, G. 2006. Speak culture! Culture in planning's past, present and future, in *Culture, Urbanism and Planning*, edited by J. Monclus and M. Guardia. Aldershot: Ashgate, 43–59.

Development, Planning and Sustainability

Edgar Pieterse

Introduction

Key policy agendas on shelter and participatory governance are woefully inadequate to address the multi-dimensional crisis of urban life in cities of the global South. This finding was the crux of *City Futures*, a book I authored regarding mainstream development opinion on the topic (Pieterse 2008). I want to build in this chapter on that work by reacting to a number of crucial mainstream policy developments that have transpired since then and extend earlier arguments about the intersections between development, planning and sustainability (Kamal-Chaoui and Robert 2009, Suzuki et al. 2010, UN-Habitat 2011, UNEP 2011a). I will not use this chapter to critique or respond to new mainstream policy discourses but rather focus on a propositional agenda that can illuminate the intersections between development, planning and sustainability. In my work I usually try to balance critique and proposition because this is the only viable, even if difficult, mode of scholarship that makes sense amid large-scale dysfunction and exclusion (Parnell, Pieterse and Watson 2009).

The chapter proceeds via a number of illustrative diagrams and in this sense also reflects an attempt to stitch together various pieces of writing on urban development, planning and sustainability that I have been crafting over the last few years. The diagrammatic form also reveals my immersion in the world of practice through an advisory relationship with various public bodies at numerous levels: metropolitan governments, national government departments, regional development agencies and international actors in the urban development arena. Diagrammatic sketches have become an indispensable mode of communication and synthesis in trying to act as an interlocutor between scholarly debates and policy search forums.

The core argument of the chapter is: Urban sustainability requires a simultaneous transformation in four 'operating systems' of cities: 1) infrastructural, 2) economic, 3) spatial, which implicates land-use, and critically 4) political prioritization, which underpins the former three. This implies a profound interdisciplinary and heterodox cross-pollination of ideas, institutional practices and systems, informed by a dynamic conceptualization of how urban systems intersect. This alternative conceptualization is the confluence of resources and energy flows that circulate and mutate through urban systems and the territorial landscapes that enfold the built environment with its embodied flow dynamics. Since sustainability cannot be delinked from resource efficiency, the 'flow and form' lens (that is, resource flows and spatial form) offers a useful perspective on how the unsustainability of urban areas can be captured, argued over and potentially remade in order to achieve a much more resilient and viable urban trajectory. At the core of both urban flow and form dynamics is politics. Without a deeply nuanced and differentiated understanding of urban politics, it is impossible to open up transformative forums to take difficult decisions to adapt to or bend the trajectories of cities. However, beyond vigorous politics, one also needs an effective series of institutional mechanisms and connectors – such as growth management strategies – to translate political intent into concrete actions. All of these dimensions are in turn embedded in culturally mediated processes and institutions as the chapters in this volume attest.

Urban Sustainability

> *A 'sustainable city' enables all its citizens to meet their own needs and enhance their well-being, without degrading the natural world or the lives of other people, now or in the future. We have to ask ourselves what specific measures need to be taken to create sustainable urban habitats, and how environmental and social concerns can be brought together into one compelling win-win scenario. (Girardet 2004: 6)*

A fundamental problem with the conventional approach to sustainability is that it reinforces the teleological assumptions of mainstream development economics. In other words, it implies that there is a particular trajectory from under-development to full development that all societies need to travel along and the question is whether the pathway can be made more or less damaging to the natural environment. In reality, the policy options, constraints and dynamics of societies are much more diverse, divergent and complex (Connell 2007); and as endogenous economic approaches suggest, the critical issue is how societies can configure their own, unique and self-defined pathways to achieve locally defined social and political goals, even in an interdependent globalizing world.

In light of this critique, it is apposite to explore a process-oriented conceptual approach to sustainability. Such an alternative is provided by the National Science Foundation Workgroup on Urban Sustainability, who argue for

> *... a definition of sustainability that focuses on sustainable lives and livelihoods rather than the question of sustaining development. By 'sustainable livelihoods' we refer to processes of social and ecological reproduction situated within diverse spatial contexts. We understand processes of social and ecological reproduction to be non-linear, indeterminate, contextually specific, and attainable through multiple pathways ... Within the terms of this definition, sustainability:*
>
> • *Entails necessarily flexible and ongoing processes rather than fixed and certain outcomes;*

> • *Transcends the conventional dualism of urban versus rural, local versus global, and economy versus environment; and*
> • *Supports the possibility of diversity, difference, and local contingency rather than the imposition of global homogeneity. (NSFWUS 2000: 7)*

This approach is compelling because it does not presuppose a particular model of modernity or economic system but also recognizes high levels of inter-dependence between contexts (or scales), which means one cannot fully escape the contemporary condition but there is also always room to work at locally specific agendas and solutions. This starting point is compatible with an open systems perspective that places economic capital in a subservient and dependent relationship to stocks of natural capital in ontological terms (Gowdy 1999). This working definition is well attuned to the scalar or relationally nested nature of urban systems. This is often missing in thinking and practice on urban sustainability. Inputs, standards, flows and design at the household scale are fundamentally different to how one thinks about sustainability at a neighbourhood, district or city-region scale. The spatialities involved are varied, irregular and seldom coincide in terms of functional geographies and unless urban sustainability thinking and interventions can be open to such spatial pluralism, it is difficult to envision the prospects of a systemic understanding of how a territory can foster more sustainable lives and livelihoods.

It is also problematic because it eschews too easily the centrality of economic development and growth for the reproduction of societies, despite the obvious benefits of the conceptual decentring of a Western imaginary of development pathways for other societies. It is absolutely vital to think through the prospects of an alternative economic model, particularly in a context where economic relations position certain economies and societies in a structurally asymmetrical relationship, that can appropriately be thought of as more durable, inclusive, respectful of non-renewable resources and connected to the meaning-making desires of people and the various communities into which they enmesh themselves. The more urgent theoretical task is to identify how one can simultaneously embrace the idea of multiple pathways to economic development while also pursuing the equally important conceptual work of figuring out alternative forms and dynamics of economic life, rooted in contemporary patterns. The economic character and functioning of urban areas must feature prominently in an enlarged understanding of sustainability, since cities and urban areas account for a disproportionate share of total economic output with the concomitant responsibility for greenhouse gas emissions. I will return to this theme later in the chapter.

The further aspect that requires a more considered treatment in the understanding of urban sustainability is the question of resource constraints. Cities depend on a number of finite resources for their economic, physical, social and ecological reproduction and for now, most of these resources are not adequately identified, understood or accounted for in conventional accounting systems. Recognizing fundamental resource limits is central to the broader goal of decoupling growth from resource use and environmental impacts. 'Decoupling means using less resources per unit of economic output and reducing the environmental impact of resources that are used or economic activities that are undertaken' (UNEP 2011b: xiii). If we consider that during the twentieth century 'the extraction of construction materials grew by a factor of 34, ores and minerals by a factor of 27, fossil fuels by a factor of 12, and biomass by a factor of 3.6' (UNEP 2011b: xiii), and recognize the dramatic growth in economic output with the rise of China, India and other emerging economies, the implications of decoupling become frighteningly clear. And since most of contemporary and future economic output will be anchored in cities, it is in those spaces that the broader sustainability question will have to be resolved.

The focus on resource consumption is also a useful way of opening public debates and discourses on the distributional dimensions of urban development. For example, research by Mark Swilling (2006) was able to capture the differential carbon footprint of the elite suburbs versus informal settlements on the periphery of Cape Town. Swilling demonstrates that the top earning 7 per cent of households have an average ecological footprint of 14.8 hectares per person versus 5.8 hectares per person for the next 9 per cent of households classified as upper-middle class. However, the footprint of the 15 per cent of households categorized below the breadline is only 1 hectare per person on average (Swilling 2006: 38–9).

A final dimension that should be weaved into an enlarged understanding of urban sustainability is the question of institutions. Since a transition to more sustainable resource consumption, more equitable access to resources and fairer allocation of the costs of degradation is a necessary passageway to more sustainable lives and livelihoods, it is vital to recognize that such processes require institutional embedding. Existing patterns of growth, resource distribution, consumption and regulation of behaviour are held in place by regulatory, discursive and organizational systems that live within the state but are also distributed through various networks of governance. We will be left with a limited understanding of sustainability unless the nature of urban sustainability and its transitional pathways are defined in terms of its institutional implications and imperatives. Most importantly, we need to explore and theorize how sensibilities such as 'diversity, difference and local contingency' invoked by NSFWUS (2000) can be enfolded into institutional frameworks, systems and cultures. This is a frequent silence that obscures the power dynamics at work in reproducing the status quo that is ineluctably wedded to the predominant patterns of extractive and exploitative capitalist reproduction.

In concluding this section, I will desist from offering an alternative working definition of urban sustainability but simply suggest that in this space of critique it is possible to identify and explore elements of a policy perspective that can build on the ideas of process, continuums, diversity, local contingency and situated political agreements that facilitate endogenous strategies to enhance sustainable lives and livelihoods. Instead, I now turn to my first diagram that seeks to evoke and elucidate four interdependent systems that propel the reproduction of cities.

Simultaneous Transitions[1]

Cities are restless animals that continuously need to feed off nature, ideas, culture, capital and rules – with legal force and tacit leverages – to reproduce urban life and aspiration. This embodies a variety of assemblages of 'people, resources, places, and mobilities' (Simone 2004: 14). 'This process of assembling proceeds not by a specific logic shared by the participants but rather can be seen as a recombination of contingency'

---

[1]   This section is an adaptation of an argument presented in Pieterse (2011).

(Simone 2004: 14). Thus, amid profound fluidity, practices are deployed to stabilize a sense of order and predictability to create the conditions for 'profitable' investment and reinvestment of time, cultural capital and economic resources; processes that are part science, part intuition and equal measure emergence. This dynamic of intense indeterminacy makes simplistic delineations of discrete urban systems and patterns rather foolhardy but unfortunately also unavoidable. Whether one likes it or not, in order to figure out what is going on, and moreover potentially intervene, typologies, taxonomies and categorizations are indispensable.

It is in this reflexive spirit that I would like to suggest that an interest in understanding pathways to more sustainable urban dynamics requires one to think about three critical meta domains of urban transition that need to be pursued simultaneously. These domains are: sustainable infrastructure; the inclusive economy; and efficient spatial form, glued by processes of democratic political decision-making. Put differently, one way of thinking about cities is that they require various 'operating systems'. Figure 14.1 highlights three critical operating systems that apply for all cities: 1) infrastructural, 2) economic and 3) spatial, which implicates land use. At the regulatory heart of these operating systems live the decision-making and regulatory force of the state and/or a plurality of powerful actors that can usurp the power of the state and/or exercise partial control.

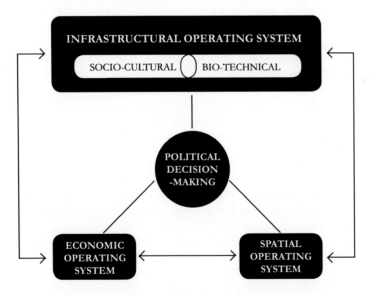

Figure 14.1   Operating systems of (sustainable) cities
Source: Author.

The infrastructural operating system can further be divided between social-cultural and bio-physical infrastructures. The latter refers to roads, transportation, information-communication technology, energy, water and sanitation, food and ecological system services that underpin the built environment and make urban life and movement possible in a concrete sense.[2] The concept 'flow management' provides a useful lens on how these infrastructures can be viewed as conductors of resource flows (Moss 2001). 'Central to the concept is the notion of flows of materials and energy, reusing resources or substituting non-renewable resources' (Moss 2001: 10).

Socio-cultural infrastructures refer to the social development investments that forge identity and community, for example cultural services, education, health, public space, libraries, food gardens, green spaces, housing and the arts. Social infrastructures by definition need to be tailored to local street-scale and neighbourhood-scale dynamics, which implies a substantial degree of community involvement and control in the execution and maintenance of these infrastructures (Rojas 2010). To ensure such local ownership and control it is important to ensure the capacity for local spatial literacy and purposive capacity, alongside practical community-organizing skills and dense institutions (Johnson and Wilson 2009, Narayan and Kapoor 2008, Sarkissian and Hurford with Wenman 2010). Recent experiences from some Latin American cities such as Medellín, Bogotá and Curitiba suggest that even though social infrastructures by definition need the fine-grain of community life to truly come to life, it is also equally important for it to articulate with a city-wide system of publicness and connectivity, especially in spatially and economically divided cities. Social infrastructure investments can send a powerful signal that public infrastructures for all class and cultural groups can and should be of the same quality, especially since the poorer citizens are much more reliant on them. The work of Alejandro Echeverri (2010) and his colleagues in Medellín is a particularly instructive.

Network infrastructures, on the other hand, often imply scale-dynamics that covers the larger urban system in all of its territorial expansiveness. This in turn holds profound implications for how political engagement is defined, structured and connected downward to the neighbourhood or community scale. In fact, as the work on splintering urbanism demonstrates (Graham and Marvin 2001), the lack of appropriate democratic oversight and engagement on 'invisible' network infrastructures produce conditions where city-wide infrastructures are tailored and routed to only service those sections of the population and economy that can contribute to the investment and maintenance of these systems. This is undoubtedly one of the primary drivers of large-scale service deficits across cities of the South (McFarlane 2010). I later return to these large-scale infrastructure systems in exploring how the form and flow of the city can be rethought, re-imagined and systematically remade.

The economic operating system involves production, consumption and market systems that underpin the exchange of goods and services. Importantly, these systems span formal and informal institutions and commonly involve their entanglement, especially in an era of intensifying globalization. However, one of the most challenging problems confronting cities in the South is that formal economic systems only absorb less than half of the labour force. The rest have to eke out an existence in the informal economy or completely disconnected from any gainful economic activities (UN-Habitat 2010). Those 'lucky' enough to engage in informal work have to put up with extremely low and often irregular income, which puts them in the category of the working poor (Chen 2008). In a broader context of ever deepening global integration of national economies

---

2   See Stephen Graham's (2010) excellent review of contemporary debates on infrastructural imperatives, pressures and transitions.

and value chains, it is becoming more difficult for national governments to protect jobs, provide support to the working poor and induce employment because such actions are, ironically, perceived as undermining competitiveness (UNRISD 2010). And as long as the intensifying financialization of economic value generation continues apace, it will be difficult to promote labour-absorptive and equalizing economic policies.

In the face of these powerful trends, it is essential that cities find creative ways of redefining and boosting local economies in order to broaden the base of those who are included in economic life. The urban development challenge is not just about enabling the generation of more formal economic jobs. On the contrary, the biggest and most urgent challenge is to target and absorb young adults between the ages of 15 and 29 in various categories of social and environmental public good activities that can reconnect them to society, nature and their surroundings. An example from South Africa may be apposite. Among the youth demographic, over 50 per cent cannot access formal jobs even though they may have completed primary and a portion of secondary schooling; at the same time, South Africa has the largest HIV/AIDS prevalence rate in the world (The Presidency 2011). In order to contain and manage the scale of the AIDS pandemic, it is vital that a national network of home-based care workers be established. These care-economy service workers need not necessarily have a formal medical training but must work with affected households to ensure that anti-retrovirals are taken in conjunction with sufficient nutritional intake, as well as various kinds of psychological support to help sufferers and their families deal with stigma and shame. Another pertinent example relates to various kinds of labour-intensive activities to restore ecosystem services. For example, river systems and canals in developing countries are often highly degraded because of upstream pollution and downstream neglect, sometimes combined with the invasion of alien species. Restoring these vital services is an essential part of improving the overall well-being of cities and communities. If done cleverly through arts-based programming, it can also be a gateway to aid young people at risk of anti-social behaviour to reconnect in more positive and enriching ways with nature and their peers. There are literally hundreds of examples that one can dream up if this logic is pursued.

The other equally important dimension of rethinking urban economic life is the imperative to confront the challenge of raising economic growth, improving the distributive aspects of growth, and improving the environmental impacts of economic processes that generate growth; or stated in policy parlance: decarbonizing growth that is closely tied to the imperative of decoupling (Suzuki et al. 2010). This is a particularly difficult challenge for economies and cities in the global South where economic growth and labour absorption cannot keep pace with social changes (for example more women joining the labour force as education attainment improves), population growth and in-migration. The easier political response is to simply welcome any kind of growth but in the context of necessary mitigation and adaptations to deal with the impacts of climate change, much more resource efficient economies need to be promoted. Put differently, traditional notions of competitiveness and productivity will now have to be redefined to reflect the imperatives of continuous decoupling (UNEP 2011b). This implies a qualitatively different conception of urban economies, various networks and how regulatory and incentive instruments are deployed to reconstitute economic patterns and dynamics. This is more difficult in contexts of rapid population growth, large-scale poverty and a lowly skilled labour force (UNRISD 2010).

Both the economic and infrastructure operating systems fundamentally depend on land and, more pertinently, land-use systems of cities. The patterns of infrastructural and economic distribution, flow and circulation add up to the spatial form of cities. If the spatial form is expansive, marked by sharp divisions between uses, functions and population groups, it is likely to be inefficient and exclusionary. In the vast majority of cities in the developing world, land-use systems further marginalize the urban poor and reinforce privilege for those who control land assets (UN-Habitat 2010). It is essential that land-use be rethought and designed to address the imperatives of greater urban efficiency and compaction, as well as access. Ideally, greater density through compaction should be linked with a much stronger emphasis on mixed-use land functions to facilitate greater efficiency and pluralism. A public-oriented approach as manifest in Bogotá, Curitiba and Medellín is encouraged in the recalibration of land uses, which informs a broader agenda to foster greater cultural and social integration. It is important to underscore that generalized shifts in consumption patterns cannot occur outside of new cultural norms and practices. Thus, fostering cities that can induce and sustain authentic cross-class and cross-cultural engagement around mundane everyday imperatives (Amin 2011) is key to unlocking social support and political commitment to push for structural transformation in the domains of infrastructure and the economy.

Finally, these three operating systems depend fundamentally on how power is distributed in society and mediated in (local) political institutions such as local governments (Pieterse 2008). If local governments are beholden to national government for revenue and resources, they will struggle to be responsive to local needs (Manor 2004). If local governments act unilaterally, or isolate themselves from the voice and actions of the organizations that represent slum dwellers, pavement dwellers, street traders, orphaned children, religious orders and so on, they are unlikely to recognize or understand the innovations that can only come from the effort and ownership of citizens themselves. However, even though there has been an unmistakable trend over the past two decades towards democratic decentralization across the world, there are still very few examples where participatory local governance is a vibrant reality. And recent research suggests that where innovative participatory instruments are introduced, they can become ensnared in over-bureaucratization, elite capture and so on (Fernandez 2010). This suggests that political reform and institutional retooling is a vital precursor to systemic change in the infrastructural, economic and spatial operating systems of cities. This can only be achieved, however, if informed by a radical democratic conception of urban politics.

Relational Power and Politics[3]

I will briefly rehearse in this section a broader argument to locate how to think about the nature of power and politics at the core of the patterning and interconnections between the various operating systems of the city. In conceptual terms, it is possible to delineate at least five domains of political engagement between the state, the private sector and civil society at various scales, ranging from the global and national to the local.

---

[3]   This section is a short summary of a much more developed argument put forward in Chapter 5 of *City Futures* (Pieterse 2008).

These are: 1) representative political forums and associated participatory mechanisms; 2) neo-corporatist political forums such as those that develop city development strategies, comprised of representative organizations – typically the government, the private sector, trade unions and community-based organizations; 3) direct action or mobilization against state policies or to advance specific political demands; 4) the politics of development practice, especially at grassroots; and 5) symbolic political contestation as expressed through discursive contestation in the public sphere. Figure 14.2 depicts these five political domains in addition to distinctions between the political and public spheres that are continuously (re)constructed through engagement in each of these five spheres and their interfaces.

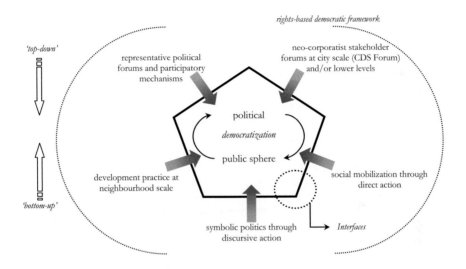

Figure 14.2   Domains of political engagement in the relational city
Source: Pieterse 2008: 89.

The value of this diagrammatic representation is that it makes visible how one can introduce a relational perspective on various moments, arenas and actors in the city. Moreover, it opens up new ground for imagining more creative progressive political strategies to undermine and subvert the oppressive functioning of dominant interests in the city. The model rests heavily on Foucault's understanding of power and therefore locates discursive and symbolic dimensions of political practice as central to re-reading political institutions and agency (Flyvbjerg 2001). I will briefly explain each domain in terms of key defining features, types of political practices, interconnections with other domains and possible pitfalls.

Political representation in domain one refers to the formal political system that characterizes national, provincial/regional and municipal government. At all levels, the main avenue of political participation in this process is through political parties that are elected. The democratic effectiveness of electoral systems depends in large measure on the democratic nature of the respective political parties along with their rootedness in their constituencies (Heller 2001). It also depends on the quality and maturity of the institutional rules and systems that structure the functioning of political chambers, council and committee meetings and associated mechanisms for transparency, responsiveness and accountability. It is important to appreciate that this represents a vital aspect of political focus to secure legal and policy commitments to urban sustainability measures, given the dramatic rise in national and local experiments with democratization and decentralization over the past two decades in most regions of the global South.

It is also true that the recent waves of decentralization occurred alongside the expansion of an international neoliberalization ideological project that firmly entrenched the idea that elected governments are either not the appropriate drivers of development or should undertake delivery in partnership with the private sector through public–private partnership institutional arrangements (Guarneros-Meza and Geddes 2010). These typically ideological currents coincide with the promotion of various types of stakeholder-based deliberative institutions – the second domain of urban politics – that provide a regulated and predictable space for negotiation and contestation between state, civil society and private sector representative organizations on urban issues of (mutual) concern. Commonly they are referred to as multi-stakeholder forums. In their book, *Local and Global*, Jorge Borja and Manuel Castells (1997) set out the case for the necessity of these kinds of deliberative spaces to co-create strategic plans for the city. While neo-corporatist institutional forms have been on the increase, more extra-formal political currents have also been on the ascendency.

Direct action, the third domain of urban politics, involves various forms of collective action by (disadvantaged) groups aimed at stretching the liberal democratic constitutional framework to its limit.[4] This implies that social movements and looser, issue-specific, social groups must claim their rights and entitlements through non-violent social action focused on concrete issues that shape the quality of life of their constituencies. In a sense, the primary function of progressive direct action is to maintain political momentum for redistribution and realization of human rights, especially socio-economic rights that include third generation environmental rights. Of all the political practices in the city, this type pushes most blatantly at the boundaries of the possible (in discursive, political and juridical terms). Direct action potentially shakes up the middle-class non-interest and disconnect in life beyond the suburb; that is, livelihood challenges in the slums and other spaces of marginalization. Street conflicts, clashes and destabilizations that spark off direct action are prerequisites for political agreements to address urban inequalities.

It is equally important to simultaneously recognize that the effective enrolment into citizenship does not only happen via the heroics and upper-case politics of direct action. As the work of John Scott (1997) suggests, the political terrain is much broader and more variegated than this. In most parts of the global South, the politics of development practice – the fourth domain of urban politics – unfolds at the neighbourhood scale (and beyond), where autonomous and state-dependent projects are undertaken to improve the quality of life and livelihoods, to protect against the vicissitudes of crime, violence and other shocks, and to deliberate future trajectories for the community in relation

---

4   There are obviously many instances where (relatively) privileged and conservative groups also embark on direct action to get their political grievances across. By focusing on disadvantaged groups, I am merely signalling an analytical preference to highlight the actions of this category of social actors but not to create an impression that other groups do not engage in this political arena.

to other communities and the larger regional economic-ecological system. It is vital to appreciate the *experiential* importance of participation in community-based associations aimed at improving the quality of life of oneself and fellow residents. The work of Arjun Appadurai (2004) on slum dweller associations in Mumbai argues for the importance of taking seriously 'the capacity to aspire' in thinking about this issue. Appadurai develops a layered argument that development, and especially its imagining, is deeply embedded in local cultures that people draw on to function in a day-to-day sense. Some of these cultural resources will be consistent with dominant societal values and norms that reproduce the acceptability of perverse inequalities. Other cultural resources may hold the germ of critique, of thinking about alternative social configurations that can lead to an improvement in quality of life and sense of self. The challenge is to use the future-shaping essence of development practice to expand 'the cultural map of aspirations' and in the process expand social citizenship and especially voice (Appadurai 2004: 69). It seems that the social struggles and debates that are implied with clarifying and programmatically effecting the three urban transitions offer novel opportunities for problematizing horizons of the future in the interest of excluded urban majorities.

If one accepts the Foucauldian frame that discourses mediate our engagement and sense of the world, our everyday spaces and ourselves, then we must bring the symbolic domain much more to the foreground in thinking through the nature and potentialities of urban politics. This is particularly relevant in an increasingly mediated world. Symbolic contestation through the deconstruction and reconstruction of dominant discourses on, for example, sustainability or the green economy, are prerequisites for achieving impact in terms of political strategies in all four of the other domains previously discussed. Symbolic politics functions through cultural resignification and therefore implies more creative practices that target the media, especially radio and popular newspapers; public spaces in the city, especially streetscapes and squares invested with symbolic meaning; and, spaces of collective consumption, such as schools, clinics, libraries. Symbolic contestation clears the ground to ask fundamental questions about given governmental discourses such as: What are the underlying rationalities of this discourse? What conditions make it possible for this discourse to pass as given and valid? What are the goals of the discourse? How can the elements of the discourse be challenged and re-arranged to turn the discourse on itself and make new meanings and imaginings possible that can be pursued through direct action or development practice or municipal policy? How can the futures of the city be re-imagined to reflect a radical openness as opposed to the conventional approach whereby there is only one alternative?

The drawback of any conceptual model is that it superimposes a false sense of structure on complex and fluid social realities. The relational conceptual model of urban politics developed here is no exception. Much of the dynamics leaks from the model to smudge the artificial boundaries between urban spaces and associated political practices. As Arjun Appadurai reminds us, cultural identities and practices are constitutively porous, relational and marked by dissensus within some aspirations for consensus. For these reasons, it is important to foreground the numerous spaces of interface between different types of political practice (see Edjabe and Pieterse 2010, Pieterse 2008).

Can Planning Bring New Awareness and Action into the World?

I want to provide in the penultimate section an argument about how we can rethink the role of planning in framing and steering the three operating system transitions, premised on an understanding of relational urban politics as sketched beforehand. But, prior to discussing the institutional implications and potentialities, it is relevant to introduce a useful policy lens promoted by the World Bank in its recent policy framework entitled *Eco² Cities* (Suzuki et al. 2010). This book suggests that cities need to develop much clearer and more accurate understandings of the form and flow dynamics of the city. Form coincides with the spatial operating system discussed in Figure 14.1 above. Flow refers to the reticulation function of urban infrastructures as various resources and energy flow through the urban system to ensure mobility, nutrition, social and economic reproduction. The flow dynamics of cities are captured through various techniques that measure and quantify the resources '… extracted from nature; processed by infrastructure; consumed by home and businesses; treated by infrastructure, and finally returned for reuse or delivered back to nature as waste' (Suzuki et al. 2010: 8).

The value of the policy frame promoted in *Eco² Cities* is that it provides a useful heuristic to explore the intersections and mutual dependencies between the infrastructural and spatial operating systems discussed above. In the logic of this particular argument it goes so far as prioritizing ecological regeneration above economic competitiveness in contrast to its sister urban policy (World Bank 2009). For example, the document argues that 'City planning is firstly about protecting and regenerating irreplaceable natural capital, especially the natural assets and ecological services throughout the urban region in which the city is located' (Suzuki et al. 2010: 31). For my purposes though it is particularly useful to extract the approach to how urban form and flow inter-relate:

> There is a fundamental relationship between a city's infrastructure systems and its urban form. Urban form and spatial development establish the location, concentration, distribution, and nature of the demand nodes for the design of infrastructure system networks. By doing so, urban form establishes the physical and economic constraints and parameters for infrastructure system designs, their capacity thresholds, and technology choices, and the economic viabilities of different options. These have tremendous implications for resource use efficiency. At the same time, infrastructure system investments (transportation, water, energy, etc.) will typically enable and induce particular spatial patterns on the basis of the market response to the investments. (Suzuki et al. 2010: 59)

These lenses and instruments offer valuable ways of representing the infrastructural and spatial operating systems of cities. Furthermore, if read in conjunction with recent UNEP (2011a, 2011b) reports on the green economy and resource decoupling, a battery of helpful policy tools become available. But, as with all policy tools, it is essential to locate them within an understanding of power relations and structural drivers of urban inequalities and injustice. So, a next step in the argument is to explore how one can institutionalize these policy approaches but simultaneously deepen the democratization and radicalization of urban politics and decision-making.

Much of the available literature on radical urban politics and the empowerment of the poor focuses on the importance of strong social movements that can champion the interests of marginal classes and groups. Even more literature then explores how this never happens and how even progressive-sounding regulatory frameworks and laws often only serve as foil for the further disempowerment, co-option and structural exclusion of powerless classes and groups. Much of this work is insightful and prescient but there is a great reluctance to identify how well-organized movements of the poor can effectively engage vested interests and formal institutions in the domains of representative politics and stakeholder-based forums. There is a deep-seated anxiety about political contamination if movements participate in formal processes, especially if there is a formal role for business and middle-class interests. In light of the earlier exploration of relational politics, it is obvious that social movements need to simultaneously occupy the five domains of urban politics and actively articulate both short-term and long-term demands across these spheres. This means highlighting everyday injustices, offering pragmatic solutions and drawing connections with long-term structural transformations that can offer a viable alternative to ecological, social and economic reproduction.

The core of this mode of 'radical incrementalism' is strategic savvy to deconstruct and redefine the institutional logics of the state and the market. Without developing this argument, I will explain one way of thinking differently about how best to engage multi-actor forums and processes. This brings me to the final illustrative diagram: Figure 14.3.

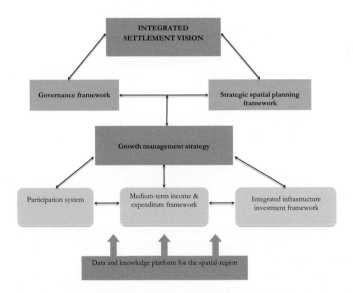

Figure 14.3   Policy and institutional dimensions of the local planning and governance systems

Source: Author.

This institutional framework is centred on the growth management strategy (GMS) for the locality or urban region. The GMS provides the baseline information about the 'state of the city' in terms of a range of development indicators, and indicates in

251

what ways it falls short of the normative ideal of a sustainable city that can address the economic, social, cultural, shelter, mobility and land needs of all its citizens. Potentially, it then proposes specific ways and targets for addressing the gap between the actual state of the city and the ideal. This could provide the starting point for sectoral and multi-sectoral plans (for example, human settlement plans, integrated infrastructure investment frameworks, Local Agenda 21, local economic development strategies, or safety and security plans and so on) that guide the detailed medium-term plans and expenditure projections of the large infrastructure sectors. In other words, an infrastructure plan cannot simply define its function in terms of its own sectoral logic, but needs to demonstrate how it contributes to closing the gap between the ideal and the current unsustainable and exclusionary city.

In addition to addressing backlogs and unmet needs, these infrastructural policies and plans must also offer *arguments* about how the technological and management approaches will shape the flow management dynamics of network infrastructural systems. Put differently, infrastructure plans can offer practical and concrete illustrations of how decoupling objectives will be met through both capital and operating investments. Most importantly, to be effective in this regard, they will have to make explicit the land-use and spatial implications of future investments and pathways. The debates about the various trade-offs and pressure points that will be prioritized in these processes is exactly what the democratic essence of growth management strategies could be. If progressive civil society organizations can colonize and steer these so-called technical discourses and forums, they can go a long way to practically advance more socially just, inclusive and sustainable pathways. What is self-evident is that spatial planning and speculation can truly come into its own relevance within this institutional configuration because it can demonstrate the effects and dynamics of contemporary processes of uneven and extractive development with vibrant tools to intimate diverse alternatives and relative consequences, depending on the paths and options chosen within the polity. Admittedly, planning becomes a sly and subversive player in this landscape. This may seem less heroic compared to traditional conceptions of spatial planning, but it is bound to be far more effective.

In Conclusion

Cities are clearly at the epicentre of contemporary efforts to redefine the 'development project' in terms of so-called low carbon futures, as a slew of recent mainstream development reports suggest. In this movement there lurks a revalorization of planning as pathways and trajectories to this 'future' has to be plotted, animated and mapped. However, with the technocratic focus on resource efficiency and optimal spatial forms, the political and cultural core of cities is in danger of being lost or obscured. I have in this chapter sought to clarify how some of the recent themes and ideas in mainstream development policy frameworks on urban sustainability are indeed useful; yet, it is equally important to recast them through a radical democratic lens in order to project a viable and creative political project that can give vitality and relevance to planning. But this is planning at the core of the pragmatic agenda of urban growth management, as multiple temporalities (2030, 2050, 2100) and spatialities come into sharper view in our increasingly inter-dependent urban worlds. This recasts planning in a more modest key but potentially with much further reach as it hijacks the terms of reference of increasingly mainstream prescripts on sustainable urbanism. However, for this sleight of hand to work, this form of 'planning' has to be done with an acute understanding of the profoundly cultural questions that lurk at the interstices of social–technical, political–economic, community–state, state–market and natural–social dialectics that shape the fortunes and risks of cities.

Acknowledgement

Research for this chapter has been undertaken with appreciated support from the Africa Centre, Mistra Urban Futures programme and the National Research Foundation. I am grateful for the editorial support of Kim Gurney and editors of the volume.

References

Amin, A. 2011. Urban planning in an uncertain world, in *The New Blackwell Companion to the City*, edited by G. Bridge and S. Watson. Oxford: Blackwell Publishing, 631–42.

Appadurai, A. 2004. The capacity to aspire: culture and the terms of recognition, in *Culture and Public Action*, edited by V. Rao and M. Walton. Palo Alto, CA: Stanford University Press, 59–84.

Borja, J. and Castells, M. 1997. *Local and Global: The Management of Cities in the Information Age*. London: Earthscan.

Chen, M. 2008. Addressing poverty, reducing inequality. *Poverty in Focus* (Brasilia: UNDP International Poverty Centre), 16, 6–7.

Connell, R. 2007. *Southern Theory: The Global Dynamics of Knowledge in the Social Sciences*. Cambridge: Polity Press.

Echeverri, A. 2010. Urbanismo social en Medellin 2004–2008. Talk at 361 Degrees Design and Informal Cities Conference, 22–23 October, Mumbai, India.

Edjabe, N. and Pieterse, E. (eds) 2010. *African Cities Reader: Pan-African Practices*. Cape Town: Chimurenga and African Centre for Cities.

Fernandez, E. 2010. Participatory budgeting processes in Brazil: fifteen years later, in *Urban Diversity: Space, Culture, and Inclusive Pluralism in Cities Worldwide*, edited by C. Kihato, M. Massoumi, B. Ruble, P. Subrirós and A. Garland. Washington, DC: Woodrow Wilson Centre and Johns Hopkins University Press, 283–300.

Flyvbjerg, B. 2001. *Making Social Science Matter: Why Social Inquiry Fails and How it Can Succeed Again*. Cambridge: Cambridge University Press.

Girardet, H. 2004. *Cities. People. Planet: Liveable Cities for a Sustainable World*. Chichester: Wiley Academy.

Gowdy, J. 1999. Economic concepts of sustainability: relocating economic activity within society and environment, in *Sustainability and the Social Sciences: A Cross-Disciplinary Approach to Integrating Environmental Considerations into Theoretical Orientation*, edited by E. Becker and T. Jahn. London: Zed Books, 162–81.

Graham, S. 2010. When infrastructures fail, in *Disrupted Cities: When Infrastructure Fails*, edited by S. Graham. New York and London: Routledge, 1–26.

Graham, S. and Marvin, S. 2001. *Splintering Urbanism: Networked Infrastructures, Technological Mobilities and the Urban Condition*. London and New York: Routledge.

Guarneros-Meza, V. and Geddes, M. 2010. Local governance and participation under neoliberalism: comparative perspectives. *International Journal of Urban and Regional Research*, 34(1), 115–29.

Heller, P. 2001. Moving the state: the politics of democratic decentralization in Kerala, South Africa, and Porto Alegre. *Politics and Society*, 29(1), 131–63.

Johnson, H. and Wilson, G. 2009. *Learning from Development*. London: Zed Books.

Kamal-Chaoui, L. and Robert A. (eds) 2009. *Competitive Cities and Climate Change*. Paris: OECD Publishing.

Manor, J. 2004. Democratisation with inclusion: political reforms and people's empowerment at the grassroots. *Journal of Human Development*, 5(1), 5–29.

McFarlane, C. 2010. Infrastructure, interruption, and inequality: urban life in the global south, in *Disrupted Cities: When Infrastructure Fails*, edited by S. Graham. New York and London: Routledge, 131–44.

Moss, T. 2001. Flow management in urban regions: introducing a concept, in *Urban Infrastructure in Transition: Networks, Buildings, Plans*, edited by S. Guy, S. Marvin and T. Moss. London: Earthscan, 3–21.

Narayan, D. and Kapoor, S. 2008. Beyond sectoral traps: creating wealth for the poor, in *Assets, Livelihoods, and Social Policy*, edited by C. Moser and A.A. Dani. Washington, DC: The World Bank, 299–322.

NSFWUS (National Science Foundation Workshop on Urban Sustainability). 2000. *Towards a Comprehensive Geographical Perspective on Urban Sustainability*. Final Report of the 1998 National Science Foundation Workshop on Urban Sustainability. Rutgers University.

Parnell, S., Pieterse, E. and Watson, V. 2009. Planning for cities in the global south: a research agenda for sustainable human settlements. *Progress in Planning*, 72(2), 233–41.

Pieterse, E. 2008. *City Futures: Confronting the Crisis of Urban Development*. London: Zed Books.

—— . 2011. Recasting urban sustainability in the South. *Development*, 54(3), 309–16.

Rojas, W. (ed.) 2010. *Building Cities: Neighbourhood Upgrading and Urban Quality of Life*. Washington, DC: Cities Alliance and IDB.

Sarkissian, W. and Hurford, D. with Wenman, C. 2010. *Creative Community Planning: Transformative Engagement Methods for Working at the Edge*. London: Earthscan.

Scott, J. 1997. The infrapolitics of subordinate groups, in *The Post-Development Reader*, edited by M. Rahnema and V. Bawtree. London: Zed Books, 311–27.

Simone, A. 2004. *For the City Yet to Come: Changing African Life in Four Cities*. Durham, NC and London: Duke University Press.

Suzuki, H., Dastur, A., Moffatt, S. and Yabuki, N. 2010. *Eco² Cities: Ecological Cities as Economic Cities*. Washington, DC: World Bank.

Swilling, M. 2006. Sustainability and infrastructure planning in Cape Town. *Environment and Urbanization*, 18(1), 23–50.

The Presidency. 2011. *Diagnostic Overview. Report of the National Planning Commission*. Pretoria: Republic of South Africa.

UN-Habitat. 2010. *State of the World's Cities Report 2010/11: Bridging the Urban Divide*. London: Earthscan.

——. 2011. *Cities and Climate Change: Global Report on Human Settlements 2011* (Abridged Version). London: Earthscan.

UNEP (United National Environment Programme). 2011a. *Towards a Green Economy: Pathways to Sustainable Development and Poverty Eradication*. Paris: UNEP.

——. 2011b. *Decoupling Natural Resource Use and Environmental Impacts from Economic Growth: A Report of the Working Group on Decoupling to the International Resource Panel*. Paris: UNEP.

UNRISD (United Nations Research Institute for Social Development). 2010. *Combating Poverty and Inequality: Structural Change, Social Policy and Politics*. Geneva: UNRISD.

World Bank. 2009. *Systems of Cities: Harnessing Urbanization for Growth and Poverty Alleviation. World Bank Urban and Local Government Strategy*. Washington, DC: World Bank.

Planning and Place Identity

Kim Dovey

Introduction[1]

Concepts of 'place' and neighbourhood 'character' have emerged in recent decades to take key roles in the discourse and practice of urban planning – deployed in the defence and protection of existing urban neighbourhoods against transformational change, but also in the design and marketing of new developments (see also Low, Paddison, Nyseth and Montgomery, Chapters 17, 18, 19 and 20, this volume). This chapter explores the relationships of place identity to urban planning. How is neighbourhood place identity experienced in everyday life? How is it defined and constructed in planning discourses and legislative practices? How is it created through urban design and protected through regulation? The potency of experiences of place identity and character makes this a crucial issue in the politics of urban planning and the production of urban cultures. One clear finding is that conceptions of place and character are fundamentally both social and spatial, and that the production and protection of place can be complicit with the production and protection of social privilege. Perceptions of place and character are multiple and flexible; it follows that the implementation of such concepts in planning legislation can be problematic. Fluidities of place identity can open opportunities for creative urban design but also the creative destruction of deregulated markets. Desires to legislate place identity can protect valuable places, but also paralyse urban development and ironically kill the very phenomenon that is to be protected.

If we explore the discourse of urban planning since about the 1990s we find an increasing use of the term 'place' in the titles of books, papers and journals – a trend that extends across the disciplines of geography, architecture, cultural studies, landscape architecture and urban design. This emerged first through applications of phenomenology in geography from the 1970s with works such as *Place and Placelessness* (Relph 1976), *Space and Place* (Tuan 1977), *The Human Experience of Space and Place* (Buttimer and Seamon 1980), *Dwelling, Place and Environment* (Seamon and Mugerauer 1985) and *Space, Place and Gender* (Massey 1994). We also see the emergence of academic fields such as place marketing and management as well as journals such as *Places*; *Health and Place*;

---

[1]   I wish to acknowledge the long-term collaboration of Ian Woodcock and Stephen Wood in the empirical and theoretical work that underpins this chapter.

Gender, Place and Culture; *Ethics Place and Environment*; and *Journal of Place Management and Development*. There has long been a 'Psychology of Place' (Canter 1977) and of 'Place Attachment' (Altman and Low 1992); followed by a more recent proliferation of planning and urban design books with the term 'place' in the title: (Beatley and Manning 1997, Corburn 2009, Cresswell 2004, Dovey 1999, Dreier, Mollenkopf and Swanstrom 2005, Eyles and Williams 2008, Healey 2010, 2005, Hubbard, Kitchin and Valentine 2004, Levinson and Krizek 2007, Madanipour, Hull and Healey 2001, Massey 1994). Research on 'place' is not a phase. The term has a currency that enables it to span disciplines and scales – from interior design to urban and environmental planning; from the room to the nation state. It spans research methods from phenomenology to empirical psychology and post-structuralism. It also encompasses popular, commercial and political discourses. Disputes over urban development often focus on questions of 'inappropriate' development that, when probed, reveal questions of urban or neighbourhood 'character' which is largely synonymous with a sense of 'place'. Yet because of their popular understanding, it becomes convenient to presume a shared understanding of place experience and to deploy it in planning discourse and practice without critical depth.

Ontologies of Place

I want to begin with a brief discussion on the ontology of place as a prelude to presenting some empirical research. First, what is the distinction between 'space' and 'place'?[2] While there is much confusion in the literature about this I suggest there is much less in our everyday use of the language. In the Wittgensteinian sense that one finds meaning in the use of language, 'place' is more than 'space' – more primary, more social, more intense. To ask 'what kind of place is New York?' may generate a variety of answers but this question has a sense that 'what kind of space is New York?' does not. When we say 'this is a great place' we mean something more social and less formal than 'this is a great space'. A large part of what distinguishes 'place' from 'space' is that place has an intensity that connects sociality to spatiality in everyday life. We can say 'do you have enough space?' but not 'do you have enough place?'. While a space may have physical dimensions it is intensity that gives place its potency and its primacy. The ways that place makes sense in everyday life is the primary understanding of the sense of place. How we make academic sense of that sense of place is a different matter. In academic literature 'space' and 'place' are often indistinguishable or are distinguished in ways that best suit the theory, abstracted from everyday life. I cannot introduce the range of theory here but Cresswell (2004) does a commendable job from a geographic perspective.

There are a range of problems in trying to understand the experience of place in everyday life, foremost among them is that it is taken for granted, it is the *doxa*, the context, the ground that we are often unaware of. Yet another problem emerges when place is interpreted in terms of deep and intrinsic meanings. This is the view that is generally accused of essentialism – to see the sense of place as deeply rooted in stabilized

---

2   For a more detailed discussion of these issues see Dovey 2010.

modes of dwelling (homeland and history) that can't be changed. This is also what is often referred to by a 'spirit' of place or 'genius loci' and related to the Heideggerian view of place as a primordial ground of being (Norberg-Schulz 1980). Such a view often conflates the sense and the ontology of place into one seamless whole, a reduction to essence that ignores social constructions of place identity.

The best case for an anti-essentialist theory of place is the avowedly anti-Heideggerian work of Massey in geography, centred on the notion of an open, global and progressive sense of place (Massey 2005, 1994, 1993). For Massey all notions of place derived from Heidegger are problematic and regressive. Against such views she proposes an open conception of place where place identity is provisional and unfixed. Massey's progressive sense of place is outward-looking, defined by multiple identities and histories, its character comes from connections and interactions rather than original sources and enclosing boundaries. Her example is a high street in London with mixed uses and ethnicities to which she ascribes character and identity without the Heideggerian primordiality. Such a sense of place is seen as primarily global rather than local, forged out of its connections with other places rather than local contingencies, privileging routes rather than roots.

There is little doubt that many Heideggerian approaches to place are regressive in the way Massey suggests, but there is an important distinction between Heidegger's argument about the spatiality of being on the one hand, and a much more spurious argument about a primordial sense of place with a singular identity, authentic history and exclusion of difference (Dovey 2010: 4–6). There is little doubt that Heidegger can be read in both these ways but the one does not imply the other. The claim that existence is spatial does not require that place experience is primordial or fully given. If we sever place from ontology then we are left with a weak theory about the relations of place to power, we have robbed place of its potency to construct ontological security and seemingly naturalized identity. The reason disputes over neighbourhood development can be so emotional is because places embody deep ontological investments in identity.

Researching Urban Character

As a means into an understanding of planning and place identity I want to discuss a project investigating the concept of place identity through the lens of urban or neighbourhood 'character' in Melbourne.[3] During the 1990s a state-led policy of urban consolidation under a neo-liberal economic and political agenda produced a substantial deregulation of urban development in Melbourne. Taller buildings were permitted in low-rise districts and apartments in suburban streets. Resident groups emerged in response to defend their suburbs against developments they saw as damaging to the 'character' of the neighbourhood. The state's response was to demand 'respect for the neighbourhood character' in residential planning codes, although 'character' was left largely undefined. Neighbourhood character studies were undertaken, physical

---

[3]   This work was undertaken as part of Australian Research Council Discovery Project DP0344105.

characteristics mapped and character areas defined, but these were generally thinly researched stereotypes.

Issues of 'place' and 'character' were also high on the agenda of the property development industry where new suburban developments were designed with the overt goal of producing an instant sense of place, community and character. On both post-industrial inner-city sites and on the urban fringe we find large tract developments, often influenced by neo-traditional new urbanist principles with a sense of 'place' and 'community' enclosure focused on a common 'green'.

In both of these cases 'place' and 'character' became key criteria for planning and urban design decisions in the absence of any clear definition or research. Our key research questions were: How is urban or neighbourhood character experienced and understood in everyday life and what is its relationship to urban morphology? What is the relationship between character and practices of urban regulation? A series of six case studies were conducted in neighbourhoods where either the creation or protection of 'character' was an issue. Three were older neighbourhoods where urban character was being defended against change while the other three were new neighbourhoods designed with a self-conscious desire to create a sense of place, character and community. In-depth interviews were conducted primarily with resident activists in order to understand experiences and conceptions that were driving both the resistance and the markets.[4] Morphological mapping was used to draw out connections between interviews and physical characteristics that were variously seen as salient – density, height, building style, pedestrian networks, streetlife, functional mix, open space, multi-unit development.[5] I will begin with the older suburbs and proceed to the new.

Defending Place Identity

Camberwell

Camberwell is an older leafy and up-market 'middle-ring' suburb about 10 km from the city centre (Dovey, Woodcock and Wood 2009a) developed initially in the 1880s as a railway suburb with detached houses on large blocks. The 'character' of the suburb was seen as under threat from multi-unit development, different housing styles and the densification of transit-oriented development. Interviews revealed a suburb characterized by its most vociferous residents in terms of 'comfort', 'consistency', 'modesty' and 'taste'. The consistency was often found in building styles, even when residents recognized they were defending a consistency that no longer exists. Descriptions of 'character' persistently slipped between the social and the physical: '[Camberwell is] old world, traditional, well preserved, and the people who make it up are of a solid base.' This 'comfort zone' was easily punctured by formal and social

---

[4] A total of 52 interviews of between 1 to 2 hours each were conducted across the six case studies with residents active in neighbourhood associations. All quotes in this paper are from these interviews.

[5] A full understanding of these studies requires graphic data that is not possible in this chapter. For illustrations see: Dovey, Woodcock and Wood 2010, or www.placeresearch.net/urban-character-place-research.htm.

differences yet its identity was also constructed from what it is not: 'Around the area [one can find] what might be described as "nice" houses, not, not new modern monstrosities, not totally derelict old places, not high-rise, low-cost type housing ....' Camberwell was defended against differences of built form and style but also, more subtly, against differences of class and ethnicity: ' ... the trouble is we're getting all these people in who all they've got is money, you see, no taste.' Urban character was seen as a legacy inherited from a largely imagined past; the values that were seen to prevail were modesty and taste. One outcome, ironically, has been that the only new architecture to be approved is in neo-traditional styles and is seen by residents to lack taste. This is in some ways a familiar view of an older suburb defended against the differences of ethnicity and class; it is a good example of a relatively closed conception of place identity and is revealing of the ways in which character can be deployed as part of the politics of place.

Hedgeley Dene

Hedgeley Dene is a neighbourhood in the wealthy middle-ring suburb of East Malvern which became the first place in Victoria to become the subject of a legislated Neighbourhood Character Overlay (Wood, Woodcock and Dovey 2006). The neighbourhood is centred on a small linear park that was developed from a creek during the early twentieth century. Building styles are mixed with a predominance of inter-war housing, generally set behind well-treed gardens but increasingly replaced with multi-unit developments that are seen to damage the 'character' of the neighbourhood as a whole and the park in particular as they cluster around to capture the view. A highly political process led eventually to the state's first neighbourhood character legislation in 2003. The legislation ostensibly enacts an agreement between residents, consultants, planners, and local and state government concerning the particular 'character' in need of protection. Yet analysis reveals that parties to this agreement were rarely talking about the same thing. For some the protection of character was about '... family homes as opposed to multi-unit developments with a transient population ...'; while for others it was about protecting the park from overlooking and preserving an illusion of being in the countryside: 'you feel like you're walking by yourself, not subject to any inspection by anyone else ... the false sense of being able to get away from the suburbs in such a small space ....' For some it was about the enforcement of building style – one key outcome was a new urban design code controlling fences, materials and roof pitch enforcing all new architecture into a mock historic style. For planning consultants defining character was an issue of professional expertise; and for the Minister for Planning it was the politics of demonstrating support for the neighbourhood. The boundaries of this character precinct were repeatedly altered as different agents and definitions of character prevailed.

Fitzroy

Fitzroy is Melbourne's oldest suburb, within walking distance of the city centre, developed from the 1850s with a mix of factories, warehouses and working-class row housing, now substantially gentrified (Dovey, Woodcock and Wood 2009b). The issue of character erupted first in 2002 when an architecturally innovative housing project was proposed for a former industrial site filling most of an urban block and was fiercely opposed by local residents primarily on the basis of height. The proposed buildings ranged from 3 to 8 storeys in a context of 1 to 5 storeys and was seen to violate the local 'character'. The architect/developer, who was also a local resident, argued that the character was comprised of a mix of heights to which he was adding with further variegation. Other residents largely shared this understanding of Fitzroy's character as mixed: layers and juxtapositions of large/small, old/new and factories/housing. This conception of the mix involved an openness to difference in both formal and social terms – to new forms of architecture and different kinds of people: 'you don't get the sense that people really care what you look like or what you say or how you act because there's so many different people doing so many different things ....' Here character was defined not only as a place with a difference from other neighbourhoods, but also as a place where embodied differences become character:

> *(Fitzroy) is different, it is … it has that 'edge' that people are interesting, that it has a good atmosphere. It has a sort of a seedy side, a sort of an underbelly that is in a way a little bit scary, but also has a community, it has character and it has depth.*

The preservation of character in Fitzroy was found in opposition to bulk and height rather than differences of building style, social class or ethnicity: 'I think you can take some buildings that are three storeys, but eight – no. I think that does start to change the village quality that we historically had about Fitzroy.' The gentrifying residents who led the defence tended to idealize the place as more socially diverse than it is; indeed part of what they opposed was the loss of diversity represented by more of their own kind. This case raises the question of how to regulate for irregularity, and how to protect the values of openness to change?

Creating Place Identity

Beacon Cove

I now want to contrast these defences of neighbourhood character with examples of market-driven instant place-identity. Beacon Cove is a new inner-city waterfront project developed in the late 1990s on a former industrial site adjacent to the former working-class suburb of Port Melbourne (Dovey, Woodcock and Wood 2010). Here the desired 'character' was a key driver of the design process, an instant place identity constructed largely from scratch using models from both the local context as well as 'new urbanism' and global prototypes. A limited number of housing types were replicated with many small variations to enclose a series of common parks or 'greens'. Minor differences of form were coupled with a strong consistency of building types, materials and landscape design, all fixed in perpetuity by detailed covenants. Residents purchase a collective identity, as the architect puts it: 'There's a family of colours … coded to type … it changes from precinct to precinct … If you purchase the house in the white precinct … It's felt that once you paint that house brown you're destroying the character.'

This sense of closure around the greens involved a spatial structure that excludes through paths as a form of 'soft gating'. In its development and marketing Beacon Cove was a self-conscious construction of place identity that depended on a bounded distinction from its surrounds. The repetition of types and the protected 'greens' led to it being labelled by outsiders 'Legoland' and 'Pleasantville'. Yet the constructions of identity in Beacon Cove are more interesting and contradictory than either the marketing or these labels indicate. Beacon Cove is not inhabited primarily by residents wishing to retreat from the context, but to engage with it – the more 'real' character of the adjacent port town is a key attraction for residents. While the project turns its back on neighbouring public housing, a remnant street of formerly working-class housing was incorporated into the project (after a long struggle to defend it) and some residents now proudly point to this much more socially and formally mixed street to show the real character of Beacon Cove. While the development is up-market it is relatively ethnically diverse and has attracted residents without the class connections of traditional Melbourne: 'people came here from (places) where they weren't accepted into the character of that area unless they'd been there 30 or 40 years'. The market turned out to be more diverse than the anticipated uniformity of empty-nesters – more tolerant of diversity and density, with a taste for contemporary rather than neo-traditional styles. Beacon Cove is a market-led development that appears to have been led more by ideology than the market.

Caroline Springs

Caroline Springs is an instant suburb on the urban fringe of Melbourne's traditionally disadvantaged western suburbs, developed into a series of 'villages' focused on waterways and parks with names like Brookside, Springlake and Chisholm Park (Dovey, Woodcock and Wood 2010). These 'villages' are semi-separated and marketed to different sectors from first-home owners to new wealth. With their keynote slogan of Creating Special Places the developers have marketed and created a vision of place identity and character in which residents invest their faith. This is a suburb that claims to escape the conformity of suburbia with a stronger sense of place, community, identity, security and home. The developers build only the public realm with detailed covenants enforcing a consistency of character within each of the 'villages' through setbacks, landscaping, height and materials. This project represents another form of 'soft gating' with elaborate gateways (that never close) framing the entrances to named places marketed to social class segments. Some villages are clearly more upmarket than others and climbing the ladder of opportunity involves moving from house to house and village to village: 'I'm Chisolm Park. It's how it's marketed to you when you buy, you have your identity ....'

Consistency of housing type and bulk is coupled with a diversity of housing styles from neo-traditional to contemporary. The marketing invites residents to *Choose your look, select your favourite façade*. This mix of different characters within a common covenant reflects a market of social and ethnic differences: 'Caroline Springs has a lot of character about it, but it's all a lot of different characters put together.' Ethnic differences are not evident in the architecture where the range of styles camouflages rather than reveals ethnic difference: 'you wouldn't walk past this one and say oh this one belongs to an Indian, this one belongs to a Maltese'. This is a place for aspirations, social mobility and identity formation. Interviewees speak of the faith they had in the developer's vision and its covenants – the swift creation of community and place identity. Here community and place identity is consumed from a brochure and then becomes a dream for the future within an ethos of individual self-reliance, self-creation and self-expression.

Kensington Banks

Kensington Banks is an inner-city housing development on the site of a former abattoir, driven by the ideal of a walkable 'urban village' with generous green spaces influenced by the ideas of 'new urbanism'. This was a public–private partnership driven by public interests such as community participation, sustainability, diversity, walkability, affordability and open space – all strongly mediated by market imperatives. A key desire was to replicate the valued character of the existing inner urban context with relatively low-rise small-grain development and a diversity of house types. The 'character' of Kensington Banks as understood by its most active residents is formed out of the intersections of these often-contradictory values and imperatives. While the project has been conceived for a mix of household types and social classes (including public housing) the market has determined that it has become dominated by young singles. The very generous provision of open space and pedestrian networks is underutilized, yet this highly walkable design has largely strangled car transport, producing traffic jams in peak hour. The open space is often lined with housing that 'faces' the park yet is entered from the 'rear'; back lanes lined with garages have become the main entries while the front becomes symbolic. The ideals of the urban 'village' have been deployed at a density too low for shops or public transport – the result is the village 'green' without the village: 'We find that in this estate particularly there's not a lot of character … it doesn't have that interaction that we're looking for … there's not that feeling.'

The naming, framing and enclosure of these new developments operates to construct and fix identity in urban form. They are designed to meet a market for a commodified community – the utopian ideal of *Gemeinschaft*, with citizens who know each other gathered in a circle around the 'green'. In each case the realities of this market are more diverse than they might seem.

What is Urban Character?

There are many themes that lace through these discourses and places: uniformity and diversity; building height and density; building type and style; community, class and ethnicity. There is not scope here to explore all of these but I want to discuss what this might mean for the ways in which we define neighbourhood character and place identity, how we might theorize it and how it meshes with practical issues of urban governance.

By far the most common definition of urban or neighbourhood character in all case studies is the 'feel' of a neighbourhood – character is 'the feel of a place, what it represents to you; the people, the buildings, the things that happen there are all part of the urban character'. The word 'feel' is generally used as a noun as if an emotion or sensation has become stabilized and objectified in built form. There is often the sense that neighbourhood character has agency: 'Just something about it. You know you drive in some places and it makes you feel at home, and you drive in other places it makes you feel – I really can't explain it, why.' While there is a common understanding at this level, it often falls apart under probing; interviewees often begin with some confidence on the topic and end confused: 'Until I really thought about [urban character], I thought I knew what it meant. Does that make sense? And then you think: Geez, what is it?' At times this produces an oscillating discourse: 'it probably is "feel", but it's probably not how I would describe it. You see I'd probably think of it more as – yeah, no, it probably is "feel", in which case it's hard to move.' The concept is often unstable and can be changed by the quest for character: 'when you've had to fight for things, and had to analyse it in such a detailed way, I think you get to a point where you think – I don't know what I saw in it in the first place.' Some residents express frustration at the need to define it: 'We've got this concept of where we live – its specialness, the choices it offers people – how come no-one else gets it, how come we've got to somehow define this?' Others become cynical and portray character as entirely subjective: '… all (character) is, is a vehicle by which someone can say, "oh, I like it" or "I don't like it".' Place identity is variously portrayed as objective and as subjective; the challenge is to understand how it flows across this divide.

Pressure for higher density, driven both by the market and sustainability imperatives, is widely seen as the most serious threat to character and it has a number of dimensions: a change in house type from detached to multi-unit; a change in the public/private interface from garden setback to direct street frontage; and an increase in height and bulk. Perceptions of density are related to a general experience of spatiality linked to sociality – an unselfconscious phenomenology of the everyday that becomes self-conscious only when threatened. The social identity that is seen as threatened by higher densities is linked to a loss of individual identity – a preference for 'a low-scale area where I have a sense of my own being and not being dwarfed'. Debates about loss of amenity due to higher density development – primarily overshadowing and overlooking – take a relatively minor role is this discourse. When they are mentioned there is often an underlying issue of social class: 'It'd be like living under the housing commission … we overlook one another here, but we sort of know one another, there's a relationship.' Thus concerns about height meld together the belittling effect of a dominant 'monolithic' form with the feeling of being juxtaposed against a social 'other'. Perceptions of density are embodied in the sense of home, the 'feel', the unselfconscious everyday *doxa* and the sense of social distance. Attitudes to density are clearly changing, and there is evidence in all case studies of some residents appreciating higher densities and the amenity and character they produce. Most residents, however, presume a right to maintain their neighbourhood at current densities: 'we have our fair share of units; to have more is not appropriate in this area.' Some of the opposition is based on building type and particularly the change from detached houses to multi-unit flats, seen also as a shift from stable to transient populations: 'I would hate to see lots and lots of flats, I think that does I think change the character of a place, it looks, would it be fair to say, more transient.'

Resident concerns about height are also linked to a fear that any increase that is conceded will become a precedent for further escalation: 'If they get that building in they'll say it's not viable then they'll make it bigger. There's no limit.' Here we find a lack of faith in the planning system largely produced by market-led planning – the 'viability' issue refers to the way the market has come to replace public interests as a legitimating ground of urban planning within a neo-liberal political economy (Dovey 2005). The argument is made in planning tribunals that a certain height and density is required by the market. Planning has become fluid, it seeks flexibility of urban regulation and concepts such as 'character' and 'place', which, while they appear to serve resident interests, can also serve market interests.

In everyday language the term 'character' is primarily applied to people and has dual meanings as both normative and descriptive, as depth and as difference. Character is normative in the sense of a 'character building exercise' or the sense of valued identity formation over time; this is what Sennett sees as threatened in his book *The Corrosion of Character* (Sennett 1998). Character is also descriptive of particular characteristics or eccentricities of people in the sense that 'she's quite a character' – whether valued or not. The term 'place' shares this dualism of meaning both a valued depth and a sense of difference. The meaning of character being ostensibly defended or created in the cases above is one of depth more than of difference, yet they are not easily separated. When urban character studies are undertaken by planning consultants we often find a list of formal characteristics that identify a neighbourhood and distinguish it from the context – building height, type, style, block size, setback, vegetation, street grid and so on. Character becomes identified with the internal consistencies of its characteristics and against its difference from other neighbourhoods. The deeper notion of character as 'feel' or 'atmosphere' is easily lost in this process as certain formal characteristics are selected and identified to stand for place identity. What is primarily established is differences between places rather than depth of character. The danger is that this selection of characteristics turns character into caricature.

The crucial distinction here is between two kinds of difference – differences between places and places of difference. 'Differences between places' are what distinguishes one building, neighbourhood or city from another. Such distinctions are firmly embedded in practices of power and they may well be driven by the quest for purification and the exclusion of difference. 'Places of difference', on the other hand, are about the degree to which an internal mix is constitutive of a particular place identity. The desire to establish differences between places can lead to the boundaries and gateways we see at Caroline Springs and Beacon Cove but also the closed sense of place pursued in Camberwell and Hedgeley Dene. Such purified places have a capacity to limit identity formation while places of difference can open up new possibilities.

This returns us to the open versus closed ontologies of place introduced earlier. The ideal of 'place' based on consistency and closure where neighbourhoods are differentiated by uniformities of character is the one being defended in Camberwell, constructed in Hedgeley Dene, and that largely drove the new morphologies of Beacon Cove and Caroline Springs. By contrast Fitzroy appears as a paradigm case of what Massey conceives as an open sense of place – progressive, globally connected, creative, multiplicitous. Yet it is not open to some kinds of change since the diversity of the mix is seen as threatened by further gentrification and greater density.

Slippages between social and physical aspects of character tend to confound attempts to operationalize it as a code of urban regulation. While planning codes and consultants' studies generally try to reduce character to a set of formal elements, the ways it is experienced in everyday life tend to resist attempts to separate the social from the physical; and this very slipperiness becomes attractive to proponents of flexible planning systems. 'Character' becomes discursively constructed in the field of politics where it comes to mean what different interests want it to mean. The problems here are threefold.

First, resident groups can use urban design codes as camouflage for social codes; struggles to prevent the wrong kinds of building can slip into the exclusion of the wrong kinds of people. When character is identified as consistency or uniformity then codes, regulations or covenants designed to discriminate against spatial difference may also discriminate against social difference. Second, lack of clear definition provides a legal loophole that can open the door to highly damaging developments. The experience in Melbourne has been that when 'character' becomes a key planning criterion then the system transforms from 'as of right' controls to site-by-site decisions in a legal tribunal where definitions of character are decided by lawyers. A further effect has been to shift debate into the field of aesthetics where height limits and amenity are traded against design quality (Dovey, Woodcock and Wood 2009b). Finally, attempts to control character through the regulation of built form has a tendency to reduce character to formal characteristics, and can turn character into caricature. In the worst examples of this new buildings are only permitted in neo-traditional style – a new character is created in the guise of protecting an old one.

In sum, moving issues of place identity to the top of the urban planning agenda can be a mixed blessing. While residents may be eager to put 'character' at the centre of the planning process this is matched by the eagerness of developers to engage in site-by-site exceptions to urban regulation. While it may appear that character-based planning may be more sensitive to the nuances of place experience, the dangers are also profound.

Re-Thinking Place Identity

The phenomenon of place is difficult to define because it is so deeply rooted – it is a site where habit fuses with habitat and social ideology takes root. To understand resident resistance to urban change requires that we penetrate beneath NIMBY stereotypes and engage with social processes and the politics of identity formation. Place and identity are dynamic social constructions that we often come to see as fixed and natural. To understand the role of place experience in urban planning requires theoretical frameworks that can connect the ontology of dwelling with the larger structures of political economy. For Harvey the politics of place can be seen in the context of global/ local tensions – at once grounded in a local phenomenology of dwelling yet subject to the appropriations of global capital (Harvey 1996: 297–8). From this view place identity attracts capital through a market desire for uniqueness and authenticity; capital seeks to retain or to produce a unique sense of place as a form of local monopoly (Harvey 2001: 394–411).

The experiences of place in everyday life, often taken for granted until threatened, surface as part of the politics of place where they are further constructed and shaped. The spatiality and sociality of place are inextricably intertwined; space is socially constructed as the social is spatially constructed (Lefebvre 1996, Massey 1993). This reciprocity is apparent in the continual slippage between social and material aspects of character as defined by residents. In this context there is a clear need for concepts and approaches that cut across the sociality/spatiality divide, a need to move beyond a false choice between place as pre-given (fixed, essential) or as entirely socially constructed. There are two conceptual frameworks that I suggest may be fertile in this regard: Bourdieu's work that connects the everyday *habitus* to the *field* of power, and Deleuze and Guattari's theory of assemblage.[6]

Bourdieu's conception of the *habitus* is a set of pre-conscious dispositions that structure the taken-for-granted *doxa* of everyday life: '... he inhabits it like a garment (*un habit*) or a familiar habitat. He feels at home in the world because the world is also in him, in the form of habitus ...' (Bourdieu 2000: 142–3). The *habitus* is described as 'a sense of one's place' but also a 'sense of the other's place' (Bourdieu 1990: 113), and as a 'feel for the game' of social practice (Bourdieu 1993: 5). The concept of *habitus* is derived in part from Merleau-Ponty's phenomenology of embodied spatiality (Carman 2008: 217–19). The resonance between *habitus* and habitat can be a useful conceptual frame here because it parallels that between social and physical character, between the feel and the form. As we turn place identity into planning codes we move from the pre-conscious experiences of place in everyday life to the production of a discourse of 'place' and 'character' within institutionally structured fields of power (Bourdieu 1993): news media, housing markets, planning tribunals. From this view the taken-for-granted 'doxa' of urban life becomes a para-dox of urban design and planning. The *habitus* is the 'feel' that is threatened by the 'form'. For Bourdieu the *habitus* is aligned with an institutionally structured *field* of power. If the *habitus* is a 'feel for the game', the *field* is akin to the game board wherein certain resources are at stake in the form of different kinds of capital – social capital, economic capital, symbolic capital. Resident groups who unite to defend a neighbourhood against change often produce and rely on such forms of power as community solidarity. But the closed sense of place is also implicated in the production and defence of privilege through networks of influence.

From such a view we can understand neighbourhood place identity as being deep-seated without being deep-rooted in any essentialist manner. It is immanent rather than transcendent; grounded in the myriad particularities and everyday practices of particular places. Another useful conceptual framework in this regard is the work of Deleuze and Guattari (1987) on *assemblage*. Assemblage theory is a loose body of ideas primarily developed from a framework of Deleuzian philosophy wherein urban places and development projects are largely seen in terms of the connectivity and flow between parts of a socio-spatial assemblage (DeLanda 2006). Flows of people, traffic, ideas, capital and goods are linked to flows of desire for profit, views, amenity, sunshine, privacy, open space and access. Such desires play out in the politics of urban planning through the interests of developers, residents, retailers, commuters and neighbours – interests that variously intersect, reinforce and contradict.

Assemblage is both verb and noun, not a collection of things or a spatial container but a socio-spatial territory wherein material forms and discursive practices become aligned. Assemblage is thus a conceptual framework that potentially connects both the 'feel + form' and the 'social + physical' dimensions of place. The concept of place can then be seen not as bounded location but as an emergent property of the urban

---

6   For a more detailed account of what follows see chapters 2 and 3 of Dovey 2010.

assemblage. 'Sense', 'feel', 'atmosphere' and 'character' can be seen as intensities in the sense that desire, love, flavour, light, colour, tone and experience have intensity (while height and bulk have extension). When place identity becomes legislated, feel is reduced to form, intensity is reduced to extension. Urban regulation is a process of coding: character is coded into characteristics; parts are made to stand for the whole; place identity becomes fixed.

Here is the deep dilemma of planning for place identity – how to create or protect urban place identity in a manner that does not kill the very dynamism that produces it in the first place. It may be useful in this regard to conceive of place as a conceptual 'plateau' – a place defined by its situation between levels. While often associated with Deleuze and Guattari's book *A Thousand Plateaus*, the term originates with Bateson (1972) where it is defined in opposition to schizmogenesis: the way a positive feedback process (like an arms race) can escalate out of control. The way that one tall building in a neighbourhood can set a precedent that triggers the right to go ever higher is an example of schizmogenesis. There are links here to Jacobs' (1965) theory of the 'self-destruction of diversity' and to Harvey's (1985) work on the circuits of capital leading to creative destruction. For Deleuze and Guattari the plateau is also a 'plane of consistency' in an assemblage that is open to change but not to suicidal escalation. The concepts of place, plateau or plane (note the shared etymology of these *pla* words) denote immanent fields of everyday practice that ground modes of thought and identity formation without transcendent ideals. The suggestion here is that place identity and character can be conceived as a socio-spatial plateau: an assemblage that is open to change but is resilient to escalation.

The Place Intensification Conundrum

In order to understand what is at stake in this issue we need to look at the larger scale of the assemblage where we find that what is being created and defended in the name of place and character is often an unsustainable low-density city. While it is quite possible to service low-density suburbs with high frequency public transport (Mees 2009) it is also very clear that low-density car-based urban assemblages generate significantly more carbon emissions (Newman, Beatley and Boyer 2009). While supporters of suburbia live in hope for green cars and public transport to save the suburbs, low-density/low-carbon suburbs are a vain hope without major technological change and massive long-term state subsidies. The best prospect to save the suburbs is to service them with an intensified network of higher-density, transit-oriented development that brings high-frequency public transport within walkable distances. It is the imperative of low-carbon cities that produces the conundrum – resident activists are often well-organized to prevent what they see as overdevelopment in their neighbourhoods and governments are fearful of losing power in a suburban backlash. If neighbourhood planning is democratic then existing densities are likely to be defended. In common interest developments where density is established by covenant, intensification is close to impossible – indeed protection from democratic planning is part of their market advantage. We need to find a way to reconcile the desire to create and protect place identity with the imperatives for low-carbon cities. The public debate becomes polarized into those who think intensification is a threat to place identity and those who think that resident democracy is a threat to sustainability. Since whatever character the city currently possesses was created by intensification in the first place, why is it that further intensification is so often seen as a threat? How can we manage urban development in a manner that enhances rather than damages character and place identity?

In this regard the very problems we have in defining and legislating place identity can be seen as opportunities. Since urban place identity is not static, fixed or pre-formed; since it has emerged from the very process of intensification – then it can also be enhanced by further intensification. High-quality urban design can create greater levels of urban amenity, access and equity along with reduced car dependency, by raising densities within a walkable distance of public transport services. The low densities of suburban neighbourhoods – open space, sunshine, generous roof spaces, trees – can also be seen as resources that can contribute to a more sustainable city through urban agriculture, solar energy production and water harvesting. In other words places can change in incremental ways that do not trigger the ontological crisis of a perceived loss of identity.

The transformation we need most is from places that are closed, purified and static towards those that are open, multiple and dynamic. The difference that makes a difference is that between places of difference and places of purity; between places that have a place for the dis-placed, and places where identity formation is fixed and finished.

References

Altman, I. and Low, S. 1992. *Place Attachment*. New York: Plenum.

Bateson, G. 1972. *Steps to an Ecology of Mind*. New York: Ballantine.

Beatley, T. and Manning, K. 1997. *The Ecology of Place*. Washington, DC: Island Press.

Bourdieu, P. 1990. *The Logic of Practice*. Cambridge: Polity.

—— . 1993. *The Field of Cultural Production*. New York: Columbia University Press.

—— . 2000. *Pascalian Meditations*. Cambridge: Polity Press.

Buttimer A. and Seamon, D. (eds) 1980. *The Human Experience of Space and Place*. London: Croom Helm.

Canter, D. 1977. *The Psychology of Place*. London: Architectural Press.

Carman, T. 2008. *Merleau-Ponty*. London: Routledge.

Corburn, J. 2009. *Toward the Healthy City: People, Places, and the Politics of Urban Planning*. Cambridge, MA: MIT Press.

Cresswell, T. 2004. *Place: A Short Introduction*. Oxford: Blackwell.

DeLanda, M. 2006. *A New Philosophy of Society*. New York: Continuum.

Deleuze, G. and Guattari, F. 1987. *A Thousand Plateaus: Capitalism and Schizophrenia*. Minneapolis, MN: University of Minnesota Press.

Dovey, K. 1999. *Framing Places*. London: Routledge.

—— . 2005. *Fluid City*. London: Routledge.

—— . 2010. *Becoming Places*. London: Routledge.

Dovey, K., Woodcock, I. and Wood, S. 2009a. Understanding neighbourhood character. *Australian Planner*, 46(3), 32–9.

—— . 2009b. A test of character. *Urban Studies*, 46(1/2), 2595–615.

—— . 2010. Slippery characters, in *Becoming Places*, edited by K. Dovey. London: Routledge, 57–78.

Dreier, P., Mollenkopf, J. and Swanstrom, T. 2005. *Place Matters*. Lawrence, KS: Kansas University Press.

Eyles, J. and Williams, A. (eds) 2008. *Sense of Place, Health and Quality of Life*. Aldershot: Ashgate.

Harvey, D. 1985. *The Urbanization of Capital*. Baltimore, MD: Johns Hopkins University Press.

—— . 1996. *Justice, Nature and the Geography of Distance*. Oxford: Blackwell.

—— . 2001. *Spaces of Capital*. New York: Routledge.

Healey, P. 2005. *Collaborative Planning: Shaping Places in Fragmented Societies*. Basingstoke: Palgrave Macmillan.

—— . 2010. *Making Better Places*. Basingstoke: Palgrave Macmillan.

Hubbard, P., Kitchin, R. and Valentine, G. 2004. *Key Thinkers on Space and Place*. London: Sage.

Jacobs, J. 1965. *The Death and Life of Great American Cities*. Harmondsworth: Penguin.

Lefebvre, H. 1996. *Writings on Cities* (translated and edited by E. Kofman and E. Lebas). Oxford: Blackwell.

Levinson, D. and Krizek, K. 2007. *Planning for Place and Plexus*. London: Routledge.

Madanipour, A., Hull, A. and Healey, P. (eds) 2001. *The Governance of Place*. Aldershot: Ashgate.

Massey, D. 1993. Power-geometry and a progressive sense of place, in *Mapping the Futures*, edited by J. Bird, B. Curtis, T. Putnam, G. Robertson and L. Tickner. London: Routledge, 59–69.

—— . 1994. *Space, Place, and Gender*. Minneapolis, MN: University of Minnesota Press.

—— . 2005. *For Space*. London: Sage.

Mees, P. 2009. *Transport for Suburbia*. London: Earthscan.

Newman, P., Beatley, T. and Boyer, H. 2009. *Resilient Cities*. Washington, DC: Island Press.

Norberg-Schulz, C. 1980. *Genius Loci: Towards a Phenomenology of Architecture*. New York: Rizzoli.

Relph, E. 1976. *Place and Placelessness*. London: Pion.

Seamon, D. and Mugerauer, R. (eds) 1985. *Dwelling, Place and Environment*. The Hague: Martinus Nijhof.

Sennett, R. 1998. *The Corrosion of Character*. New York: Norton.

Tuan, Y. 1977. *Space and Place*. Minneapolis, MN: Minnesota University Press.

Wood, S., Woodcock, I. and Dovey, K. 2006. Contesting characters at Hedgeley Dene, in *Contested Terrains*, edited by S. Basson, Proceedings of the 23rd annual international conference of SAHANZ, Curtin University, Perth.

Case Study Window – Culture in International Sustainability Practices and Perspectives: The Experience of 'Slow City Movement – Cittaslow'

Tüzin Baycan and Luigi Fusco Girard

Culture and Planning

'Cultural Turn' and Cultural Planning

In recent decades, a shift from a more traditional concept of culture as linked to the classical 'fine arts' towards an understanding of 'cultural resources', 'cultural and creative industries', 'cultural diversity' and 'a way of life' has been observed (Baycan 2011, Baycan-Levent 2010, Bianchini 2004, Cooke and Lazeretti 2008, Cunningham 2001, Evans 2009, Ghilardi 2001, UNCTAD 2004). While knowledge, culture and creativity have been increasingly recognized as key strategic assets and powerful engines driving economic growth, the high concentrations of heterogeneous social groups with different cultural backgrounds and different ways of life have made cities incubators of culture and creativity (Baycan-Levent 2010, Merkel 2011, UNCTAD 2008). Besides knowledge and innovation, culture and creativity have become the new key resources in urban competitiveness. Cultural production in itself has become a major economic sector and a source for the competitive advantage of cities (Florida 2002, Merkel 2011, Miles and Paddison 2005, Musterd and Ostendorf 2004, Zukin 1995). Knowledge, culture and creativity have also become the new keywords in the understanding of new urban transformations (Hall 2004). The existing literature shows that cultural and creative industries are deeply embedded in urban economies (Foord 2008, Pratt 2008, Scott 2000). The role of cultural production in the new economy has radically changed the patterns of cultural consumption (Quinn 2005), and cities have transformed from functioning as 'landscape of production' to 'landscape of consumption' (Zukin 1998). In parallel to this transformation or 'cultural turn', 'culture-led regeneration' and 'cultural planning' have become the main strategies of cities.

The 'cultural planning' approach was first used and recommended in 1979 by economist and city planner Harvey Perloff as a way for communities to identify and apply their cultural resources to society's objectives. However, the roots of cultural planning can be found in nineteenth-century amenity planning, the City Beautiful Movement, the Works Progress Administration cultural jobs creation programmes of the 1930s and the pioneering work of the community arts movement of the 1940s (Dreeszen 1998). Cultural planning has been discussed since the early 1990s in North America, Australia and Europe (Bianchini, 2004, 1996 and 1990, Bianchini and Parkinson 1993, Grogan, Mercer and Engwicht 1995, Hawkes 2001, McNulty 1991, Mercer 2006, 1996 and 1991a, Stevenson 2004) as a possible alternative to traditional cultural policies as well as cultural policy-led urban regeneration strategies.

Against the traditional cultural policies which are based on definitions of 'culture' as 'art', cultural planning adopts as its basis a broad definition of 'cultural resources' (Bianchini 2004, Ghilardi 2001, NSW Ministry for the Arts 2004). Cultural resources include: 1) history; 2) heritage including archaeology, gastronomy, local dialects and rituals; 3) diversity of local people: the cultures of youth, ethnic minorities and communities of interest; 4) diversity and quality of leisure, cultural, eating, drinking and entertainment facilities and activities; 5) arts and media activities and institutions; 6) natural and built environment including public and open spaces; 7) the repertoire of local products and skills in crafts, manufacturing and services, including local food products, gastronomic and design traditions; 8) local milieux and institutions for intellectual and scientific innovation, including universities and private sector research centres. Mercer (1991b) has argued that while putting cultural resources at the centre, compared to traditional cultural policies, cultural planning has intrinsically become more democratic, more conscious of the realities of cultural diversity and more aware of the intangible features of cultural heritage and patrimony. Another distinctive feature of cultural planning is its territorial remit. While traditional cultural policies tend to take a sectoral focus, cultural planning adopts a territorial remit to cultural and community development (Bianchini 2004, Ghilardi 2001, see also Bianchini, Chapter 22, this volume). The purpose of cultural planning is to see how cultural resources can contribute to the integrated development of a community. Mercer (1991a) has defined cultural planning as 'the strategic and integral planning and use of cultural resources in urban and community development'. According to Dreeszen (1998: 9), 'Community cultural planning is a structured, community-wide fact-finding and consensus-building process to assess community needs and develop a plan of action that directs arts and cultural resources to address those needs.' Cultural planning has been described in *Cultural Planning Toolkit* published by 2010 Legacies Now and Creative City Network of Canada (2010: 1) as 'a process of inclusive community consultation and decision-making that helps local government identify cultural resources and think strategically about how these resources can help a community to achieve its civic goals'. However, Bianchini and Ghilardi Santacatterina (1997) have argued that cultural planning is not the 'planning of culture', but a cultural (anthropological) approach to urban planning and policy. Cultural planning has many common features with other planning disciplines, thus it can be seen as a process which should give rise to vision and leadership on local and cultural development (NSW Ministry for the Arts 2004). For example, the *Cultural Planning Handbook: An Essential Australian Guide* argues that cultural planning needs to achieve three main objectives: firstly, to utilize well-analysed high-quality data; secondly, to work with an agreed community vision for an area; and thirdly, to seek to integrate cultural planning (through new and existing linkages) with other forms of planning including social, economic and environmental planning (Grogan, Mercer and Engwicht 1995).

The cultural planning approach has been widely applied in both the United States (since the 1970s) and Australia (since the mid-1980s) as a successful way of enabling policy makers to think strategically about the cultural resources of localities and the delivery of policies capable of responding to local needs in a flexible, bottom-up way. However, in Europe, cultural planning has had little application due to the tendency of policy makers in interpreting the notion of local cultural resources in a rather narrow way, mostly as heritage, thus overlooking potential synergies between sub-sectors of the local cultural economies (Bianchini 2004, Ghilardi 2001).

Cultural Planning Versus Globalization: 'Slow City Movement – Cittaslow'

The idea of 'cultural planning' can be seen as a possible response to the problematic cultural implications of globalization for cities (Bianchini 2004). By recognizing the value of local cultural resources, cultural planning can be seen as a 'manifesto' in response to globalization that promotes standardization and homogenization, with the risk of losing specific identities, as well as an attempt to challenge traditional approaches to urban development. When culture is described as 'a way of life', the central characteristics of cultural planning consider the integration of the arts into other aspects of local culture and into the texture and routines of daily life in the city (Bianchini and Ghilardi Santacatterina 1997, Bianchini and Parkinson 1993). As underlined by Mercer (2006), '… what is being planned in cultural planning are the lifestyles, the texture and quality of life, the fundamental daily routines and structures of living, shopping, working, playing'. Therefore, against the implications of globalization, cultural planning must address the issues of identity, autonomy and sense of place and must be able to establish and maintain a real and effective policy equilibrium between 'internal' quality and texture of life and 'external' factors (Mercer 2006) relating to the demands raised by globalization such as tourism and attractiveness to potential residents and visitors including businesses.

The 'Slow City Movement' is one of the best examples of cultural planning in a globalization context as it aims to ensure sustainable development at the local level and to promote it in the global scene. The 'Slow City Movement' has emerged as a cultural 'manifesto' as well as a response to the problematic cultural implications of globalization for cities. Against the 'fast world' generated by globalization in which the more places change, the more they seem to look alike and are less able to retain a distinctive sense of place, the 'Slow City Movement – Cittaslow' began in 1999 in Italy as an alternative urban development strategy to the fast world. The purpose of the 'Slow City Movement' is to emphasize the local distinctiveness in a context of globalization and to improve the quality of life locally. The 'Slow City Movement' is the resistance of people and local governance to the threats to cultural difference and the standardization of everyday practice imposed by so-called modern living. With this feature, the 'Slow City Movement' is also one of the best examples of 'cultural resilience'. The 'Slow City Movement' has expanded around the world at an increasing pace since 1999. Today, the network has reached 149 cities in 24 countries connecting administrators, citizens and slow food partners.

This chapter aims to investigate the experience of the 'Slow City Movement – Cittaslow' as an international sustainability practice from the perspective of cultural resilience and cultural planning. We will first address sustainability, creativity and resilience with a special focus on cultural dimension, then examine the experience of the slow city network on the basis of 'Cittaslow policies' and their applications.

Sustainability, Creativity and Resilience

The Four Pillars of Sustainability

Sustainability, as it has become formally adopted, has three pillars: 'ecological sustainability', 'social sustainability' and 'economic sustainability'. With its three dimensions – economic, social and environmental – sustainable development has become the mantra of contemporary planning. However, it has been argued that there should be four pillars, the fourth being 'cultural sustainability'. Hawkes (2001: 25) has described the four pillars of sustainability as: 1) cultural vitality (wellbeing, creativity, diversity and innovation); 2) social equity (justice, engagement, cohesion, welfare); 3) environmental responsibility (ecological balance); and 4) economic viability (material prosperity). He has emphasized that cultural vitality is as essential to a sustainable society as social equity, environmental responsibility and economic viability, and that sustainability can only be achieved when it becomes an enthusiastically embraced part of our culture.

Culture in its widest sense is about relationships, shared memories and experiences as well as identity, history and a sense of place. On the other hand, culture is our way of connecting the present with the past and the future and it is about the things we consider valuable for passing on to future generations. In this sense, culture is linked to sustainability. By linking culture and other aspects of economic and social life, cultural planning can be instrumental in creating development opportunities for the whole of the local community (Ghilardi 2001). Cultural planning can help local governments tackle social exclusion, contribute to urban regeneration, create employment opportunities, build safer communities, improve community well-being and encourage healthier lifestyles. However, cultural planning is not a matter of directing people's values and aspirations, but about providing opportunities and removing obstacles to people's cultural expression, creativity and sense of place (NSW Ministry for the Arts 2004).

The critical condition for real success in implementing sustainable development is to invest in city creativity and resilience. Sustainability, creativity and resilience are closely intertwined as revealed by best practices of creative cities in implementing sustainable development (Fusco Girard and You 2006). The image itself of a creative city reflects the interdependences among sustainability, resilience and creativity. Creativity and innovation enhance the capacity to face new risks and perturbations (that is, the resilience of the ecological, economic and social systems). In other words, creativity enhances sustainability as it guarantees a higher resilience capacity to urban systems.

Multiple Perspectives of Resilience

Resilience has multiple perspectives. Innovations can involve, for instance, a new circular metabolism (with reuse, recycling and regeneration of materials) in urban ecosystems (ecological resilience), new economic competitiveness with the identification of original development trajectories in wealth production (economic resilience), and the opening up of new social bonds and community relationships (social resilience). The intensity of the resilience depends on specific innovations that are introduced into the urban system. They improve the comprehensive city self-organization and thus sustainability.

If resilience is the capacity of a system to maintain over time its original organizational structure (its identity and unity), absorbing shocks from outside, thanks to its self-organizational capacity, it is possible to identify some common elements in ecological, economic and social resilience such as: the notion of memory, conservation, stability, correlations and feedbacks.

Social resilience depends on the density of formal and informal social networks, which are able to preserve over time a certain organizational order: it depends on the existing community and the sense of place.

Economic resilience is the capacity to produce wealth, business and profits by changing and experimenting with innovative production technologies, organizations and strategies. New connections among networks and new relationships are strongly improved through ICT.

Ecological resilience reflects the health and robustness of the system, that is, the density of connections/relations between its different components that allow for circular processes (with a reduction of consumed materials, energy, water and so on).

Cultural resilience is the internal energy, the inner force (or vitality) that allows the city or the society to react to external forces, adapt to them and conserve their specific identity in the long run, in spite of turbulent transformation processes and to design new win-win solutions. Cultural resilience stresses the notion of the cultural memory of the community as a formative strength of collective consciousness, foundation of continuity, and engine of the future and new actions in order to improve trust, cooperation and coordination of actions and to promote a sense of community.

Cultural resilience depends on: 1) the capacity to think and choose in a systemic, multidimensional, open and relational way; and 2) linking short, medium and long-term perspectives while paying attention to the 'memory' of the system in achieving common interests. If these common elements/goals are recognized, cooperation and coordination of actions is possible. The approach becomes constructivist (not 'or ... or', but 'and ... and'), and characterized by gradualism. Cultural resilience is built on the basis of a way of thinking, founded upon a critical approach (that is, on evaluation capacity), able to recognize and compare tangible and intangible elements – values, objectives, goals – by considering all existing interdependences and distinguishing a hierarchy or priority. Critical thinking is the capacity to learn from experiences and to select appropriately not only means but also objectives that have values and meanings, that are reference anchors in orienting innovations and creativity. In other words, city cultural resilience depends on each inhabitant's capacity to transform data and information into critical knowledge and wisdom, and to adopt a new way of life that rejects the current 'more, bigger, faster' culture.

Sustainable, Creative and Resilient City

A resilient city is also a creative city, able to reinvent a new equilibrium against destabilizing external pressures. It multiplies the potential of people to build new opportunities/alternatives. The notion of the creative city and its various components has been extensively analysed in the literature (Baycan 2011, Baycan-Levent 2010, Florida 2005, 2002, Fusco Girard, Baycan and Nijkamp 2011, Hall 1998, Kunzmann 2004, Landry 2006, Markusen 2006, Scott 2006, 2000, Simmie 2001).

The concept of 'creative cities' is fuzzy (Kunzmann 2004) and can be interpreted from many different perspectives. Peter Hall (1998) identifies in history some types of creative cities: technological-innovative; cultural-artistic; artistic and technological; and artistic, technological, and organizational. Technological-innovative creativity of the city that proceeds through sharp breaks and discontinuities leads to the introduction of new strategies and to the rapid obsolescence of pre-existing ones. There is also a creativity of cities that feeds itself continuously and starts from the *status quo*, through successive incremental improvements (adaptive creativity). But, in both these cases the empirical evidence shows that innovative elements are deeply rooted within history and tradition (path dependency). They have a bond with the city's past history, its soul and spirit. They have been able to 'metabolize' the urban history. In both cases, to promote an 'innovative milieu' in order to valorize existing skills and talents is necessary. This 'innovative milieu' allows cities to be creative in the accelerated change: to be resilient from the inside (not only because they receive exogenous resources or adapt the best practices developed elsewhere). An essential element of this milieu is represented by a better knowledge that provides the possibility to think different subjects in a new way, thus to identify new alternatives, new solutions and new choices.

Today, cities all over the world are investing in cultural infrastructures (research parks, cultural districts, hubs and so on) as a catalyst to sustain local development. Schools, universities and research institutions are becoming the main investments in order to develop new knowledge and to transform this knowledge into actions. They are cities' real wealth and replace the traditional urban economic industrial base. Culture shapes relations among economic, ecological and social systems; promotes 'circularization' of the city's economy, and thus resilience. Ancient heritage as well can shelter incubators of innovations and resilience. Strategic local planning for culture (for example Barcelona Local Agenda 21 for culture) is the general tool to better correlate all city policies into an effective network and to promote human sustainable development.

On the other hand, cities are faced with: 1) increasing urbanization processes that are not economically, socially and environmentally sustainable; and 2) an increasing standardization and homogenization promoted by globalization, with the risk of losing specific identities. This process increases vulnerability and reduces resilience. In order to stress their identity, many cities are investing in culture, heritage and historic landscape. They are, in particular, investing in maintaining the 'places' that represent the signs and symbols of their creativity and the relations between people and the built environment: the 'spirit of places'. Investments in the regeneration of ancient centres, dismissed areas and waterfronts are well known examples.

Spatial planning increasingly becomes culture-oriented: cultural districts, cultural industries, cultural incubators are proposed to catalyse economic development, sustaining urban cultural policies (see also Montgomery, Chapter 20, this volume). Urban planning aims to improve, in particular, the cultural resilience that constitutes the inner energy that allows the city to react to external forces and to conserve its specific identity. Planning means not only reducing the ecological foot-print of urban areas, stimulating a 'circular economy' in the city systems, and embellishing the urban scene with good architecture and restoration of cultural heritage, but also regenerating the life in the streets, squares and neighbourhoods while valorizing all local resources and rebuilding the social bonds, sense of community, social capital and thus, cultural resilience. The new urban planning or cultural planning focuses on priorities like the production and regeneration of public spaces as specific areas of identity, exchanges, life. The regeneration of 'places' (historical centres, cultural landscape, waterfronts and so on) is often included in a strategy for the 'beautiful city'. Many experiences are now spreading all over the world. 'Arts plans', 'heritage plans', 'culture plans' are examples of the tools to build actions towards creative and beautiful cities that stimulate investments in creative sectors as a priority in the new city economy.

European Capitals of Culture is an interesting example of city development strategy based on culture. Glasgow, Turku, Liverpool, Lille, Linz, Vilnius, Essen, Istanbul are some of the examples of cities that experimented with this culture-led model (Palmer 2004). Slow cities is another interesting example of culture-led urban development and cultural planning. The following section examines the 'Slow City Movement' and its policies.

'Slow City Movement – Cittaslow' and Cultural Resilience

The experience of 'Slow Cities – Cittaslow', an international network of small cities, offers an alternative approach and a set of interesting practices towards sustainable development. The good practices of slow cities have a very high benchmarking value from which we can learn many lessons including how to implement the culture of 'savings', 'reuse', 'recovery', 'recycling', 'regeneration', the 'renewable' and how to transform the ecological, territorial and landscape values into economic as well as cultural/civil values. The experience of slow cities also offers interesting interpretations of 'good governance', in particular from an environmental perspective. The practices of slow cities are good examples of concrete creativity and of increased ecological, economic and socio-cultural resilience of the city to promote its sustainability.

The 'Slow City Movement – Cittaslow' started as a cultural and a new lifestyle proposal inspired by the 'Slow Food' movement. The 'Slow Food' movement was founded in 1986 by an Italian food writer who was alarmed by the opening of a McDonald's restaurant next to the Piazza di Spagna in the heart of Rome. The movement's goal is to protect the 'right to taste' (Slow Food International, http://www.slowfood.com) by preserving almost-extinct traditional food products, raising the awareness of the pleasures of eating (including the social aspects of sharing a meal), taste education, and paying attention to traditional agricultural methods and techniques among other initiatives. The 'Slow Food' movement touches on important aspects that keep local community economies vital. In particular, 'Slow Food' is locally grounded through its goal of maintaining the viability of locally owned businesses such as restaurants and farms. At the core of the movement is the concept of 'territory'. 'Slow Food' emphasizes local distinctiveness through the connection to the specificity of a place as expressed by traditional foods and ways of producing and growing produce such as wine, cheese, fruits and vegetables. In the words of Carlo Petrini (2001), the Italian food critic who spearheaded the resistance against McDonalds, the concept of territory is a 'combination of natural factors (soil, water, slope, height above sea level, vegetation, microclimate) and human ones (tradition and practice of cultivation) that gives a unique character to each small agricultural locality and the food grown, raised, made and cooked there' (2001: 8). The understanding of territory in the 'Slow Food' movement connects the environmental aspects of a place to the culture and the history of people who inhabit the territory and have utilized it for generations for traditional food production. While the 'Slow Food' programmes address the notion of place through the concept of 'territory', the 'Slow City Movement – Cittaslow' provides an explicit agenda of local distinctiveness and urban development. The purpose of the 'Slow City Movement' is to emphasize the local distinctiveness in a context of globalization and to improve the quality of life locally. The word 'slow' is used in contrast with the 'fast' lifestyle that characterizes the big cities (Cittaslow 2011, http://www.cittaslow.org).

At present, living and managing a 'Cittaslow' is just a particular way of carrying on an ordinary lifestyle rather than following today's trends. 'Cittaslow' is both an urban social movement and a model for local governance. Basically, the 'Slow City Movement' is the resistance of people and local governance to the threats to cultural difference and the standardization of everyday practice imposed by so-called modern 'fast' living (Miele 2008, Parkins and Craig 2006: 82). According to 'Cittaslow' philosophy, good living means having the opportunity to enjoy solutions and services that allow citizens to live in their town in an easy and pleasant way. Living slow means being slowly hasty; '*festina lente*' latins used to say, seeking every day the 'modern times counterpart', or in other words, looking for the best of the knowledge of the past and enjoying it thanks to the best possibilities of the present and of the future (Cittaslow 2011). All of this will result in technological opportunities – modern solutions in communication, transportation and production.

The 'Cittaslow' Association, an international network of small and medium-sized towns all over the world where quality of life is important, aims to promote and spread the culture of good living through research, testing and the application of solutions for the city organization. 'Cittaslow' is looking for

> ... *towns where men are still curious of the old times, towns rich of theatres, squares, cafes, workshops, restaurant and spiritual places, towns with untouched landscapes and charming craftsman where people are still able to recognize the slow course of the seasons and their genuine products respecting tastes, health and spontaneous customs* .... (*Cittaslow 2011*)

'Cittaslow' is a membership organization. Full membership of 'Cittaslow' is only open to towns with a population under 50,000. To be eligible for membership, a town must normally score at least 50 per cent in a self-assessment process against the set of

'Cittaslow' goals, and then apply for admission to the appropriate 'Cittaslow' national network. There are six 'Cittaslow' membership goals:

1. *environmental policy* – pollution control; municipal plan for saving energy, with reference in particular to the use of alternative sources of energy (renewable sources, green hydrogen, mini hydroelectric power plant);
2. *infrastructure policy* – protection of cultural and historical values; arrangement of transportation systems (safe mobility and traffic); supporting new facilities and local activities (centre for medical assistance, quality green areas and service infrastructures); new commercial activities (commercial centres for natural products, programme for urban restyling and upgrading);
3. *technologies and facilities for urban quality policy* – window for bio-architecture; equipping the city with cables for optical fibre and wireless systems; dissemination of municipal services via internet;
4. *safeguarding autochthonous production policy* – plans for the development of organic farming; programmes for the safeguarding of artisan and/or artistic craft products in danger of extinction; safeguarding traditional methods of work and professions at a risk of extinction;
5. *hospitality policy* – training courses for tourist information and quality hospitality; preparation of 'slow' itineraries for the city (brochures, websites, home pages and so on);
6. *awareness policy* – campaign to provide the citizens with information on the aims and procedures of what a Slow City is, preceded by information of the intentions of the Administration to become a Slow City.

A detailed list of 'Cittaslow' goals and policies can be found on the Cittaslow website (http://www.cittaslow.org).

The 'Slow City Movement – Cittaslow' has expanded to 149 towns in 24 countries around the world since 1999, connecting administrators, citizens and 'Slow Food' partners (Cittaslow 2011). The countries (and number of towns) in the network are Australia (2), Austria (3), Belgium (4), Canada (2), China (1), Denmark (1), Finland (1), France (2), Germany (10), Hungary (1), Italy (71), Netherlands (3), New Zealand (1), Norway (3), Poland (6), Portugal (4), South Africa (1), South Korea (8), Spain (6), Sweden (1), Switzerland (1), Turkey (5), the UK (9) and the United States (3). Although the 'Slow City Movement' is increasingly spreading all over the world, it can still be seen as a European movement as most of the slow cities are located in Europe and approximately half of them, 71 of 149, are in Italy.

The 'Slow City Movement – Cittaslow' has some important practical consequences in terms of more balanced regional polycentric development. The practices of slow cities show that they manage to reduce the depopulation trend and the exodus of activities towards the larger sized centres as well as costs of congestion, agglomeration and overuse of resources both in the areas of concentration and the inner cities. The following section examines the implemented policies and practices of slow cities.

'Slow City Movement – Cittaslow': Policies and Practices

In this section we evaluate the implemented policies and practices of the 'Slow City Movement – Cittaslow' on the basis of our two previous studies: 1) a website-based analysis of slow cities' policies (Baycan, Akgün and Erginli 2012) and 2) a survey (conducted by Baycan and Fusco Girard in 2010 and 2011) based on observation on site as well as an in-depth interview with the mayor or vice-mayor of 19 slow cities in Italy.

Our first study (Baycan, Akgün and Erginli 2012) aimed: 1) to investigate the slow city movement and to compare and evaluate its policies and actions towards sustainable local development; 2) to highlight the most important and common slow development policies through the 'Cittaslow' network; and 3) to evaluate the geographical and regional differences in implemented policies. The study analysed the implemented policies of and events organized by slow cities and identified the priorities in slow development policies by deploying the well-known multi-criteria analysis, Analytical Hierarchy Process (AHP).

The data and information used in our first study was collected from the website of each 'Cittaslow'. While investigating the websites, we used a couple of indicators such as organized events and their characteristics; implemented projects and their characteristics; and related 'Cittaslow' policies that we summarized in the previous section. Although we visited all 149 slow cities' websites, due to the limited availability of data and information from the websites, we could only include 46 slow cities from 16 countries: Australia (1), Austria (2), Belgium (2), Canada (1), Germany (2), Italy (13), Netherlands (1), Norway (3), Poland (1), Portugal (4), South Korea (6), Spain (1), Switzerland (1), Turkey (1), UK (5) and the United States (2), which organize events and/or implement projects in our sample.

Our website-based analysis revealed 90 events organized by 46 slow cities. The results of our analysis show that organized events and activities are usually unique to the city. While one event can correspond to more than one 'Cittaslow' policy, events can be local, national or international and organized usually by the slow city itself. The results of our study show also that among 'Cittaslow' policies, those focusing on the protection of local production, or in other words, 'safeguarding autochthonous production policy', are the most supported ones (37 per cent) by events and activities. 'Awareness policy' (29 per cent) and 'infrastructure policy' (16 per cent) follow local product policies, respectively. On the other hand, the results of our analysis highlighted that events and activities organized by or in slow cities are not always related to 'Cittaslow' policies (37 per cent). In addition, the available data and information show that slow cities do not primarily focus on technological improvement; their focus seems to be more on local products, their protection and increasing their economic value. A greater focus on local products is understandable as it is easier to manage both financially and in terms of 'tacit' knowledge, especially in the beginning of becoming 'slow'. Technological improvement, on the other hand, involves larger investments, higher economic costs and long-term planning activities. Therefore, less focus on technological improvement (0 per cent) should not be understood as less willingness to improve the technology, but a need for time to find necessary financial resources as well as to establish financial models and instruments (Baycan, Akgün and Erginli 2012).

The results of our comparative analysis show also that while small cities in the network have fewer tendencies to apply 'Cittaslow' policies or any other policies via events or projects, relatively big cities have a higher interest in slow policies. Our comparative analysis highlights a regional differentiation as well: North European slow cities seem to give a higher priority to the 'Cittaslow' policies than other slow cities in other regions, whereas North American cities follow policies other than the described 'Cittaslow' policies (Baycan, Akgün and Erginli 2012).

Our second study considers a survey conducted by Baycan and Fusco Girard in Italian slow cities. As almost half of the slow cities are located in Italy, in order to better understand the main features of slow city development strategies, the survey was conducted in different regions in Italy since September 2010. The survey was based on observation on site as well as in-depth interviews with mayors and/or vice-mayors of slow cities. Nineteen slow cities were visited and qualitatively analysed. These 19 slow cities are, in alphabetical order: Abbiategrasso, Amalfi, Caiazzo, Cerreto Sannita, Chiavenna, Cisternino, Francavilla al Mare, Fontanellato, Giffuni Valle Piana, Giuliano Teatino, Orsaia di Puglia, Orvieto, Penne, Pollica, Positano, San Potito Sannita, Teglio, Tirano, Trani.

Our observations and interviews in these 19 Italian slow cities have clearly highlighted that the development strategy of a slow city starts from the recognition and promotion of local identity, of what is specific to the territory (Figure 16.1), culture and geography of places that determines a competitive advantage. Slow cities we have visited describe themselves from the perspective of their local identity: Positano as the summer fashion city in a unique landscape (Figure 16.2); Giuliano Teatino as the household recycling/composting town; Cerreto Sannita as the artistic ceramics town; Penne as the tailoring town; Francavilla al Mare as the Cenacolo City; Fontanellato as the fairs city; Amalfi as

Figure 16.1   Orvieto wine
Source: Tüzin Baycan 2010.

Figure 16.2   Positano: The Vertical City
Source: Tüzin Baycan 2010.

the historic maritime city; Teglio as the city of Pizzocchero Academy; Tirano as the city of Red Bernina Train; Chiavenna as the Crotti City and so on. These resources become a lever to promote original development trajectories based on 'relationality'.

The slowness culture that we observed during our survey assumes the notion of the centrality of the person that lives in 'relationship' with other people (Figure 16.3) as well as in relationship with the ecosystem – therefore living in a bundle of relationships/ interdependencies. The public spaces (squares, avenues, parks and so on) become central elements to improve the exchange of relationships. The natural/ecological ecosystems, but also the social ones are considered 'commons' to be preserved and enhanced. Commons and communities are strongly interdependent: one affects the other and vice versa, causing a whole range of consequences in terms of fighting against standardization and homogenization.

The slowness culture is based on the reconstruction of the local 'identity' in the globalized society of standardization. It promotes a regeneration of identity for the territory, enhancing agricultural production (especially organic), handmade production, gastronomy and so on, and preserves, at the same time, the landscape and makes it attractive for agro-touristic demand. However, slowness culture not only refers to the specificity/identity, but also promotes the 'processes of circularization', or in other words, the transition towards the 'green economy' that the experiences of, for example, Giuliano Teatino (recycling) and Penne (cultural tourism and bio-agriculture) have successfully demonstrated. In these experiences, cities have aggregated creative ideas, interests and goals around specific 'common goods' and, therefore, have promoted their

Figure 16.3   Mayor and residents, public square Penne

Source: Tüzin Baycan 2010.

conservation of natural and cultural landscape as well as their management system in parallel with the production of new common goods such as landscape, biodiversity, historical/artistic/cultural heritage and cultural diversity. These experiences have provided the starting point for a cultural transformation of the inhabitants through the reconstruction of a link between these new experiences and the territory. Thus environmental, historic, artistic values have been transformed into cultural/civic values.

These experiences also show the economic convenience of conserving common goods. The beauty of cities has been considered as having an economic value as it has an 'attractive power' affecting tourist demand as well as entrepreneurship, investment and specialized jobs. The beauty of cities also has cultural, social and civil value. Here, art has been considered important in terms of the capacity to live and to work together, to produce spillovers on the art of life, and on the capability to transform each inhabitant into an 'artist of citizenship' who is able to assess and combine conflicting elements in a creative way. These experiences reveal the concrete capacity of the slow cities network to combine choices in beauty and utility, beauty and fairness, rights and duties, private interests and general interests/common goods.

Around the 'commons', the experimental projects that synthesize tradition and modernity have been developed in order to activate new creative practices in which relations and connections are reconstructed while increasing the intensity of social relations. The carried out experiences have been able to enlarge the temporal and spatial horizon of the participants in their decisions due to the recognized importance of the role of the third sector: co-operative, social, civil economy. These experiences have also issued an interesting message, from a cultural perspective, that it is possible: 1) to construct a combined ranking of priorities with respect to multiple, heterogeneous and conflicting objectives shared by inhabitants; 2) to overcome the fragmentation and to stimulate cooperation processes, social networks and communities; 3) to invest successfully in the capacity for self-organization and self-government as the cooperation as well as the promotion of the common good is convenient for everyone; and 4) to integrate intrinsic and instrumental values into each other in the short, medium and long term. These experiences verify that a future can be built that 'goes beyond' the traditional perspective of the real estate economy and that particular and general interests can be woven together successfully.

The strategic approach of the slow city network can be interpreted as a pattern based on promotion of an increasing 'density of relations' and 'relational exchanges', starting from the enhancement of endogenous resources. This approach has improved the ability to be competitive with other cities, an ability based on the optimization of local identities and specialization of certain products, expertise and resources, all of which play an important role in: 1) increasing attractiveness to the outside; and 2) improving the capacity to export goods and services.

Concluding Remarks

Our study has investigated the experience of the 'Slow City Movement – Cittaslow' as an international sustainability practice from the perspective of cultural resilience and cultural planning. Our investigation has shown that the 'Slow City Movement' is one of the best examples of cultural planning in a globalization context as well as a cultural 'manifesto' to deal with the problematic cultural implications of globalization for cities. With this feature, the 'Slow City Movement' also offers one of the best examples of 'cultural resilience'.

On the basis of two surveys, our study has focused on policies and practices implemented by slow cities in order to better understand the 'Slow City Movement' and the critical success factors towards a sustainable local development. Although the availability of data and information is rather limited – due to the lack of information on websites and the limited number of slow cities that we could visit – the results of our surveys reveal a general framework of the culture constructed in these experiences. The results show that the most remarkable Cittaslow policy to emerge is a focus on local product. This can be seen as typical slow city behaviour, explained by the 'slowness' philosophy that emphasizes the protection and spread of local product, in other words, the 'uniqueness' of localities. The results also show that 'slowness' philosophy and policy help inhabitants to think in a different way – a systemic and a holistic way – and to change the traditional way of making choices. The culture of the 'Slow City Movement' stresses critical thinking to help improve the capacity for finding balance when making choices and promotes an evaluative culture among all people to consider not only instrumental but also intrinsic values. This culture extends the time perspective towards the long term and increases the central role of 'commons' such as the cultural and natural landscape and environment. On these 'commons', a pact has been built by people that stimulates a sense of belonging, participation and self-organization. This culture provides the critical energy to increase city resilience and thus to maintain sustainable development from the bottom up. Furthermore, this culture is based on relationships and on the principle of 'relationality' (Ravetz 2011). It promotes a balanced and creative relational approach to economic wealth production and wealth distribution as well as ecological wealth conservation. Moreover, the 'relationality' principle, stimulating a relational way of thinking (a relational rationality), promotes circular patterns/models. Thus it is possible to maximize synergies and synergy economies. The aim is to learn how to integrate differences, transforming them into complementarities via a win-win perspective. Different priorities are identified to integrate economic criteria with social and environmental criteria.

Although the 'slowness' strategy has been developed for small cities, it is transferable to medium-sized cities as well under certain conditions. This strategy may reduce the unsustainable impacts of urbanization in bigger cities by promoting intelligent/smart regional development that celebrates diversities and specializations and maintaining small scales in activities. In other words, these experiences can suggest original perspectives for cities of a larger size.

References

Baycan, T. 2011. Creative cities: context and perspectives, in *Sustainable City and Creativity: Promoting Creative Urban Initiatives*, edited by L.F. Girard, T. Baycan and P. Nijkamp. Farnham: Ashgate, 15–53.

Baycan, T., Akgün, A.A. and Erginli, B.E. 2012. Slow cities: a comparative policy framework analysis. *European Planning Studies*.

Baycan-Levent, T. 2010. Diversity and creativity as seedbeds for urban and regional dynamics. *European Planning Studies*, 18(4), 565–94.

Bianchini, F. 1990. Urban renaissance? The arts and the urban regeneration process, in *Tackling the Inner Cities*, edited by S. MacGregor and B. Pimlott. Oxford: Clarendon Press, 215–50.

——. 1996. 'Cultural planning': an innovative approach to urban development, in *Managing Urban Change*, edited by J. Verwijnen and P. Lehtovuori. Helsinki: University of Art and Design Helsinki, 18–25.

——. 2004. *A Crisis in Urban Creativity? Reflections on the Cultural Impacts of Globalisation, and on the Potential of Urban Cultural Policies*. Paper presented at the International Symposium: The Age of the City: The Challenges for Creative Cities, Osaka, 7–10 February 2004.

Bianchini, F. and Ghilardi Santacatterina, L. 1997. *Culture and Neighbourhoods: A Comparative Report*. Strasbourg: Council of Europe.

Bianchini, F. and Parkinson, M. (eds) 1993. *Cultural Policy and Urban Regeneration: The West European Experience*. Manchester: Manchester University Press.

Cittaslow. 2011. *Cittaslow International Network*. [Online]. Available at: http://www.cittaslow.org/ [accessed 4 April 2013].

Cooke, P. and Lazzeretti, L. 2008. *Creative Cities, Cultural Clusters and Local Economic Development*. Cheltenham: Edward Elgar.

Cunningham, S. 2001. From cultural to creative industries: theory, industry, and policy implications. *Culturelink Review* (special issue), 19–32.

Dreeszen, C. 1998. *Community Cultural Planning: A Guidebook for Community Leaders*. Washington, DC: American for the Arts.

Evans, G. 2009. Creative cities, creative spaces and urban policy. *Urban Studies*, 46(5–6), 1003–40.

Florida, R. 2002. *The Rise of the Creative Class*. New York: Basic Books.

——. 2005. *Cities and the Creative Class*. New York: Routledge.

Foord, J. 2008. Strategies for creative industries: an international review. *Creative Industries Journal*, 1(2), 91–113.

Fusco Girard, L. 2010. Creativity, resilience and sustainability: toward new development strategies of port areas. *International Journal of Sustainable Development*, 13(1/2), 161–84.

Fusco Girard, L., Baycan, T. and Nijkamp, P. 2011. *Sustainable City and Creativity: Promoting Creative Urban Initiatives*. Aldershot: Ashgate.

Fusco Girard, L. and You, N. 2006. *Città attrattori di speranza*. Milano: Franco Angeli.

Ghilardi, L. 2001. Cultural planning and cultural diversity, in *Differing Diversities: Transversal Study on the Theme of Cultural Policy and Cultural Diversity*, by T. Bennett. Strasbourg Cedex: Cultural Policy and Action Department, Council of Europe Publishing, 123–34.

Grogan, D., Mercer, C. and Engwicht, D. 1995. *The Cultural Planning Handbook: An Essential Australian Guide*. St Leonards: Allen and Unwin.

Hall, P. 1998. *Cities in Civilization*. London: Orion Books.

——. 2004. Creativity, culture, knowledge and the city. *Built Environment*, 30(3), 256–8.

Hawkes, J. 2001. *The Fourth Pillar of Sustainability: Culture's Essential Role in Public Planning*. Melbourne: Common Ground Publishing with Cultural Development Network (Vic).

Kunzmann, K. 2004. An agenda for creative governance in city region. *Disp*, 158, 5–10.

Landry, C. 2006. *The Art of City Making*. London: Earthscan.

Legacies Now and Creative City Network of Canada. 2010. *Cultural Planning Toolkit* [Online]. Available at http://www.creativecity.ca [accessed January 2012].

Markusen, A. 2006. *Cultural Planning and the Creative City*. Paper presented to the Annual American Collegiate Schools of Planning Meetings, Fort Worth, TX, 12 November 2006.

McNulty, R. 1991. *Cultural Planning: A Movement for Civic Progress*. Paper presented at the Cultural Planning Conference, Mornington, Victoria, Australia.

Mercer, C. 1991a. *What is Cultural Planning?* Paper presented to the Community Arts Network National Conference, Sydney, Australia, 10 October 1991.

——. 1991b. *Brisbane's Cultural Development Strategy: The Process, the Politics, and the Products*. Paper presented at the Cultural Planning Conference, Mornington, Victoria, Australia.

——. 1996. By accident or design: can culture be planned?, in *The Art of Regeneration*, edited by F. Matarasso and S. Halls. Nottingham and Bournes Green: Nottingham City Council and Comedia, 57–65.

——. 2006. Cultural Planning for Urban Development and Creative Cities. [Online]. Available at: http://www.culturalplanning-oresund.net/PDF_activities/maj06/Shanghai_cultural_planning_paper.pdf [accessed 4 April 2013].

Merkel, J. 2011. Ethnic diversity and the 'creative city': the case of Berlin's creative industries, in *The Ethnically Diverse City: Future Urban Research in Europe*, Vol. 4, edited by F. Eckardt and J. Eade. Berlin: BWV Berliner Wissenschafts-Verlag, 559–78.

Miele, M. 2008. CittàSlow: producing slowness against the fast life. *Space and Polity*, 12(1), 135–56.

Miles, S. and Paddison, R. 2005. Introduction: the rise and rise of culture-led urban regeneration. *Urban Studies*, 42(5–6), 833–9.

Musterd, S. and Ostendorf, W. 2004. Creative cultural knowledge cities: perspectives and planning strategies. *Built Environment*, 30(3), 189–93.

NSW Ministry for the Arts. 2004. *Cultural Planning Guidelines for Local Government*. Sydney: NSW Ministry for the Arts – Department of Local Government.

Palmer, R. 2004. *European Cities and Capitals of Culture*. Study prepared for the European Commission, Palmer-Rae Associates, Brussels.

Parkins, W. and Craig, G. 2006. *Slow Living*. Oxford: Berg.

Perloff, H.S. 1979. Using the arts to improve life in the city. *Journal of Cultural Economics*, 3 (December), 1–21.

Petrini, C. 2001. *Slow Food: The Case for Taste*. New York: Columbia University Press.

Pratt, A. 2008. Creative cities? *Urban Design*, 106, 35.

Quinn, B. 2005. Arts festivals and the city. *Urban Studies*, 42(5–6), 927–43.

Ravetz, J. 2011. *Shared Intelligence for the Urban Century*. Paper presented to the Informed Cities Forum 2011, Naples, Italy, 26–27 October 2011.

Scott, A. 2000. The cultural economy of Paris. *International Journal of Urban and Regional Research*, 24(3), 567–82.

—— . 2006. Creative cities: conceptual issues and policy questions. *Journal of Urban Affairs*, 28(1), 1–17.

Simmie, I. 2001. *Innovative Cities*. London: Spon.

Stevenson, D. 2004. Civic gold rush: cultural planning and the politics of the third way. *International Journal of Cultural Policy*, 10(1), 119–31.

UNCTAD. 2004. *Creative Industries and Development*. Washington, DC: UN.

—— . 2008. *Creative Economy Report 2008: The Challenge of Assessing the Creative Economy Towards Informed Policy-making*. Washington, DC: UN.

Zukin, S. 1995. *The Cultures of Cities*. Oxford and Cambridge, MA: Blackwell.

—— . 1998. Urban lifestyles: diversity and standardization in spaces of consumption. *Urban Studies*, 35(5–6), 825–39.

PART 5
Cultural Practices

Preface to Part 5

Deborah Stevenson

The chapters in this Part of the book are concerned what are amongst the most critical challenges facing contemporary cities and urban planning – how to support, understand and manage the cultural practices of urban populations. At stake is the creation of environments where different social groups can live, work, mingle, and express their cultures and values. Also important is the equitable distribution of urban resources. When contemplating such issues, attention frequently turns to public space because this is where the texture, complexity and diversity of urban life is most readily revealed. And so it is with the chapters here, which in differing ways acknowledge that the quality, use and presence of urban public space are central to the way in which people experience and relate to the city as resident or visitor, and whether encountering it in the context of work, education or leisure (to name but three). Public space is also linked to the development and maintenance of civic culture. Parks, footpaths, beaches, the verges of rivers and waterways are all important in this respect, as are roads and thoroughfares – the sites of intersection, movement and transportation.

What is understood as public space also embraces publicly funded institutions, including museums, libraries and galleries, and town halls and civic centres, as well as leisure spaces, such as swimming pools. These are the places that are owned collectively by residents and managed by local councils, state or national governments and which are entwined with ideas of citizenship and urban culture. Too often, however, far from seeking to encounter diversity in public space, people actively try to avoid it and, increasingly, urban policy and planning are used to support this avoidance. Different social groups inhabit different areas of the city for work, home and leisure; they trace different paths through cityspace often in an effort to avoid encountering the 'other'. This is a city of watchfulness and suspicion comprised of enclaves of homogeneity where the mingling with strangers is eschewed. For many contemporary city dwellers, therefore, the urban environment is one that is controlled, homogeneous and predictable – a space of managed diversity.

In the first chapter in this Part of the book, Setha Low examines the use, planning and policing of public space in order to gain an understanding of urban diversity and its consequences for social equity. She argues that in the United States in particular, as a result of economic restructuring and the hegemony of neoliberal urban policies, cities have become highly segregated with people living in different parts of the city having inequitable access to a range of services and opportunities. Public space has also been transformed because of the withdrawal of the state from its provision and maintenance. The result is the deprivation of public space in many poorer areas and its privatization and subsidy in others. With reference to two case studies, Low suggests that in order to promote cultural diversity, social inclusion and increased equity what is needed is an expanded concept of social justice grounded in a detailed understanding of how public space is used and regarded by locals and visitors. Also important to her analysis is to probe the tensions that have emerged between the contradictory forces of globalization and 'venacularization', which is the inscribing of culture and meaning on local landscapes in the context of global flows of labour and capital.

Public space, its uses and fragility are also central themes in the chapter by Ronan Paddison who regards it as being central to the structures of feeling which residents have towards the city, which means that changes in public space have considerable consequences for the way in which people relate to their city. He argues that the increased privatization of urban space has affected the ways in which public space is produced, reproduced and used and explores the consequences of such changes to public space. The chapter focuses on considering the repercussions of two practices of 'aesthetic intervention' which have become commonplace in cities in the global North and, increasingly, in the global South – the use of iconic architecture and public art. He suggests, that contrary to what is officially endorsed or expected, the use of space and the consumption and interpretation of art and architecture can be sources of resistance, meaning and the assertion of the local.

Picking up on some of the issues probed by Low and Paddison, Torill Nyseth's chapter examines city imaging and the place identity as central elements of pervasive global competitions between cities to attract capital and people. Nyseth draws a nuanced distinction between three different strategies of place promotion – place marketing, place branding and place reinvention – suggesting that where place marketing treats place as a commodity and place branding is focused on management and place-making, place reinvention captures the complexity of place in a way that engages with its symbolic and imaginative dimensions. Nyseth explores these themes and distinctions with reference to Norwegian examples, suggesting that place reinvention undertaken in response to economic and industrial change can involve the renegotiation of local identity and a celebration of local cultures. It can be the result of both planning as well as of serendipity.

Concluding this Part of the book is a 'case study window' by John Montgomery which examines the trend of establishing cultural quarters as part of city imaging and regeneration strategies. Montgomery argues that successful cities are those with a strong place identity and a rich cultural life and where there is a commitment to developing the 'creative economy', and in this regard he argues that a well designed and managed cultural precinct or 'quarter' can be a useful element. He suggests that a thriving cultural quarter will have three key intersecting features – activity (use, venues, events, workshops), built form (public space, architecture and buildings, urban design) and meaning (gathering spaces, place identity and its markers, heritage). He then goes on to illustrate how these elements in different combinations have been pivotal to the success of the cultural quarters in Dublin, Manchester and Sheffield.

Themes of equity, diversity, place meaning and identity recur in this section of the book. Although differing in their concerns and emphases, the chapters all contribute to an engaged and probing understanding not only of the diversity and complexity of urban cultural practices but also of their significant spatial dimensions and embeddedness. Public space, its quality and accessibility, is important in this context because it is in public space in all its configurations that local cultures and democracy can be shaped, expressed and affirmed.

Public Space and Diversity: Distributive, Procedural and Interactional Justice for Parks

Setha Low

Introduction: Neoliberal Changes in New York City Public Space

Public space offers an empirical means for thinking about diversity in the creation of a more just city (Fainstein 2000, Low, Taplin and Scheld 2005). It is here that race, class, gender, age, sexual preference, ethnicity and ability differences are experienced and negotiated in a safe forum for political action, communication and democratic practice (Low 2000, Young 2001). Difficulties encountered in defining and studying what constitutes an equitable distribution of public space necessitate employing a broader framework of justice to utilize the lessons learned from planning and design practice and to encourage the use of public spaces for democratic practices (Fincher and Iveson 2008, Low and Smith 2006, Mitchell 2003).

In the United States, particularly in New York, there have been major transformations in public space due to neoliberal urban policies and economic restructuring. First, there are more immigrants and ambulatory vendors using public spaces due to increased immigration and expansion of the informal economy sector. Second, there is greater heterogeneity in the city due to the global relocations of labour and capital, while at the same time there is more social segregation in neighbourhoods. Third, even with increased urban heterogeneity, individuals may not have an opportunity to interact face to face because the economics and politics of home ownership and rental restrictions produce ghettos for the poor and secured communities for the wealthy. Fourth, consequently there are escalating tensions between the processes of globalization and vernacularization–particularly in the ways in which people reinscribe culture and meaning on the local landscape (vernacularization) while also participating in global flows of capital and labour (globalization). (See also Newman and Thornley, and Searle, Chapters 4, and 8, this volume.) Fifth, economic restructuring and globalization create a 'dual city' with social and economic disparities in economic opportunity, goods and services available to people residing in different sectors of the metropolitan region. And finally, concurrently the city and state have less money for the operation and maintenance of public spaces and only those that are subsidized by private conservancies and individuals continue to prosper, while playgrounds and parks in poorer and marginal neighbourhoods suffer from neglect.

New York has responded to these changes in a number of ways. A new structure of feeling and a fear of others has emerged which is symbolized by City Hall Park in New York City to the extent that it is now gated off with a 'do not enter' sign. Parks and plazas are being privatized and run by corporate managers so that places like Herald Square in front of Macy's Department Store on 34th Street is now surrounded by a high fence with a gate. Rules and regulations are listed including a bronze sign that reads 'Do not sit' next to a sitting wall designed for just this purpose. Herald Square is closed at 6:00 pm each night so that the teenagers and young adults, who used to hang out there, are kept out. Video surveillance cameras are also trained on all publically available spaces. For example, 265 surveillance cameras focus on Union Square, a very popular small park at 14th Street. Post 9/11 there has also been increased policing and zero tolerance arrests. And there are new historic preservation practices in parks and other public spaces that reinsert elite class values into the landscape (Kayden 2000, Low, Taplin and Scheld 2005, Shepard and Smithsimon 2011).

Access and use of these public spaces has suffered accordingly. People of colour avoid policed and secured squares, thus, there is an absence of young men of colour in these spaces (Fiske 1998). Low-income individuals read the symbolic cues in privatized spaces that may have long lists of rules, physical deterrents on benches to limit resting or ornate landscape features that make these users uncomfortable and thus they avoid these newly restrictive landscapes. Middle-class individuals and tourists are happier with highly controlled and managed public spaces, while teenagers avoid these same places because of their long lists of rules and regulations. Homeless individuals are locked out of gated public spaces at night so they have no place to stay. Individuals who are perceived as 'Middle-Eastern' or 'Arab' in appearance worry about police profiling and surveillance cameras and are stopped for questioning making them more hesitant to spend time there (Miller 2007, Shepard and Smithsimon 2011). Immigrant vendors are having a difficult time because of both the new regulations and the degree of policing and surveillance (Chesluk 2008, Miller 2007).

These dramatic changes started over 20 years ago, but accelerated during the post-9/11 years as part of the growth of neollberalism characterized by liberal governance, entrepreneurship and placing responsibility for social reproduction on the individual and local community (Harvey 2005). In the United States, neoliberalism is accompanied by homeland security practices, such as surveillance, policing and restrictive urban design, which is creating a new form of civil society called civil militancy in that it protects the home and the homeland, but abandons public space and the public sphere (Ruben and Maskovsky 2008). It celebrates the private city. The new structure of feeling based on fear mongering, war and terrorism underlies these changes (Sorkin 2008). Thus in this historically and politically fraught moment, public space in the United States has become even more important for marginalized groups and for assuring the continuance of democratic practices and urban social justice.

Of course these changes have not occurred only in the United States. Similar changes have been observed in public spaces throughout the world including in South Africa

(Banks 2011), Latin America (Low 2000), Australia (Fincher and Iveson 2008), Canada (Blomley 2004) and many Western European countries (see Low 1999 for various cases). (See also Paddison, Chapter 18, this volume.) In Costa Rica, where I had previously completed a 15-year study of plazas (Low 2000), similar public space changes were observed, but handled in a different manner. When municipal planners decided that Parque Central, the original plaza mayor – the main square – of downtown San Jose, the capital city of Costa Rica, should become a space that would attract middle-class US tourists and residents, they closed it down, remodelled, and then policed it upon its reopening so that the lower-class users, shoeshine men and pensioners could not reoccupy their places. The new design created a more barren space where homeless or men from the countryside could not sleep because of small or curved benches, and cut down most of the shade trees that had made the plaza comfortable. All of the original users were forced to relocate to less desirable spots during the renovation, and subsequently the police kept them from re-entering. As this example illustrates, the details of how public space is transformed to become less diverse are specific to each context and worth exploring comparatively, however, in this paper I focus on ethnographic cases drawn from urban parks in the northeastern United States.

The Just City: Distributional, Procedural and Interactional Justice

With the corresponding changes in public space due to neoliberalism and a climate of fear, it is necessary that social scientists, managers, planners and designers develop clearer arguments about the substantive basis by which these changes can be considered unjust. Two models have been suggested: the first is Susan Fainstein's (2004) model of the just city based on Lefebrve's (1991) 'right to the city' and Nussbaum's (2000) concept of capabilities. Rather than arguing for diversity as the principle goal of urban planning, Fainstein (2004) focuses on the inevitable trade-offs among notions of equity and diversity as well as sustainability and growth. The strength of her argument is that it moves urban planning from a normative to a utopian framework, and asserts that we should focus on under 'what conditions can conscious human activity produce a better city for all citizens' (Fainstein 2005: 121).

A second model is offered by Ruth Fincher and Kurt Iveson (2008) on the role of planning in the struggle for spatial justice in the city. They assert 'the need for planning frameworks to take adequate account of *diversity* of cities, by identifying and working with the many publics which inhabit them' and thus agree with the goals of Fainstein (2005), Dear (2000) and Campbell (2006). But they proceed by defining the three kinds of intersecting diversities that must be addressed including differences in wealth, status and hybridity, that is, the range of possible identities available to any one group. This thoughtful analysis of difference forms the framework for their three goals for socially just planning: 1) redistribution of space, services and facilities, 2) recognition of the limitations of redistribution in the valuing of cultural differences and contribution, and 3) encounter of people and identity group. Their approach argues for a more relational approach to diversity (Fincher and Iveson 2008).

My work on social justice derives from ethnographic research on urban parks, plazas and gated communities (Low 2003, Low and Smith 2006, Low, Taplin, and Scheld 2005), the just city concept as proposed by Fainstein (2005), and my critique of Fainstein's contention that diversity should not be at the centre of a social justice analysis of public space. A further criticism of Fainstein's proposal, however, is that she utilizes too narrow a definition of justice such that her utopian aims are only partially fulfilled. Also, like Fincher and Iveson (2008), but developed from the social psychological and industrial-organizational literature, I argue that three dimensions of justice – distributive, procedural and interactional – are essential to address the multiple kinds of perceived injustice.

In the urban planning literature of the United States and more recently Western Europe, advocating for a just city was based solely on distributive justice. Distributive justice refers to questions of how the wealth, rewards, benefits and burdens of society should be distributed to achieve an economically just city. The discussion revolves around whether the economic benefits and burdens should accrue to individuals equally, according to need, according to merit, or to those who are the least well off (Lefkowitz 2003, Rawls 1971). 'Equity theory argues that fairness means that people's rewards should be proportional to their contributions' (Tyler 2000: 118) in the hopes that by making fair allocations fewer conflicts will arise.

In terms of public space, distributive justice based on equity would ensure that public space was available to all people and that everyone would have some degree of access. Of course by distributive justice standards, New York is not a just city in that many poorer and low-income neighbourhoods have no public space at all, and the spaces that exist are inadequate for local needs or privatized and therefore expensive to use. While at the same time, wealthier parts of the city have a broad range of public space available as well as full access and the financial resources to enter more privatized locations.

Even if New York City could distribute public space more fairly, a distributive justice model in and of itself would not create the just city, in that there are other issues of citizenship related to fairness. Specifically, there are two additional kinds of justice that must be considered: procedural and interactional.

The term 'procedural justice' refers to the way that the processes of negotiation and decision making influence perceived fairness by individuals. Psychologists have found that distributional outcomes are not the only relevant issue when determining fairness (Tyler and Blader 2003). The way that a person is treated is equally important. While early research on social justice supported the findings that people felt most satisfied when outcomes were distributed fairly, subsequent research found that distributive justice outcomes were often biased, and that the favourability of an outcome is less crucial when the underlying allocation process is perceived as fair (Cropanzano and Randall 1993, Tyler and Blader 2003).

An example of procedural justice is the US legal system and the formal procedures related to decision-making processes in court settings (Thibaut and Walker 1975). If a person perceives the legal system and its procedures as fair, then the complainant is more willing to accept the outcome regardless of whether it is the outcome initially desired. The case of Prospect Park, a large urban park in Brooklyn, New York, offers a public space illustration. Frederick Law Olmsted had designed Prospect Park in 1867–73 to resemble an English landscape garden using large-scale earth moving techniques to form so called 'natural features' such as the pond and wetlands area that over time had deteriorated due to poor park maintenance and neglect. The administrator of the park applied for funding to restore this area without consultation with some local residents. There are two neighbourhoods surrounding the park, one populated by mostly African Americans and Caribbean Americans in Crown Heights, and one populated by mostly White professionals and Hispanics in Park Slope. Snow fences (light heavy wooden fences used to mark locations that should not be snow ploughed or damaged under snowy conditions) were put up by the park managers to allow for an ecological reconstruction of this Olmsted-designed water feature, and were not a problem for the Park Slope residents. But in our interviews with residents of Crown Heights we found that these snow fences were perceived as a way for the park managers to keep 'Black' people out of the 'White' part of the park. This sense of injustice was created by the lack of procedural justice for Crown Heights residents – they were not consulted or informed before this reconstruction project started, while Park Slope residents, many of whom were members of the Prospect Park Conservancy, participated in the decision making. Not including Crown Heights residents in the process produced charges of racism even though the fences kept everyone out. In other words, from a distributive point of view, no one could use these spaces while they were being reconstructed, but the difference in the procedure by which the project was communicated and decided influenced the reaction of residents in both neighbourhoods.

The concept of interactional justice is about the quality of interpersonal interaction in a specific situation or place. Psychologists find that to a large extent, individuals make justice appraisals based on the quality of interactional treatment they receive (Cropanzano and Randall 1993). Attributes of interactional fairness include truthfulness, respect, propriety and justification (Bies and Moag 1986). In terms of public space, interactional justice is reflected in whether people are being treated in a discriminatory way, or whether they are targets of harassment, insults, or other rude behaviour. In the case of Union Square, a square in the middle of New York City, young men of colour and people who look Middle Eastern or 'Arab' by US phenotypic stereotypes are followed by the police, spoken to rudely and asked inappropriate questions. It creates a hostile environment for them so they often avoid the place. Similarly, teenagers in public markets and in semi-public malls are often followed by private guards and asked whether they intend to buy anything. This pattern also influences their desire to use the malls as public space, even when these are sometimes the only spaces in the suburbs available for them to hang out, meet their friends, go to a movie or go out to eat.

If we want to talk about fairness in public space, we need to consider all three of these types of justice and ask the following questions:

- Distributive justice – Is there equal public space for everyone? Is there a fair allocation of public space resources?
- Procedural justice – Is there a way to gain access to public space? Is there a fair system for applying to use the park grounds for games or picnics? Can all people apply for these rights? Do park managers consult equally with diverse residents before deciding on a plan of action? Are park managers sure that all residents understand the reason for any particular action, especially those that restrict the use of park resources?

- Interactional justice – Does the public space allow for all individuals to interact safely? What about women interacting with men? Can children interact with everyone freely? Can teenagers? Can young men of colour? Are people treating each other in a way that promotes a sense of citizenship, equality and social justice?

An Ethnographic Approach to the Study of Social Justice and Diversity

One way to work towards this expanded concept of social justice, is to collect better information about how public space is being used and thought about by local residents and visitors with diverse ethnic, class, age, ability, racial and gender identities. (See also Watson, Chapter 1, this volume). William H. Whyte (1980) set out to find out why some New York City public spaces were successes, filled with people and activities, while others were empty, cold and unused. After seven years of filming small parks and plazas in the city, he developed 'his rules for small urban spaces' based on what he had observed. He found that only a few plazas in New York City were attracting daily users and saw this decline as a threat to urban civility. Instead he began to advocate for viable places where people could meet, relax and mix in the city. His analysis of those spaces that provided a welcoming and lively environment became the basis of his famous rules of how to make small urban spaces work. And these rules were used by the New York City planning department to transform the public spaces in the city.

The Public Space Research Group, in response to the problems that we have observed in larger parks, has expanded Whyte's initial ideas and developed a series of principles that encourage, support and maintain cultural diversity. They include some planning guidelines that are similar to Whyte's rules for small urban spaces that promote their social viability, but in this case, we call them 'lessons learned' based on our ethnographic fieldwork, and suggest that their application will promote and maintain cultural diversity. The lessons from our studies are not applicable in all situations, but are meant as guidelines for park planning, management and design. They can be summarized in the following six statements and examples.

First, if people are not represented in urban parks, historic national sites and monuments, and more importantly if their histories are erased, they will not use the park. For example, one of the reasons that African Americans do not visit Independence National Historic Park in Philadelphia is that their institutions and public spaces were torn down with the building of the colonial park and are not marked in the contemporary site.

Second, access is as much about economics and cultural patterns of park use as it is about circulation and transportation, thus income and visitation patterns must be taken into consideration when providing access for all social groups. An illustration is the explanation of why poor African Americans living in Jersey City just 400 yards from the Ellis Island Museum do not visit even though they say they would like to: the cost of the ferry ride ($10 per person). This high fare is complicated by the fact that most family activities include the extended family of more than ten people, so that the cost and pattern of use together make it financially impossible for a low-income family to visit, even when there is no entrance fee.

Third, the social interaction of diverse groups can be maintained and enhanced by providing safe, spatially adequate 'territories' for everyone within the larger space of the overall site. Prospect Park, a culturally and socially diverse New York City park is successful in maintaining its diversity because there is adequate room for a variety of activities and groups from open playing fields for soccer and Frisbee as well as baseball diamonds, picnic areas, secluded groves for lovers, semi-secluded areas for Muslim mothers and children who like to stay away from men, and a loop road for walkers, runners, bicycle-riders and roller-bladers where everyone can mix.

Fourth, accommodating the differences in the ways social class and ethnic groups use and value public sites is essential to making decisions that sustain cultural and social diversity. For example, being aware of different user values – from flower lovers, family-focused activities, naturalists and bird-watchers – allows park managers to accommodate all users in their decision-making process and provide for each.

Fifth, contemporary historic preservation should not concentrate on restoring the scenic features without also restoring the facilities and diversions that attract people to the park. The case of Central Park's restoration of Olmsted's vision illustrates how elements of the park such as the grounds and bandstand can be restored without the social programmes Olmsted intended. In Prospect Park the boathouse no longer has boats and food is no long sold in the places where it used to be. So part of the historic vision of what it would mean to recreate a vibrant social park is eliminated. Unfortunately only restoring the historic design elements and reconstructing a classic turn-of-the-century park is seen as elite and exclusionary by some groups of people who do not share this historic vision, but look to parks for the provision of outdoor activities and family programming.

Sixth, symbolic ways of communicating cultural meaning are an important dimension of place attachment that can be fostered to promote cultural diversity. For example, playing salsa music every Saturday night in the summer at Pelham Bay Park in the Bronx has encouraged Latino immigrants and local Spanish-speaking residents to enjoy the space and claim it as their own.

These principles for promoting and sustaining cultural diversity in urban parks and heritage sites are just a beginning. More research is required to understand the importance and difficulties of maintaining a vibrant public space. (See also, Ashworth, Chapter 11, this volume.) But at the very least, they demonstrate how diversity can be an essential component of evaluating the just city. In the remainder of this chapter I illustrate how our ethnographic approach uncovered distributive, procedural and interactional injustice in two National Park Service parks and heritage sites that the Public Space Research Group studied.

Independence National Historical Park

In 1994, Independence National Historical Park began developing a general management plan that would set forth a basic management philosophy and provide strategies for addressing issues and objectives over the next 10 to 15-year period. The planning process included extensive public participation including a series of public meetings, televised town meetings, community tours and planning workshops. As part of this community outreach effort, the park wanted to work cooperatively with local communities to find ways to interpret their diverse cultural heritages within the park's portrayal of the American experience. This study was designed to provide a general overview of park-associated cultural groups, including an analysis of their concerns and the identification of cultural and natural resources used by and/or culturally meaningful to the various groups.

Park-associated groups including African Americans, Jewish Americans, and Italian Americans, whose recent ancestors previously lived in the general area, were identified as the initial groups for contact. These groups were selected because the area has had special importance for them. Other cultural groups such as Asian Americans and Hispanic Americans were included because they were identified as rapidly growing communities who use the park grounds for ceremonial and recreational purposes, and thus would be affected by the proposed changes to Independence National Historical Park.

Four local neighbourhoods were selected for study: Southwark for African Americans, Little Saigon for Asian Americans, the Italian Market area for Italian Americans and Norris Square for Hispanic Americans based on the following criteria: 1) they were within walking distance from the park (excluding Norris Square); 2) they had visible spatial and social integrity; and 3) there were culturally-targeted stores, restaurants, religious organizations and social services available to residents, reinforcing their cultural identity. We selected the Vietnamese American community to represent the Asian American cultural group because of its proximity to the park, and its recent population growth.

A number of methods were selected from the Rapid Ethnographic Assessment Procedure (REAP) methodology to generate data from diverse sources that could then be integrated to provide a comprehensive analysis of the site. Behavioural maps recorded people and their activities located in the park throughout the day and early evenings on weekdays and weekends. Transect walks recorded what identified community consultants described and commented upon during guided walks across the site. Individual interviews based on the study questions were completed in Spanish, English or Vietnamese. Expert interviews were collected from individuals such as religious leaders, local historians, historic preservation specialists and tour guides identified as having special expertise to comment on the cultural significance of Independence National Historical Park. Focus groups were set up with major religious institutions in the neighbourhoods – churches and synagogues – as well as with active community organizations. Historical and archival work accompanied all phases of the study, and newspaper clippings, articles in local magazines and other media-generated materials were collected. The data were then organized by coding all responses from the interviews, and then analysed for content by cultural/ethnic group and study question.

From the findings we learned that African Americans are the most concerned about their lack of cultural representation in the park's colonial history, that Asian Americans and Hispanic Americans are less directly concerned but would like to see their stories integrated as part of the American experience, while Italian Americans and Jewish Americans are at best ambivalent about presenting themselves as distinct from other Americans. Three of the cultural groups, African Americans, Hispanic Americans and

Jewish Americans, mentioned places they would like to see commemorated or markers they would like to see installed to bring attention to their cultural presence within the park boundaries. And many of the cultural groups – particularly the Hispanic Americans, the African Americans and the Asian Americans – were anxious to have more programming for children and activities for families. The Hispanic Americans were particularly interested in the recreational potential of the park, and their sentiments were echoed by at least a few consultants in each of the other cultural groups.

Overall, there are distinct messages from each cultural group, as well as general preferences that relate to the majority of the groups. They include: 1) most of these local residents do not use the park except to take visitors, and although many have fond memories of the past some now find the park unsafe; 2) most cultural groups feel that the park's meaning is related to the struggle for freedom and relate this history to their own histories; 3) some cultural groups have appropriated the symbol of the Liberty Bell and given it their own cultural meanings; 4) many cultural groups feel excluded from the park because of the lack of cultural representation and identification; and 5) most of the cultural groups would like more participation in the park.

The case study of Independence National Historical Park demonstrates that cultural representation of cultural groups is critical to their use of and relationship to the park. The erasure of history documented for the African American and Jewish American communities, and the exclusion of the Hispanic American and Vietnamese American communities through monolingual programmes and signage illuminates how cultural/ethnic groups respond to the cues and affordances of the physical and social environment. If we want culturally diverse groups to participate in designed public spaces, then it is the responsibility of the designers and planners, as well as the federal, state and municipal governments, to take seriously the words of these consultants: 'Design places that erase our history, and/or create places that exclude us in subtle ways, we will not come.' Cultural representation in urban space is material evidence of its history and local politics of exclusion. Urban parks provide social and environmental mnemonics that communicate who should be there, and historical buildings and places, markers and monuments set the stage for human behaviour.

This also provides evidence of how planning and design practices of historic preservation can disrupt a local community's sense of place attachment and disturb expressions of cultural identity for local, ethnic populations. New ethnic and immigrant groups can be excluded because of a lack of sensitivity to cultural barriers such as an inability to read or speak English, non-verbal cues of formal furnishings and dress, as well as signs of cultural representation. Understanding the intimate relationship between ethnic histories, cultural representation and park use is critical to successful design and planning in any culturally diverse context.

Jacob Riis Park: Conflicts in the Use of a Historical Landscape

Jacob Riis Park is a beach with a boardwalk and playground and concession facilities located in Gateway National Recreation Area in southwestern Queens. The park is named for Jacob Riis, a late nineteenth-century reform figure famous for his photographic documentation of life among immigrant children in the Lower East Side slums of New

York. The National Park Service inherited Jacob Riis Park from the New York City Parks Department in 1974 to be incorporated within Gateway National Recreation Area. The National Park Service has made some improvements, but deterioration rather than upkeep is the dominant note. Part of the problem with park maintenance is the difficulty of obtaining funds and federal support because there is no political constituency for urban parks at the national level. Further the majority of new park users are recent immigrants who do not necessarily speak English or have the citizenship status or knowledge to demand facilities or services. Even more problematic for the National Park Service is that the new users' recreational and social needs do not fit the historical landscape design of the park. Because of park deterioration, lack of funding for new facilities, and changing visitor use patterns, attendance fell off sharply during the 1990s.

The Public Space Research Group was asked to conduct a Rapid Ethnographic Assessment Procedure (REAP) study at Riis Park for the National Park Service as part of an effort to understand this decline in park use. Specifically, park managers wanted to know more about their new immigrant users and how to meet their social, recreational and cultural needs. Our goal was to find ways to enhance their park experience with an idea to increasing the number of users and accommodating their activities within their limited budget and historic landscape constraints.

At Riis Park, the boardwalk, the bathhouse and other buildings along the boardwalk, the parking lot and the landscaped grounds constitute the Jacob Riis Park historic district. The effort to tether the park to a 'period of significance' leaves management less able to be flexible in adapting to changing needs. While a non-historic playground can be rebuilt or eliminated as needed, the historic 72-acre parking lot, which is more than half empty on even the hottest summer days, must be preserved. As of 2000, the Park Service had spent $15 million to reconstruct portions of the bathhouse structure in keeping with historic preservation standards, yet the public still has no place to shower and change clothes.

Picnicking – and people and cultural groups who take part in this activity – is the point of sharpest conflict at Riis Park between contemporary park use and a management policy based on historic preservation. The Rapid Ethnographic Assessment Procedure helped to understand the new immigrant users so that the park could begin to find ways to accommodate their picnicking needs while at the same time providing the necessary services for long-term beach users and protecting the historic landscape.

A number of REAP methods were used during the various phases of the research process. Individual interviews were completed in Spanish, Russian or English, depending on the preference of the interviewee. The interviewers had a map of the park available for noting any specific site and to stimulate discussion about Jacob Riis Park. Expert interviews were collected from park staff and volunteers identified as having special expertise to comment on current park problems.

Overall, Jacob Riis serves a diverse population of users, from recent immigrants with limited economic means who are coming for the first time, to wealthy professionals who have visited the park for over 20 years. Household income, education and occupation is evenly distributed, so that from a socioeconomic point of view, it is truly a park for everyone. Even the distribution of first-time and once-a-summer users (32 people out of 131, or 24 per cent) and frequent users – people who come one to four times each week (46 people out of 131, or 35 per cent) – is somewhat balanced. There are more long-term users than first-time users (25 people have been coming for more than 20 years compared to 17 first-time users), but the park still attracts an influx of newcomers while retaining strong ties to long-terms users who come from local neighbourhoods.

At the same time there are distinct territories claimed by their users, and these territories and users have distinct needs and desires. To develop a plan of action for renovation and change, it is important to understand that the different areas of the park – the bays and back beaches – have clearly articulated, yet distinct concerns. From the perspective of planning and design, it is difficult, therefore, to develop one proposal that will meet the needs of all constituents.

For instance, the back beach picnickers are the newest visitors to the park. These visitors – many of whom speak only Spanish and have recently come to the United States from Central and South America – are the poorest and least able to provide services and resources for themselves. They need more picnic areas equipped with tables, shade (trees, tents or cabanas), and grills. Bathrooms for the large number of children and elderly in these families, safe playgrounds so that children can play nearby and be supervised, and adjacent beach lifeguards are all necessary for their visit to be optimally successful and satisfying. These newcomers prefer to visit as large groups of families and friends. They prefer shade and are accustomed to hanging fabric for shade and hammocks for seating.

Currently the historic landscape does not accommodate these visitors' desire for shade and large gatherings. The limited number of dying trees must be replaced with the historic black pines that provide little shade, and there are very few picnicking areas in the original 1930 plan. Robert Moses intended that visitors would sit on blankets on the beach and never envisioned (and never wanted) large groups of people barbequing at what he designed as a 'middle-class' park. In order to accommodate the new visitors' needs, creative designs and better communication between the park and these users will be necessary to solve this culture/landscape dilemma.

Further, immigrant visitors are concerned that their children learn as much as possible about their new life in the United States, so programmes including swimming, safety, as well as educational and recreational programmes are desirable. But many of the parents cannot read or write English, so letting them know about the programmes requires specifically targeted and innovative forms of outreach. Finally, these groups enjoy music and dancing – especially Latino rhythms and salsa – and would enjoy summer afternoon concerts that remind them of home (and bring a bit of home to their new beach). Because these newcomers are the poorest visitors, and a population that Jacob Riis can effectively serve, their needs should be the focus of the first intervention made in the park.

On the other hand people who use the beach bays would prefer not to see limited park resources distributed in this way, and have no interest in picnic areas and educational or cultural programming. Instead, short-term changes should focus on lifeguards, bathrooms and garbage cans. Lifeguards are a top priority for the safety of visitors and because visitors identify with just this one particular part of the beach. For a variety of reasons – gay identity, tolerance and safety issues – Bay 1 visitors have less flexibility of movement at the beach. Therefore, when there is no lifeguard these visitors are at risk. While immediate funding for another lifeguard may not be feasible, an arrangement could be made whereby a lifeguard would be posted at Bay 1 several times a week.

People using the beach bay – rather than the picnic areas – felt strongest about the condition of facilities behind the beach: the rough concrete on the boardwalk with weeds coming through the cracks and sand drifts on the surface, the deteriorated ball courts, the closed bathhouse. The basic amenities are most important: a smooth, resurfaced boardwalk, free of sand; a railing in good condition, and basketball and handball courts in good working order. More showers – ideally in the bathhouse, with some changing facilities were their highest priorities.

Most visitors to the beach bays complain about the lack of restrooms or showers or type of food concessions. Perhaps there are not enough visitors to justify such amenities, but even some improvements would likely attract new visitors. For example, users like the quiet, secluded atmosphere located at the farthest beach bay, but providing the basic amenities – bathrooms and showers – would change the atmosphere too much, and there are more bays to the west for people who want real seclusion. Almost all the users of the beach bays were not concerned with picnic tables, grills, playgrounds or music. These visitors come to Jacob Riis for the beach and swimming, so their concerns focus on the cleanliness of the beach, the availability of bathrooms and showers, and having a lifeguard nearby. The focus is on individual activities rather than family oriented activities. Other than bathrooms, the needs of the different cultural groups are distinct.

The REAP at Jacob Riis Park, thus, uncovered the conflicts that arise when cultural and social groups compete for very limited resources in a restricted, historical landscape. But at the same time, Jacob Riis has been extremely successful in attracting a wide variety of users by offering diverse niches that allow people to enjoy the park. Even though visitors would chose different improvements, they all agree that it is a wonderful park that accommodates their activities and cultural patterns of park use without conflict with others. And the boardwalk provides a place for everyone to come together and experience the diversity of users in a safe and supportive way.

The contradictions in the National Park Service policy towards urban parks, particularly beaches such as Jacob Riis, can be read in the neglect of the contemporary landscape. Even the good intentions of management and staff and the local efforts of committed users can not reverse the lack of funding and attention to the park's deterioration. Yet at the same time, Jacob Riis Park works as a beach that serves recent immigrants and poor to middle-class residents of Brooklyn and Queens. The spatial organization of beach bays and back beach areas creates territories that encourage a strong sense of stewardship and place attachment. And at the same time, these 'territories' promote social tolerance and cultural integration at the level of the site.

Social Justice and Diversity in Public Parks

These two case studies illustrate how a social justice analysis that includes distributive, procedural and interactional dimensions promotes cultural diversity and a sense of inclusion and fairness in urban parks. From a distributive justice perspective there was not equal public space for everyone. At Independence Historical National Parks the remains of African American historical presence were destroyed and not recorded in this national vision of US colonial history. Similarly, new immigrants such as the Vietnamese and the Puerto Rican communities did not have any claim through language or symbolic marking for use of this space. They were not overtly excluded but the lack of representation and programming defined this National Park Service park as 'White' and 'for tourists'. At Riis Park the unequal distribution of space is also determined historically through a system of historical preservation of a design that gives priority to the beach and little thought or space for the picnicking or play needed by new immigrants. Further, many of the beach bays did not have lifeguards, bathrooms or concessions while other, whiter or heterosexual, parts of the beach did. Thus, in both parks some users and residents did not have accessible or welcoming public space.

Procedural justice is a problem especially at Jacob Riis Park where the immigrant users are not able to participate in the designation of park design and management. In a sense, the REAPs that we undertook are one step toward remediation of this lack of fairness, and certainly an indication that the park managers are aware that they do not have open communications with many of their resident and user groups. The problem of communication with residents and citywide users is an ongoing one, since park staff is often limited in terms of their language skills and time restrictions in these underfunded urban parks.

Finally, interactional justice is also not evident in these parks because of the barriers to interaction discussed. Although the absence of interactional justice is subtle in Independence, that is, there is no visible discrimination or incidents like in Riis Park, nonetheless, residents interviewed did talk about how they felt treated by tourists when in the park environs. Many of them talked about how they felt afraid in the park, which could be explained in a number of ways, but could derive from their ill treatment because of ethnicity or colour. Interactional injustice is most evident, however, at Riis Park where the needs and demands of Central and South American immigrants were overlooked and when the staff cannot speak the language of their largest user group. We were not able to study some of the subgroups as thoroughly as we would have liked, but we did learn that there was physical violence and verbal abuse of people of colour who entered Bay 14 that was territorialized by White local males. There were also conflicts between the gay beach users and families over appropriate behaviour and dress.

Overall, by examining these incidents of inequality, inequity and unfairness we were able to articulate the social justice problems we uncovered throughout the research. Although it is only a first step, it is one that we were able to take in prompting better relationships and fairer use of resources in these large public urban areas.

References

Banks, L. 2011. *Home Spaces, Street Styles: Contesting Power and Identity in a South African City*. Johannesburg: Pluto Press.

Bies, R.J. and Moag, J.S. 1986. Interactional justice: communication criteria of fairness, in *Research on Negotiation in Organizations*, edited by R.J. Lewicki, B.H. Sheppard and M. Bazerman. Greenwich, CT: JAI Press, 43–55.

Blomley, N. 2004. *Unsettling the City*. New York and London: Routledge.

Campbell, H. 2006. Just planning: the art of situated ethical judgement. *Journal of Planning Education and Research*, 26(1), 92–106.

Chesluk, B. 2008. *Money Jungle: Imaging the New Times Square*. New Brunswick, NJ and London: Rutgers University Press.

Cropanzano, R. and Randall, M.L. 1993. Injustice and work behaviour: a historical review, in *Justice in the Workplace: Approaching Fairness in Human Resource Management*, edited by R. Cropanzano. Hillsdale, NJ: Lawrence Erlbaum Associates, 3–20.

Dear, M. 2000. *The Postmodern Urban Condition*. Oxford: Blackwell.

Fainstein, S.S. 2000. New directions in planning theory. *Urban Affairs Review*, 35(4), 451–78.

——. 2004. Cities and diversity: should we want it? Can we plan for it? *Urban Affairs Review*, 41(1), 3–19.

——. 2005. Planning theory and the city. *Journal of Planning Education and Research*, 25(2), 121–30.

Fincher, R. and Iveson, K. 2008. *Planning and Diversity in the City: Redistribution, Recognition and Encounter*. Basingstoke: Palgrave.

Fiske, J. 1998. Surveilling the city: whiteness, the black man, and democratic totalitarianism. *Theory, Culture and Society*, 15(2), 67–88.

Harvey, D. 2005. *A Brief History of Neoliberalism*. Oxford: Oxford University Press.

Kayden, J.S. 2000. *Privately Owned Public Space: The New York City Experience*. New York: John Wiley and Sons.

Lefebvre, H. 1991. *The Production of Space*. Cambridge and New York: Blackwell.

Lefkowitz, J. 2003. *Ethics and Values in Industrial-Organizational Psychology*. Mahwah, NJ and London: Lawrence Erlbaum Associates.

Low, S. 1999. *Theorizing the City*. New Brunswick, NJ: Rutgers University Press.

——. 2000. *On the Plaza: The Politics of Public Space and Culture*. Austin, TX: University of Texas Press.

——. 2003 *Behind The Gates: Life, Security and the Pursuit of Happiness in Fortress America*. New York: Routledge.

Low, S. and Smith, N. 2006. *Politics of Public Space*. New York and London: Routledge.

Low, S., Taplin, D. and Scheld, S. 2005. *Rethinking Urban Park: Public Space and Cultural Diversity*. Austin, TX: University of Texas Press.

Miller, K.F. 2007. *Designs on the Public: The Private Lives of New York's Public Spaces*. Minneapolis, MN: University of Minnesota Press.

Mitchell, D. 2003. *The Right to the City: Social Justice and the Fight for Public Space*. New York: The Guilford Press.

Nussbaum, M.C. 2000. *Women and Human Development: The Capacities Approach*. Cambridge: Cambridge University Press.

Rawls, J. 1971. *A Theory of Justice*. Cambridge, MA: Harvard University Press.

Ruben, M. and Maskovsky, J. 2008. The homeland archipelago: neoliberal urban governance after September 11. *Critique of Anthropology*, 28(2), 199–217.

Shepard, B. and Smithsimon, G. 2011. *The Beach Beneath the Streets: Contesting New York City's Public Spaces*. Albany, NY: State University of New York Press.

Sorkin, M. 2008. *Indefensible Space: The Architecture of the National Insecurity State*. New York and London: Routledge.

Thibaut, J.W. and Walker, L. 1975. *Procedural Justice: A Psychological Analysis*. Hillsdale, NJ: Lawrence Erlbaum Associates.

Tyler, T.R. 2000. Social justice: outcome and procedure. *International Journal of Psychology*, 35(2), 117–25.

Tyler, T.R. and Blader, S.L. 2003. The group engagement model: procedural justice, social identity, and cooperative behavior. *Personality and Social Psychology Review*, 7(4), 349–61.

Whyte, W.H. 1980. *The Social Life of Small Urban Spaces*. Washington, DC: Conservation Foundation.

Young, I.M. 2001. Equality of whom? Social groups and judgments of injustice. *The Journal of Political Philosophy*, 9(1), 1–18.

Public Spaces: On Their Production and Consumption

Ronan Paddison

The Levels of City Living

For the visitor a late evening stroll along the riverfront in Guangzhou might bring a surprise, the sight of dancing in a public space. The music, a live band, and the evident skill of the dancers might suggest that it is a commercially organized event taking advantage of the warmth of the summer evening and the availability of a suitable space. Yet, the scale of the performance – the size of the public space in which the dancing is taking place – suggest otherwise. Equally, the obvious enjoyment of the participants, their familiarity with one another, the extent to which partners are being alternated as the evening progresses, the conviviality of the scene suggest that this is far from being a commercially organized event. A regular occurrence, many of the dancers are from the local community, the performance being a way to give alternative, and enjoyable, expression to living in a Chinese mega-city with its attendant pressures.

The space in which the performance is organized, and its meanings, is all the more remarkable for its setting. The dancers, seemingly unselfconscious of the location, have appropriated a space whose backdrop is the impressive skyline of the city, the music having to compete with the endless drone of the nearby arterial highway. The intimacy of the event appears to unfold in spite of its background, a bubble in which living the city socially is made possible, the evident enjoyment of the dancers testimony to its place. Against the dynamic backcloth of a rapidly emergent global city, the performance expressed the importance of the local in everyday life.

Cities are a mosaic of different spaces whose meaning is defined by their function as part of the overall city and the experience and imaginations citizens have of them. For most of its inhabitants much of the city is unknown territory, imagined rather than experienced through habitual movements. Mundane urban life incorporates a fraction of the city, limited to certain rhythms. The scale of the Asian mega-city, and the speed with which the urban form is changing, might be supposed to accentuate the atomism within which city living takes place (Forrest 2009). In the otherwise alienating environment of the urban the importance of the local neighbourhood becomes accentuated, the space within which not only is much of daily life transacted but the space which gives meaning to living within the city.

As much as the local neighbourhood is central to both the experience and meaning of urban living, citizens too feel a sense of belonging to their city. Part of urban culture is rooted in the sense that its citizens have of being part of the city. Commonly, then, regardless of the specific city being considered, or the world region to which it belongs, citizens will be protective of their city, particularly when it appears to be threatened by the activities of another city (or the state or a powerful corporate entity seeking to disinvest or reinvest in a competing city). Predictably, such emotions will vary between citizens; those with a deeper investment in the city, by virtue, say, of their length of residence within it or the economic or cultural or social capital which they have tied up in it, may be more defensive of what they consider to be their city than those who consider themselves more transitory within it. Yet, for most (if not all) of its residents there is rationale to a sense of belonging to the city. (See also Dovey, Chapter 15, this volume.)

The emotive ties to the city become played out at different scales and through the imaginations and experiences we have of it. Key here is the role of public space, where and through which so much of urban life is both experienced and imagined. The particularity embraced by the term 'public' courts the binary of public and private which has its own conceptual dangers, added to which in the contemporary restructuring of cities, and linked to the growing use of surveillance and control technologies, much public space is being re-created as semi-private space. These qualifications aside, the recognition of the role public space plays, how it is imagined and used, is central to the construction of the meanings residents give to city living. (See also, Watson, Low, and Nyseth, Chapters 1, 17 and 19, this volume.) The ability of community residents in the Guangzhou neighbourhood to appropriate a local space, and the defence that is commonplace to local activism aimed at protecting local amenity – both demonstrate the importance of public space.

Globalization accompanied by, and fuelled by, the rise of urban neoliberalism has accelerated the rate of urban change, demonstrated most visibly by the ways in which cities have sought to reconfigure themselves in order to be more competitive. Through what McCann and Ward (2011) have defined as policy mobilities the forms of urban development and redevelopment become standardized – waterfront developments, the spread of planned shopping centres to cities in the global South, the adoption of iconic architecture as a means of branding the city, the spread of gentrification from the global North to South, the increasingly widespread use of culture as the means of fostering the economic base of the city and repositioning it. (See also, Watson, Chapter 7, this volume). How these policy discourses become mobilized may be contextually dependent were their transferability needs to take into account local circumstances. Yet, the physical transformation of the city under deepening globalization displays strong commonalities.

The spread of such physical change has direct consequences on the public spaces defining cities. Early discussions of such changes became associated with the debate on the 'end of public space', not in any absolute sense but through the progressive appropriation and redefinition of public to semi-private space through, in particular, the spread of the planning shopping centre and the spread of disciplinary practices accompanying the regeneration of the city. Even where the term 'end' was contested and needed qualification, there is a common understanding that in different ways cities have become increasingly privatized and that this has had significant consequences on the ways in which public spaces have been produced and reproduced, and consequently on the ways in which they are consumed. The contradictions and tensions to which such changes give rise may not be new: the development of cities in early industrial urbanism gave rise to not dissimilar conflicts, as did the development of colonial cities. It is their re-emergence under more globalized and competitive conditions that helps to define the present round of urban restructuring.

This chapter explores the ramification of such changes on public space. Taking the simplified distinction outlined earlier of the two scales into which city living can be divided, the chapter centres on the implications of two forms of aesthetic intervention which have become commonplace in cities in the global North and increasingly so in the global South, the use of iconic architecture and of public art. Focusing the argument in this way is important for a variety of reasons – both are linked to the cultural appreciations and imaginations of the city and of local neighbourhoods, both potentially link to the emotional understandings that residents have of city spaces at different scales, both are connected to the re-aestheticization of the city, both are connected to the recapitalization of the city and the extraction of surplus value through the redevelopment of city spaces. Critically, too, they tend to operate at different scales, the production of iconic architecture to the city as a whole and how it is (and should be) imagined both externally and internally for its residents, public art as a means of (re)aestheticizing (though not exclusively) local neighbourhood spaces. For both the processes of production and how they are consumed are inter-connected, though as the chapter will argue how such inter-connections unfold differ to the extent to which the interventions are contested or not. Key here – and bringing into focus again the dancers on the riverfront space in Guangzhou – is the accessibility of public space, its possible appropriation, if only temporarily, for collective good, in effect the degree to which city residents are able to claim a sense of ownership over a local space. The argument revisits the definitions that surround public space to which the chapter turns before looking at iconic architecture and public art.

Defining Public Space

What constitutes public space, what is its nature, what have been its histories in different cities, and how should it be defined has been the subject of considerable debate. First, it is clear that public spaces are both ubiquitous and fall into a large number of types. Taking Brown's (1990) understanding of public space as the public sector expressed in physical terms, it includes not just parks, squares and streets – those spaces traditionally associated with the public realm – but also a diverse range of other facilities meeting common needs through collective provision, including churches, marketplaces, sports centres as well as public libraries. Some, such as major civic spaces, define the city as a collective entity and are an appeal to *external* audiences, others, by virtue of their location, as well as often their function, become defined more around the local spaces into which the city is divided and which are more closely bound with everyday urban living. It is in the latter in particular that Mean and Tims (2005) identified a range of public spaces that were defined by their relatively high levels of social interaction and sometimes conviviality, but which perhaps were not thought of as public spaces in the first instance, including libraries, car boot sales and allotments. Their list begins to define spaces which may be to a degree privatized – allotments, for example – while car boot sales reflect a commodified, if open, use of space. Listing in any exhaustive sense the different types of public spaces, while able to show their diversity besides in many cases their obvious importance to the conduct of everyday life in the city, begins to demonstrate the problems in defining what is meant by public space and in maintaining the public/private binary.

Most definitions of public space begin by emphasizing its accessibility and how this can be connected normatively to the notion of a democratic public space. Young (1986) offers a description of such spaces as 'accessible to anyone, where people engage in activity as individuals or in small groups ... [where] people are aware of each other's presence ... [where] the diversity of the city's residents come together and dwell side by side' (1986: 22). Accessibility, inclusion tolerance of difference – the openness to 'unassimilated otherness' – define the ideal Young (1990) terms the 'unoppressive city'. For Arendt (1958) public space is part of what she terms the public realm and again it must be accessible to all, but added to this it should be used by all and historically durable. The emphasis given to accessibility is based on the inclusivity of public space; by definition, public spaces should avoid being exclusive to particular groups or individuals.

The normativity of such arguments reflects their origins from political philosophers, just as their aspirations for public space are challenged by reality. One consistent observation of any city – and probably approximating a universal truth in the study of cities – is that public space, its production and reproduction, is deeply contested. The origins of such contestation varies, but if unequal power relations and the consistent role of dominant forces, capital and/or elites in particular, linked with appropriation of public space and its resistance, is commonplace, it is not the only source of such contest. Intra-community differences are also a common enough source of contest, where how public space should be used (for example) highlights different interests, setting in tow possible conflicts. (See also Low, Chapter 17, this volume.) Within these too, however, differences in the power relations between groups will tend to be critical to how conflict unfolds, as well as potentially to outcomes.

Where power and power relations become central to understanding the 'operation' of public space, Mitchell (2003) has argued that increasingly our public spaces are being produced *for* us rather than *by* us. The claim is in some ways contestable, particularly in its apparent downplaying of the historicity of such relations – in pre-industrial, early urban-industrial cities in the global North, and colonial cities, the production of public spaces and how they were managed invariably reflected elite interests. Public spaces that are produced by us are the exception. Nevertheless, the argument helps to reaffirm arguments about how dominant interests are able to appropriate public space, and how this in turn reflects the fragility of public space and the extent to which it can be captured by elites. As a significant vein of urban scholarship has sought to show, it is through these arguments that we can begin to understand how and why in the contemporary city there is an apparent assault on *public* space.

Such claims provide an important backdrop to the arguments here. But what is even more important to emphasize is how public space, its production and reproduction, is closely linked to inclusion and accessibility. Both help to define a sense of (collective) ownership over space within which (arguably) how such spaces are produced may play a part. Expressed alternatively and in Mitchell's terminology, public spaces made *by* us are the ones more likely to be considered democratic, inclusive spaces. Whether and how reality matches such an expectation is a moot point, but what is critical here is that the premise underpins much of the endeavour to incorporate public participation into the planning process – that, even if it is articulated somewhat opaquely, public participation will be able to assist the processes of designing cities and result in outcomes that are closer to the preferences of its residents. Clearly, the argument is deeply contestable, as well as possibly naïve, but what is salutary to bear in mind is that its converse, the production and reproduction of public spaces which are imposed, and therefore reflective of particular interests, is linked to current unease among scholars over the trends apparent in public space. We can begin to explore these ideas through the use of iconic architecture and public art which is undertaken through looking at specific examples. The emphasis here will be on that class of cities, largely within the global North, that as a result of deindustrialization and attendant processes of decline, have had to redefine themselves in terms of their economy, a project which in turn has become closely associated with their imagineering.

Interpreting Iconic Architecture

The use of iconic architecture as a means of projecting the image of the city has become a hallmark of post-modern urbanism just as it was to be employed in the modernist, and indeed, earlier cities. Debates around the meaning of iconic architecture and its rapid rise in the urban townscapes of a growing number of cities globally have sought to unravel the processes through which its production has been achieved and to offer an explanation of its increasing use (Jencks 2004, Sudjic 2005). As Kaika and Thielen (2006) have argued, during industrial urbanism cities sought to showcase their economic and political prowess through the building of impressive town halls (as in Brussels, Manchester, the major German cities and many other examples). To these could be added new cultural facilities catering for the interests of the burgeoning middle classes (concert halls, art galleries) and less commonly, though more spectacularly, through architectural interventions linked to mega-events (Crystal Palace to the holding of the Great Exhibition of 1851 in London, for example). The spread of contemporary iconic building is not dissimilar to its nineteenth-century counterpart in European cities in a number of respects; in particular the ability of the iconic building to be a visual marker of distinction. The reasoning behind the Burj al-Arab hotel in Dubai in this sense is not unlike that of the Eiffel Tower completed more than a century earlier.

The use of iconic architecture is contentious not least because of how it is produced as well as what it represents. Architecture itself is a 'discourse', the projection through ideas and material construction of a normative appreciation of the city. By this token iconic architecture becomes an emphatic type of architectural discourse, one in which the production is aimed at ensuring that the intervention is visually arresting, memorable as well as often challenging. Underpinning its production are unequal power relations, an argument equally true of pre-industrial as well as modern cities.

Thus in the pre-modern city religion was a powerful force around which elite interests coalesced reflected in the building of cathedrals (Kaika and Thielen 2006) while in the industrial city the local state became dominant, reflected in the development of town halls and other municipal buildings aimed at making a visual statement reflecting the prowess of the city. Contemporary iconicity for Sklair (2009, 2005) is rooted in the power of the international capitalist class, although in many cities its adoption may be more immediately linked with the aspirations of local political and economic elites striving to rebrand the city (Evans 2003, see also Searle, Chapter 8, this volume).

Where the contemporary revival of iconic architecture differs from its earlier counterpart is in its cooption with the commodification of the city. Iconicity has become drawn into the wider neoliberal project in which the city is a product to be marketed and as city marketing becomes through policy emulation a widely adopted strategy, so iconicity becomes the means by which to create distinction (DCMS 2004). As Jones (2011: 117) has argued

> *the incorporation of iconic architecture and the respective brands of the famous architects responsible for the design of such buildings into the place marketing projects so central to UK urban regeneration strategies represents a rich research agenda for sociologists interested in the ways in which regeneration agencies seek to mobilize urban space and culture to produce surplus value.*

Nor, as it has been argued, is such an analysis to be restricted only to the UK where the quest for the architecturally iconic has become globalized.

Besides the production and extraction of surplus capital value iconic architecture is also able to replicate the processes through its symbolic value and it is through this that its ramifications for citizens will become apparent. Where iconic architecture has become contested often results from the different audiences it attempts to address, external and internal, and the extent which the former is apparently privileged over the latter. As part of the wider project of reimaging the city iconicity becomes entangled with strategies aimed at making cities attractive, of drawing them into the network of urban tourism destinations. More ambitiously, iconicity becomes drawn into the discourse of the attractive city (Florida 2004, Landry 2004, 2002). Such strategies tend to be aimed, initially at least, at the external market – at drawing in external investment to the city, at attracting tourists to the city. How it addresses the city's own population, and how it is consumed by them, may raise different issues than it does for external audiences.

These issues were to become apparent in the ill-fated 'Fourth Grace' commissioned as part of Liverpool's status as Capital of Culture in 2008. Commissioned from a key architect, Will Allsop, the Fourth Grace was to complement the existing three graces of iconic status on the city's waterfront. Jones (2011) details its origins leading through to its abandonment which reflected the problems realization of the intervention encountered. Among these were problems arising from its cost and funding as well as concerns as to whether the building would be completed in time. Significant, too, and reflecting the audacious ambitions of the architect, there was considerable opposition to its installation from Liverpudlians. Drawing from the discussion aired in the local press the building was variously described as an 'eyesore', a 'monstrosity' and a 'cow pat'. Allsop's design was considered intrusive to the city and the cityscape and, however much the architect sought to stress that the structure reflected the city's history, one which was perceived to be out of place. Public reaction was to be just one factor in the abandonment of the project, and may have been less significant than the other reasons, but the lack of popular support did counter what initially had been a project that was considered by the local elites as central to the wider project of culture-led regeneration.

The story of the Fourth Grace demonstrated that, where iconic architecture can be visually contentious, its installation within public space may become contested and that contestation in itself is able to contribute to the abandonment of the project. Yet, iconic buildings, insofar as they reflect elite visions for the marketing of the city, the interests of capital combined with the professional expertise of 'starchitects' embrace powerful agents whose ambitions become key to how cities are regenerated (McNeill 2009). In attempting to attract the attention of external audiences the appeal of installing an iconic building tends to become taken for granted in what has been defined, on the precedent set by Bilbao, as the 'Guggenheim effect' – tourists and, more generally, investment will be drawn to the city because of how iconicity contributes to place-making. Such projects need to be sold to the city too – ensuring that the development is able to attract popular support in the city, the 'internal audience' becomes part of the marketing process. Exploring how this was undertaken in another city – Glasgow with the opening of the Riverside Museum in 2011 – helps to show how what can be a contentious process, can be depoliticized.

The recent transformation of Glasgow, its re-instatement as a post-industrial city, has achieved almost paradigmatic status in the urban literature. It is frequently cited as a textbook example of an old industrial city that has sought within the last 30 years to engineer a new image for itself as part of a wider project to rebuild the city's economy. From the 'dark days' of the early 1980s, when unemployment rose to over 15 per cent, well in excess of twice the UK rate, the city council initiated a city marketing campaign – 'Glasgow's Miles Better' – that sought to project a more positive image of the city, nationally and internationally as well as for citizens themselves (Paddison 1993). Based on ideas from the marketing of New York the initiative was aimed at countering the poor external image of the city linked to its faltering economy based on heavy industries that had largely collapsed accompanied by high levels of social deprivation, poor housing and high crime rates. While some 30 years later the city has made inroads into redefining itself as a post-industrial city – in which the establishment of new financial industries, emergent cultural industries, the development of international tourism and the hosting of conferences have played a particular role – Glasgow remains in many respects a dual city in which high levels of social deprivation persist alongside the ostentatious consumerism reflected in the city's new logo, 'Glasgow, Scotland with Style'.

The process of transformation and how it has been achieved by the city's development agencies has not gone unchallenged. Much of the opposition has centred around the issue common to many cities undergoing physical and economic restatement of 'whose city'. During the milestone year in the city's reimaging, 1990, in which it had been awarded European Capital of Culture, there was considerable opposition as to what and whose culture was being represented in the reimagineering of the city (Boyle and Hughes 1991). The conflict has been persistent, reflecting how the city was actually developing and its growing inequalities, reflected in a major and more recent report by Demos that presented alternative imaginations for the future city to those being implemented by its development agencies. Central to the project of the city's restructuring has been its physical transformation dominated by waterfront regeneration and the gentrification of parts of the inner city, both of which have been contributory to the growing inequalities in the city and to what Mooney (2004) and others have characterized as the dual city.

Part of the process of re-presenting the city as a dynamic and attractive place that had shed its old image centred on the quest for architectural iconicity. The quest has had a chequered history. Initially, attention focused on the installation of a tower providing an architectural stop at one end of the city's most prestigious shopping thoroughfare, though the idea was abandoned in favour of a high tower to accompany the then newly built Science Centre on the riverside. As part of a larger set of developments taking place on either side of the river, the Science Tower was to incorporate a lift enabling visitors to appreciate the changes taking place in the city and particularly along the Clyde. Since its opening the Tower has been plagued by successive mechanical problems all but making it defunct – it is not even included in the current edition of the official map of city sights published by Glasgow City Marketing Bureau. These ill-fated attempts to establish an iconic building for the city have been answered through the recent (2011) opening of the Riverside Museum. Designed by the starchitect Zaha Hadid, the Museum is a flagship development within the ongoing redevelopment of the riverfront which, as the Leader of the City Council put it, is 'a building which itself would be a major statement and attraction' (Jamieson 2011).

A significant addition to the cityscape, what is of importance to the argument here is that, while it has had its detractors, the museum, occupying a prominent space that addresses the city as a whole rather than the specific neighbourhood in which it is located, appears to have been enthusiastically accepted by the city's residents (as well as becoming an attraction to the city's 'external audiences'). Part of the reasoning no doubt as to why it has become a popular local attraction stems from what the museum contains – replacing the former popular transport museum, the contents centre on exhibits many of which play to the memories citizens have of the city. In the immediate sense of the number of visitors attracted to Riverside the museum has rapidly become a major cultural attraction, the vast proportion of those visiting from the city itself or its immediate hinterland. More concealed in its influence is the role marketing has in persuading the city's residents that the project was an essential part of the regeneration process – in effect, a 'natural' addition to the city's public spaces expressed through rhetoric that effectively depoliticizes the intervention. Through this argument an iconic tourist attraction is not only important to attracting tourists, but beneficial to the urban economy and, therefore, to its residents. At any rate, by comparison with earlier junctures in the city's harnessing of culture, opposition to the museum has been muted, in spite of the cost and the more hidden benefits of the project in the gentrification of the city.

Public Art in Local Neighbourhood Spaces

Iconic architecture might be considered as public art distinguished by scale and its aspiration to be a memorable (and attractive) visual statement. Though public art itself embraces different types of intervention and is not limited to being a visual intervention, its renaissance is widely associated with the installation of artworks which not only aspire to engage with different audiences but also to be instrumental in the process of place-making. Typically, the installation of public art operates at a more modest scale than iconic architecture, though some public art has achieved iconic status, not just because of its scale or because of what it seeks to represent, but also because of its location. In the United Kingdom Anthony Gormley's *Angel of the North* (a major sculpture located on a hilltop alongside a major road welcoming visitors to the approaching Tyneside conurbation) has achieved international iconic status within a relatively short period of time. Nor are such examples limited to the present day (re) inscription of the city – earlier examples of monumental public art, Nelson's Column in London and its ill-fated counterpart in Dublin, Nelson's Pillar, demonstrate how such interventions can occupy key public spaces besides becoming, through habituation, integral parts of the everyday urban landscape (Whelan 2003).

The ubiquity of public art as part of the reaesthecization of the city is linked to its spread within the city beyond its central public spaces to more local spaces within the urban fabric. Here its emergence is found as decoration to public institutions, such as hospitals, to recapitalized spaces including waterfront regeneration, to gentrified spaces, to the mixed neighbourhoods associated commonly with urban regeneration and to a host of other spaces comprising the city. Its visual ubiquity has been the source of some opprobrium where it has even been described in terms of applying 'lipstick to the gorilla' (Evans and Shaw 2004). Such criticisms reflect not just the negative attitudes the product itself can engender – attitudes that are rooted in taste – but because of its apparent disjuncture with the local environment in which it has been placed and the imposition it represents. Invariably, public art is contentious, its meaning contested between different audiences, contestation which in some cases has – as in the case of Richard Serra's *Tilted Arc* in Federal Plaza, Manhattan (Senie 2002) – passed into the urban folklore surrounding its use.

What the *Tilted Arc* controversy highlighted was the relationship between the artist and the space in which (s)he was working and their responsibilities to the local publics. As Gablik (1995) argued

> What the Tilted Arc controversy forced us to consider is whether art that is centred on notions of pure freedom and radical autonomy and subsequently inserted into the public sphere without any regard to the relationship it has to other people, to the community, or any consideration except the pursuit of art, can contribute to the common good. (quoted in Miles 1997: 90)

In realizing their project the artist – as no less the starchitect – needs to recognize the power their agency has to reinscribe public spaces and the tensions to which this can give rise.

The broader lesson pointed to by the controversy – and one in which there is growing evidence of it having been learnt in a wide range of cities – is that the processes through which public art reinscribes public space should be inclusive, drawing the local community into the process of production. Complementing the physical restatement of Barcelona as a world city centred on major developments such as the waterfront has been a less-publicized but widespread use of public art as part of the process of local neighbourhood regeneration. Though the claims made for how public art has been

319

drawn into local neighbourhood spaces in Barcelona has not proved uncontentious (Balibrea 2001), how art has been used to complement public spaces has sought to work with local populations rather than simply impose projects on them, developments that have become widespread practice in North American and European cities.

In its attempts to be more inclusive, public art has spread to urban neighbourhoods it has been a stranger to in the past, including in British cities, localities that are characterized by high levels of social deprivation. (See also Montgomery, Chapter 20, this volume). In Gateshead, sister city to Newcastle in the north-east of England, public art has been deliberately extended to deprived neighbourhoods as part of a wider project aimed at social inclusion (Sharp, Pollock and Paddison 2005). Similarly, in Glasgow public art has been installed in a variety of neighbourhood spaces, in estates at the edge of the city characterized by high levels of social deprivation and poverty as well as in inner city mixed neighbourhood regeneration. The use of participatory techniques represents an alternative interpretation of inclusion – by working with local communities, how public art is to be installed is less imposed than it is a more cooperative process between the artist and the local population.

In practice, the inclusion of public participation as the means by which the dedication of art can enhance local public spaces raises its own problems and conflicts. In common with the problems of public participation elsewhere in urban development, its practice may be more tokenistic than real. Similarly, the mechanics of how participation is undertaken is problematic, as (for example) in ensuring that the multiple publics constituting local residential populations are represented effectively in the process as well as their preferences reflected in the final installations. Nor does local participation resolve the tensions that can arise between local community aspirations and the autonomy of the artist. Added to these problems are continuing uncertainties professed by some as to the purpose of public art, the value of it, and where it is being used in mixed neighbourhood regeneration who is benefiting from it. These difficulties may not be resolvable, yet there is sufficient case practice to indicate that, however problematic, working with local populations is potentially productive in that it is able to generate public art that is able to enhance local public spaces. Even if as a conclusion it needs qualifying to take account of contingency, it begins to provide support for Mitchell's argument introduced earlier, distinguishing between public spaces made by, rather than for, us.

Conclusions: The Entanglement of Production and Consumption

Democratic public space is fragile, its imprint in the city consistently being challenged. The restructuring of cities in the interests of capital and of local elites, the impress of professionalism operating through city planners and others, the emergence of new technologies – the motor car and surveillance – and their spatial ramifications, each has had erosive effects on the maintenance of democratic public spaces. Yet, simultaneously public space plays a key role in giving meaning to urban life, materially and symbolically. Public spaces are the fora in which social interaction is played out, in which the diversity of the city is encountered. Materially public space is fundamental to meeting everyday needs and, more symbolically, to contributing to identities, both of the self and collectively. Its centrality to the cultural as well as the economic and social life of the city gives added meaning to its vulnerability arising from its ability to be appropriated by particular (and powerful) interests.

Contemporary trends to reaestheticize the city as part of the wider project of creating cities that are attractive and able to compete with other cities in the global market-place are linked to processes that are eroding democratic public spaces (Harvey 1989). Privatized spaces, and the spread of gentrification are achieved through neoliberal policies in which the reconfiguration of public spaces is produced through the efforts of local development agencies. Integral to these processes has been the re-recognition of the power of architecture and of public art in reinscribing place, contributing to the creation of attractive and liveable cities in projects that are targeted particularly at external audiences. The dystopian view is that the arts and the mobilization of spectacles – including iconic architecture – are being mobilized in the interests of capital and of specific class fractions, the middle classes in particular.

Such a view has been challenged, centring around the ways in which art can be used to critique hegemonic power and furnish the basis for resistance. In her analysis of cultural capitals Johnson (2009), following the lead provided by Jacobs (1998), argues that 'the activation of creativity in the form of cultural capital can and does have material effects on cities which may trouble dominant views of that city … and offer alternative representations' (Johnson 2009: 43). Art, then, can challenge dominant representations. Thus, in London the Docklands Community Poster Project interposed local community narratives showing aspects of local history and identity otherwise airbrushed out by the reimaging projected by the development agency. Further, the practice of installing public art, as has been discussed earlier, has sought to be more inclusive of the multiple voices comprising local communities, linking the production, reinscribing, of public spaces through art in ways which aim to be more inclusive of local preferences. The premise common to these developments is that both production and consumption *should* become entangled in the reinscribing of public space.

Such a premise is not intended as a panacea, nor could it be, given the examples introduced in this discussion. The impromptu dancing on the Guangzhou waterfront with which the chapter began – though endorsing a different artistic form from the visual intervention – demonstrated how local communities can appropriate a space for its own ends. How a space becomes consumed can result in alternative uses to those that had been originally intended. How public art is read can be contrary to the reasoning attached to its production; participatory practices drawing in local opinion as to what public art is to be installed are necessarily selective in who becomes consulted and when it takes place. Yet, much public art is more enduring, communicating with citizens that were not involved in its production and with subsequent generations. Even where public art is imposed by elites into urban public spaces, habituation to its presence may result in its acceptance as a 'natural' component of a public space; when it was proposed to replace the monuments in a major public space in Glasgow the proposal met considerable opposition and was abandoned, even though the edifices represented elite members important to the city and Scottish nation from the nineteenth century.

Nor does the production-consumption process necessarily unfold linearly – that is in ways the premise would predict – suggesting that other explanations need to be sought in explaining how the hegemonic appropriation of public spaces, the imposition of iconic architecture may not be resisted, but rather be appreciated more consensually. In a critique of the recent changes in Barcelona, the physical regeneration of the waterfront and inner city neighbourhoods, Balibrea (2001) cautioned that the absence of any significant dissent to the flagship projects may not necessarily be read as support for change, but rather that through the marketing of change development agencies have been able to generate a sense of false consciousness in which it has been possible to convince even potential critics that their interests are equivalent to those of the dominant economic classes. Yet, the spaces Balibrea is referring to, as also the new museum in Glasgow, are all at the city rather than the neighbourhood scale. It is in the more local spaces that (re)aestheticization has been the more inclusive, those spaces in which it is more likely that there is a sense of ownership.

Acknowledgements

Parts of this chapter draw on Paddison and Sharp (2007), Sharp, Pollock and Paddison (2005), and on an unpublished manuscript by the author 'The Politics of Architectural Iconicity'.

References

Arendt, H. 1958. *The Human Condition*. Chicago, IL: University of Chicago Press.
Balibrea, M.P. 2001. Urbanism, culture and the post-industrial city: challenging the 'Barcelona Model'. *Journal of Spanish Cultural Studies*, 2(2), 187–210.
Boyle, M. and Hughes, G. 1991. The politics of the representation of the 'real' discourses from the left on Glasgow's role as the European City of Culture 1990. *Area*, 23, 217–28.
Brown, D.S. 1990. *Urban Concepts*. New York: St. Martin's Press.
DCMS. 2004. *Culture at the Heart of Regeneration*. London: Department for Culture, Media and Sport.
Evans, G. 2003. Hard-branding the cultural city: from Prado to Prada. *International Journal of Urban and Regional Research*, 27(2), 417–40.
Evans, G.L. and Shaw, P. 2004. *A Review of Evidence on the Role of Culture in Regeneration*. London: Department for Culture, Media and Sport.
Florida, R. 2004. *Cities and the Creative Class*. New York: Routledge.
Forrest, R. 2009. Managing the chaotic city-social cohesion: new forms of urban governance and the challenge for East Asia, in *Changing Governance and Public Policy in East Asia*, edited by K.H. Mok and R. Forrest. London: Routledge, 287–302.
Gablik, S. 1995. Connective aesthetics; art after individualism, in *Mapping the Terrain: New Genre Public Art*, edited by S. Lacy. Seattle, WA: Bay Press, 74–87.

Harvey, D. 1989. From managerialism to entrepreneurialism: the transformation of urban governance in late capitalism, *Geografiska Annaler*, 71B, 3–17.

Jacobs, J. 1998. Aestheticisation and the politics of difference in contemporary cities, in *Cities of Difference*, edited by R. Fincher and J. Jacobs. New York: Guilford Press, 252–78.

Jamieson, T. 2011. A transport of delight as architect tours museum. *The Herald* (Glasgow), 10 June, 3.

Jencks, C. 2004. *Iconic Buildings: The Power of Enigma*. London: Francis Lincoln.

Johnson, L.C. 2009. *Cultural Capitals: Revaluating the Arts, Remaking Urban Spaces*. London: Routledge.

Jones, P. 2011. *The Sociology of Architecture*. Liverpool: Liverpool University Press.

Kaika, M. and Thielen, K. 2006. Form follows power. *City*, 10(1), 59–69.

Landry, C. 2002. *The Creative City: A Toolkit for Urban Innovators*. London: Earthscan.

——. 2004. *The Art of City Making*. London: Earthscan.

McCann, E. and Ward, K. (eds) 2011. *Mobile Urbanism: Cities and Policy Making in the Global Age*. Minneapolis, MN and London: University of Minneapolis Press.

McNeill, D. 2009. *The Global Architect: Firms, Fame and Urban Form*. London: Routledge.

Mean, M. and Tims, C. 2005. *People Make Places: Growing the Public Life of Cities*. London: Demos.

Miles, M. 1997. *Art, Space and the City*. London: Routledge.

Mitchell, D. 2003. *The Right to the City: Social Justice and the Fight for Public Space*. New York: Guilford Press.

Mooney, G. 2004. Cultural policy as urban transformation? Critical reflections on Glasgow, European City of Culture 1990. *Local Economy*, 19(4), 327–40.

Paddison, R. 1993. City marketing, image reconstruction and urban regeneration. *Urban Studies*, 30(3), 339–50.

Paddison, R. and Sharp, J. 2007. Questioning the end of public space: reclaiming control of local banal spaces. *Scottish Geographical Journal*, 123(2), 87–106.

Senie, H. 2002. *The Tilted Arc Controversy: Dangerous Precedent?* Minneapolis, MN: University of Minnesota Press.

Sharp, J., Pollock, V. and Paddison, R. 2005. Just art for a just city: public art and social inclusion. *Urban Studies*, 42(5/6), 1001–23.

Sklair, L. 2005. The transnational capitalist class and contemporary architecture in globalising cities. *International Journal of Urban and Regional Research*, 29(3), 485–500.

——. 2009. Commentary: from the consumerist/oppressive city to the functional/emancipatory city. *Urban Studies*, 46(12), 2703–11.

Sudjic, D. 2005. *The Edifice Complex: How the Rich and Powerful Shape the World*. London: Allen Lane.

Whelan, Y. 2003. *Reinventing Modern Dublin: Streetscape, Iconography and the Politics of Identity*. Dublin: University of Dublin Press.

Young, I.M. 1986. The ideal of community and the politics of difference. *Social Theory and Practice*, 12(1), 1–26

——. 1990. *Justice and the Politics of Difference*. Princeton, NJ: Princeton University Press.

The Reinvention of Place:
Complexities and Diversities

Torill Nyseth

Introduction

This chapter relates to the discourse on reinvention that is affecting most cities or towns these days. In the postmodern, global society it has become more important than ever for places to be attractive to newcomers, investors and tourists, but also in the eyes of its own inhabitants. In the global competition between places local actors are fighting to attract industrial investors and offer them the best possible terms to convince them to invest in their specific place. Identity, sense of place and local distinctiveness are key features of the competitive success of places (Murray 2001). This expanding discourse on place branding and selling is addressed in this chapter through a critical and analytical social science perspective. The rationale behind ideas such as place branding and promotion is questioned. While branding is an active strategic and deliberate policy for changing the image of a place, place reinvention is underpinned by more contingent and discrete processes of change. Place reinvention goes beyond place branding and directs attention towards the relationship between symbolic and imaginative change and planned regeneration or place development. In this chapter I will elaborate further on the concept of place reinvention and its dimensions and practices through a theoretical reflection illustrated with examples form the Nordic periphery.

The Rediscovery of Place

Our era has been described as globalized and distinguished by increasing mobilities and forms of 'placelessness' (Augé 1995) and the widely held perception that place has lost power as an identity platform and marker as 'routes' has taken over from 'roots' (Friedman 2002). Yet, at the same time we see a rediscovery of the importance of place. (See also Dovey, Chapter 15, this volume.) Place is, according to Ulrich Beck, being rediscovered in the age of globalization (Beck 2000). Beck insists that 'you cannot even think about globalization without referring to specific locations and places' (2000: 23).

This rediscovery of place is, among other patterns of development, linked to global economic processes and to the diminished role of the welfare state. According to Florida (2002), place is one of those remaining mechanisms that stabilize the postmodern society. He argues that characteristics of economic development, such as a growth in high-tech, knowledge-based and creative content industries clustering within specific places, reveal that 'Place and community are more critical factors than ever before' (Florida 2002: 219). Place is not a victim of globalization that leaves the actors within specific places with no option but to 'adjust' to circumstances (Castree 2003). Rather, places are moments through which the global is constituted, invented, coordinated and produced – they are agents of globalization. Every place is a global space according to Doreen Massey. Contemporary geographers therefore argue that a concept of place fit for our times is one that sees place differences as both cause and effect of place connections. Far from heralding the end of place, the argument is that globalization is coincident with new forms of place differentiation (Massey 2004). Globalization both homogenizes and differentiates; the more linked places become, the more differences endure and are remade (Castree 2003: 176). Gillian Rose argues that 'the specificity of place is continually reproduced, but it is not a specificity which results from some long internalized history, there are a number of sources of this specificity – the uniqueness of place' (Rose 1995). Places are in this sense 'individualized', left on their own to survive in the global competition (Harvey 1989). Massey argues that places are open, fluid and dynamic. Places never reach a position of completion; they are always an ongoing affair, always in the making (Massey 1994).

This also affects how we understand place identity. As we recognize that people have multiple identities, then the same can be said in relation to places; places do not have single, unique identities; they are full of internal conflicts (Massey 1994). Increasing flows of ideas, commodities, information and people are constantly challenging senses of place and identity which perceive themselves as stable and fixed (Rose 1995). Places therefore have multiple identities (Holloway and Hubbard 2001) and they are continuously changing.

Place identity is also contested and linked to power. The recognition of multiple flows, identities and interests implies that what is valued about a place is always contestable. This *dynamic and relational perception of place*, that place not only is geographical in a material essential sense, but must be understood as dynamically and relationally constructed, is particularly relevant in my understanding of place reinvention. A place then is a never-ending story; always in the making. The change of 'place image' or 'place identity' is linked to such place reinvention in various forms. An image may represent flexibility and many different frames of interpretation that can express clarity and distinctness at the same time (Vik and Villa 2008). Understanding the way places are told and read, constructed and de-constructed is complex (Jensen 2007: 233).

Place Promotion and its Limits

A standard approach to place transformation in postmodern society is through some form of place marketing. The meaning and importance of self-presentation and place promotion is intertwined with the increased significance of experience, entertainment and a new economy, based on economic and political regionalization. These factors are increasingly evolving through the construction of regional or local identities. Hence, place branding more specifically might be seen in relation to the growth of culture and experience-based industries and their dependence on a symbolic significance of place (Jensen 2007). The development of place promotion as a competitive advantage is also related to a highly mobile and global society. Overall globalization processes consist of apparently paradoxical trends, both the homogenization of cultures, and new forms of place differentiation mechanisms that support place uniqueness. Place promotion occurs in this area of tension. Place branding and place marketing have become central aspects in the practices of both local and regional development. Small places as well as cities, regions and nations make diverse efforts to be interpreted as attractive future arenas for finance, business, housing and tourism as well as public institutions (Anholt 2007, Hall and Hubbard 1998).

Analytically we may distinguish between at least three quite different strategies that place promotion could be linked to:

- place marketing strategies (boosterism).
- place making and place management.
- place reinvention.

While the first strategy represents the entrepreneurial and place-as-commodity-oriented form, in the second understanding, place branding is treated as a form of place making and place management (Kavaratzis and Ashworth 2005). The third strategy, place reinvention relates to a more complex and broader understanding of place transformation (Nyseth and Viken 2009), and therefore, as will be argued later, is potentially more inclusive than in the first case. Place reinvention goes beyond place branding and may involve both economic and symbolic transformations constituting a changed sense of place. Place reinvention is less strategic, a more analytical approach that tries to understand how different processes are linked together in place transformations. In place branding, all these strategies could be performed, either in combination, or with an emphasis on only one of them.

Place Marketing

Place marketing is the use of publicity and marketing to relate selective images of specific localities targeted at specific populations (Gold and Ward 1994). Place marketing is primarily an outward-oriented activity, but it is also a matter of generating local pride, identity, self-confidence and counteracting negative perceptions (Bramwell and Rawding 1996). The practice of selling and promoting place is tightly linked to entrepreneurial strategies. As Brenner and Theodore (2002) point out, it is a part of the true neo-liberal vision that a place should be branded and marketed. In the marketing literature for instance expressed by Philip Kotler et al.: 'places are, indeed, products, whose identities and values must be designed and marketed' (1993: 11). Symbolic expressions of place identity are communicated through campaigns and other marketing tools in order to be competitive, and to be competitive they have to present themselves as unique. Branding means narrowing down a place's identity into fancy logos and slogans – it is selective story-telling (Sandercock 2003), a form of collective impression management which may not be in tune with local collective identifications. Product marketing deconstructs identity, whereas in place marketing, the identity is the product, which cannot be broken down easily into simplistic elements. This stands in deep contrast to the view of place as an extremely complex and multi-faceted concept, not easily reducible, represented or reflected through traditional marketing practice (Murray 2001: 8). It is as if marketing and promotion are disconnected from development, regeneration and renewal processes that are energizing many localities – the overall 'place-development' (Murray 2001). Places are packaged and sold as a commodity (Murray 2001, Ward 1998). Place marketing and place promotion has therefore been labelled the carnival mask of late capitalist urbanization (Hall and Hubbard 1998).

In order to create a more attractive place image, social and cultural meanings are selectively appropriated and conflicts are played down. Place images tend to be characterized by simplification, stereotyping and labelling (Shields 1991: 47). Branding places is a way of inscribing a certain logic in space – both symbolically through logos, slogans and so on, and materially through the construction of buildings, infrastructure and landmarks. Place branding activities are based on an understanding of demand patterns and images of place consumers, and on identifying the position of the place in the view of competitors. As with place myths, the branding process is a process of creating an evocative narrative with a spatial referent through selective narration – the act of representing the place in a favourable light (Ward 1998: 1). Place marketing is, according to Kearns and Philo (1993), characterized by making use of 'imageneering' concepts or symbols to construct competitive place images. Place marketing involves quite specific interpretations of what the symbolic attributes of a place are, and implies a symbolic communication of these interpretations. Such symbolic communication is not only directed towards the target group of an external market, it also demonstrates locally how 'our place' should be understood. 'Place marketing has become much more than merely selling the area … it can be viewed as a fundamental part of guiding the development of places in a desired fashion' (Fretter 1993: 165).

Place promotion and city branding initiatives are often based on strong local actors, and the exclusion of ordinary citizens. Exclusion is not only a result of formal power relations, institutional cleavages or differences in social capital, but also results from differences in knowledge and culture (education, professional discourses, identity, codes of conduct and so on). Different forms of cultural capital are activated in the reimagining of a place, and played out and structured in place promotion processes. Place promotion in the form of branding asks everyone to speak with one, coordinated voice, a common language to express the brand's identity, a shared commitment to the brand's promise (van Ham 2002: 266). According to the critics, place promotion reduces the complexity involved in local stories and does not justify the richness and diversity of places and people. As noted in the previous section, there is not one identity in a place, never only one story to tell. The symbolic representations (that is, images) of place involved in branding and re-imaging strategies is a power play, as these representations are often contested. A construction of image is also a construction of identity that tends to be based in favour of the social groups in power (Hall and Hubbard 1998: 28). This raises the question of inclusion and exclusion in official image-building or branding, and of how such processes relate to the identities of local residents. (See also, Stevenson, Chapter 9, this volume.)

If, however, such projects encourage local participation and are embedded in collective strategies, they might stimulate increased engagement locally, create enthusiasm and improve the place attractiveness for locals and outsiders. Such projects may in other words contribute to turning around or slowing down a negative spiral of economic and social development.

Place Making

In the second form of place promotion – 'place making' and 'place management' – strategies have moved from the most simplistic advertising campaigns and 'boosterism' towards more comprehensive strategies (Moor 2007). Attempts to construct a new place image here are not limited to advertising campaigns alone but often linked to other strategies, for instance the fabrication of a new urban landscape, flagship projects, mega-events and other cultural attractions (Healey 2010, Hubbard 1996). Also, programmes of economic development are coming to be driven more and more by image-enhancing initiatives (Hall and Hubbard 1998). It is possible to identify a transition from a more disembedded promotion practice towards a toolbox of planning instruments based on a broader understanding of places (Kavaratzis 2004, Kavaratzis and Ashworth 2005). Celebrating diversity is, for instance, present in urban branding profiles, particularly in multicultural contexts (Johansson and Cornebise 2010). A brand could also be a designation of heritage, and as such anchored in local history and identity, and to 'authenticity' (Knudsen and Waade 2010, Skogheim and Vestby 2010, see also Ashworth, Chapter 11, this volume).

Place promotion of this second type is more anchored in local identity and a more complex understanding of place. Promotion activities, however, need to find ways to open up awareness of diverse conceptions of place and to work out what is broadly shared and where deep conflicts lie (Healey 2010). Place promotion is more than 'fine words' and should be more closely linked to place making and place development. Stephen Ward invites us to a much broader understanding: 'Yet marketing, narrowly defined, is not enough. Behind the fine words and images there has to be at least some physical reality of buildings, public spaces and activities that give some genuine promise of a re-invented city' (Ward 1998: 193). This leads us to the third form – place reinvention.

Place Reinvention: Bridging the Gap Between Place Promotion and Place Development

Place reinvention invites a radically different approach, from seeing places merely as products or destinations to be promoted, towards a practice which views places (metaphorically) as living, breathing, cultural entities. In this perspective we move away from marketing as a decoupled project that is not linked to anything else, towards a more integrated approach that links place marketing to place development, urban renewal, community development and urban planning. The symbiotic relationship between people and place is a point of departure; places are shaped and made by people's actions and by their perceptions. At the core of the identity of place are local people as well as businesses, facilities and local landscapes. Without the involvement of their knowledge, concern and imagination, place marketing will be unable to move away from fictitious stereotyping towards a more diverse set of messages about places. Place reinvention switches and broadens the perspective on place development and place promotion towards underlying processes of identity transformation and a sense of place that is always at stake. Place reinvention addresses the numerous ways places are being produced and reproduced. Such strategies direct attention towards the relationship between symbolic and imaginative change and planned regeneration. Place reinvention relates to the transformations resulting from the interplay between actors such as industries, authorities and the public, between projects of construction, promotion and consumption, and processes related to information, identity and imagery. Place reinvention involves a complex dialectic between material space and discursive representation, and the complexity involved in place transformation which the branding literature seems to ignore or simplify.

Place reinvention relates to both economic and symbolic transformations constituting a changed sense of place; for instance changes in the place's industrial base are accompanied by changes in how the place is represented. The term 'reinvention' indicates that something has to be recreated, renewed or redefined, indicating that something else is left behind or forgotten. Place reinvention is a concept that focuses both on inventions and interventions as vehicles for change (Robinson 2006: 251). Inventions are the more continuous changes going on all the time, while interventions are linked to those more direct, planned and intentional processes attempting to achieve change.

Places change for a number of reasons and often as a necessity; industries and businesses adapt to changing circumstances to survive, and authorities urge change to keep up employment and settlement patterns. Over the years the character of most places change, some radically, others more modestly. Some processes change the *raison d'être* of a place or its *genius loci* as for instance with changes to a town or region's industrial basis or nature. Other processes are more related to changed landscapes and townscapes although often these two processes merge with changes due to shifts in their industrial bases. Changes in the modes of production followed by an ongoing restructuring of the local economy may lead to changes in place identities and place images. Place reinvention links place making to both material and symbolic processes of change, and to the discourses and narratives that are associated with place images. Material and symbolic production of place are intimately linked and not two different processes. Materiality as landscapes and industrial base, and their representations, are tied together in the construction of narratives that create activities and reinvent places. Changes in the mode of production followed by an ongoing restructuring of the local economy may lead to changes in place identities and images; the symbolic representation of a place.

The symbolic production of place is not only obvious in place promotion and the selling of place, but also physical regeneration and the construction of flagship projects of particular architectural value can be powerful symbolic representations of place (Hubbard 1996). (See also Paddison, Chapter 18, this volume.) Re-imaging processes then involve physical reconstructions as well as semiotic work. Even more important than semiotic changes, are shifts related to narratives and discourses. Through symbolic expressions, places also communicate their identity (Kearns and Philo 1993), an important aspect in their struggle to be attractive. The idea of a 'cultured' economy lays the ground for the semiotic focus in efforts to develop place. Signs, images and symbols of place are carrying place-specific cultural values, and must be understood as cultural expressions. Through the construction of symbolic boundaries global flows allow people to construct their 'locality' in a range of ways (Appadurai 1990).

Reinvention of places is both haphazard and a matter of intention; it is both planned and something that just happens as a more or less unintended consequence of other ongoing processes. Thus, new place images are not only the results of strategic development processes aimed at profiling and promoting place, but also products of people's everyday life. Thus, reinvention takes place in the encounters between different types of actors; it is relational, dealing with the complex and multi-layered identities of place. But place reinvention is also guided by traditions, norms and values. And at the same time, different actors tell different stories about the same place.

Inventions are not only labels that fit the larger cities. Even places in the rural periphery can be defined as experimental, innovative, open, fluid and dynamic places concerned with re-imaging in order to adapt to a new global context. This is well illustrated in the example of the Norwegian town of Narvik. I draw on the content of Granås (2009) as a case study illustrating the dynamics of place marketing within place reinvention.

Dynamics of Place Marketing within Place Reinvention

Narvik's ice-free harbour led in 1905 to the construction of a 42km railway track for the transport of iron ore from Kiruna in Sweden via Narvik for further transportation on ships. Since then, the transportation of iron ore has been the main industry of the town. The recent history of this town located in mid-northern Norway tells a story of how a place image as an industrial place has been broken down and a more heterogeneous economy is taking over. A recession started around 1980 leading to a massive drop in employment in the railway industry and related industries. When the recession led to the phasing out in the 1990s of traditional industries, more professionalized marketing processes were introduced by actors within a new governance regime of the town, focusing on the uniqueness of the place. In the Narvik Brand, 'a city of extreme experiences', the town is presented as unique, based on perceptions of 'spectacular' nature and an emphasis of how this nature is suited modern lifestyles. According to Granås: 'The marketing of Narvik has, in a remarkable way, given the "wild north" a twist that turns its meaning away from an epithet and towards something very modern and widely appreciated' (Granås 2009: 121). The exoticism of nature is fully displayed through strong illustrations and textual descriptions. The 'spectacular' is directly connected to the lifestyle of extreme sports and hence something very modern and 'civilized'. The extreme sports culture is a global one, displayed through international TV channels, internet sites and icons, for example, snowboarding stars. The modern is only one symbolic aspect of the anti-periphery statements made here. Another is that of a playful, speedy, powerful and dangerous nature-based excitement. These are representations that confront the periphery as relaxing, dull and quiet. In this case, place marketing relates to an enduring political struggle against the political and cultural marginalization of a northern town. The place marketing initiatives in Narvik have contributed to the reassertion of the cultural dignity of inhabitants, including their identities as northerners and proud inhabitants of a town that 'still is modern' (Granås 2009).

The case also illustrates that strategic place promotion includes power constellations as marketing representations also reflect interest management among local actors. Place promotion 'also involve[s] the promotion of meaning categories that are embodied with power and have homogenizing effects on place discourses, where some voices are silenced and others are emphasized' (Granås 2009: 123). It is within this terrain of intentional and strategic place reinvention that place marketing operates. But the task of place marketing is not only that of selling the place; it includes imprinting on identity formations in a wider sense; to sell a place is also to be involved in the formation of the product that is put on the market.

Reinvention of place may also involve renegotiation of local identities. This is illustrated, below, through a discussion of place reinvention and cultural celebration of indigenuity, which draws on the work of Pedersen and Viken (2009).

Place Reinvention through the Cultural Celebration of Indigenuity

In the 1990s the indigenous perspective was seriously put on the international agenda and in Norway, Sami self-governance was institutionalized through the establishment of the Sami Parliament (Minde 2005). At the same time it became increasingly accepted that northern Norway was a multicultural society and that this was something to celebrate and not suppress as had been the case in the 1960s and 1970s. The process from suppression to celebration is well illustrated by the Riddu Riđđu in Gáivuotna (Kåfjord), a yearly indigenous music and art festival. This small community located in the county of Troms in northern Norway, with a population of not more than 2,200, appeared in the early 1980s as a mono-ethnic Norwegian municipality, and an introverted and moderately developed community. In 2008 the same municipality was marketed as a multi-ethnic, outward-oriented and modern place, and the municipality has become a model for how a minority culture can be reinvented (Pedersen and Viken 2009). Thus, the image of Gáivuotna changed greatly over a 15-year period. There are many reasons for this change, one of the most important being the decision to join the Sami Language Act in 1992 and the profiling of Sami heritage (Nyseth and Pedersen 2005).

However, from the middle of the 1990s the ethnic music festival Riddu Riđđu stands out as the most important development initiative concerning the regained position of ethnic minorities. Festivals are accepted as arenas for social and cultural innovation and Riddu Riđđu has been an arena for experimenting with ethnic emblems and expressions. The festival has developed into an international indigenous festival with 3,000–4,000 participants at this annual gathering of indigenous people in the circumpolar areas. Few festivals become societal institutions to the extent Riddu Riđđu has achieved.

Emerging within a particular historical setting, the festival was created in a period of economic crisis which, in retrospect, gave rise to the revitalization and modernization of coastal Sami culture. Behind the entrepreneurial beginnings of the festival was a group of well-educated and politically conscious young people. The name of the festival means 'storm from the coast', which describes well the character of the festival in the 1990s. Indeed, the message of the festival, its form and the social behaviour of those attending it provoked the local community, particularly the Leastadians, members of a religious sect with a strong position in the municipality.

The festival has also provided a space for identity negotiations. Riddu Riddu has made visible a coastal Sami identity as something separate and different from the inland Sami identity. Two points are important in this respect: one, the festival revitalized coastal Sami traditions within a number of life spheres: traditions which many thought were Norwegian; and two, it had an innovative and modernizing effect, opening a space for different forms of art (music, dance, theatre, movies) and for other indigenous peoples' cultural expressions. But at the same time Riddu Riddu brought ethnic antagonisms to the surface. In the 1980s Gáivuotna was about to become a fully accepted Norwegian community (Viken 2008), though based on suppression and historical rejection. The festival, however, cleared the air. Today, there are vital ethnic differences, but people have learned to live with them, and the either/or identities have changed to multiple identities; people are both Sami and Norwegians, or Kvens[1] and Norwegians. Moreover, the festival has been integrative by unveiling people's ethnicities. Today, people are open about their ethnic roots and priorities. The generation which is growing up in Gáivuotna today is not ashamed of being of Sami descent, but regards such an association with pride. They are a part of Gáivuotna, but they are also a part of the indigenous people of the world. The municipality's multi-ethnic status is accepted. In fact, in everyday life ethnicity is not a problem but a foundation for a vital community.

The innovative aspects of the Riddu Riddu festival are overwhelming, building networks all over the region as well as outside the circumpolar world. Riddu Riddu is also an example of political innovation. In political rhetoric Riddu Riddu became a prime example of how the resources in a multi-ethnic environment can be mobilized, and a model for how ethnicity can be used and converted into cultural, industrial, societal and political development. Gáivuotna today understands and promotes itself as a multi-ethnic, extroverted and modern place (Pedersen and Viken 2009). The branding and marketing of the municipality is based on its Sami origins. What the municipality is selling is an ethnic-based cultural property, their Coastal Sami identity (Nyseth and Pedersen 2005).

Summing Up

Place reinvention has in this chapter been used to capture how places change meaning as a consequence of continuous and strategic processes of place making and identity

---

[1]　　Kvens are people of Finnish heritage, immigrating to northern Norway from Finland in the 1800 and onwards.

building. Place reinvention directs attention towards the relationship between symbolic and imaginative changes, and the planned regeneration or place making initiatives. Place reinvention is a concept that perhaps can only be addressed at a very abstract and aggregated level. At the same time, it involves processes that are very concrete, physical and material in the sense that places actually are being changed through these processes. Places are reinvented through continued practices. Making places attractive to investors, as well as inhabitants, newcomers and tourists, depends on traditional economic systems and market trends, but is also influenced by discourses and meta-narratives, for instance discourses concerning alternative economies, symbolic aspects of place, and place branding discourse. Fishing villages in the northern part of Norway for instance are not only places where people make a living out of fishing – increasingly such places have become integrated in an experience economy, and as a consequence, such places reorient themselves through forms of branding as unique localities for visitors. Where cultural industries are replacing manufacturing industries, for instance tourism, signs and images of place are decisive. (See also O'Connor, Chapter 10, this volume.) The cultural power to create an image may become more important as traditional institutions have become less relevant mechanisms for expressing identity (Zukin 1995).

The forces that drive place reinvention then are numerous. Some of them are related to economic crises in the local economy. At a more profound level, overall changes in the economy, for instance the cultural shift, such as a shift from Fordist mass production and industrial manufacturing towards a cultural economy, are also important. Among the deeper changes are those related to globalization and increased competition expressed, for instance, through different processes such as global tourism at one end of the spectrum, and at the other end the revitalization of local identity. All places are more or less inscribed into the global discourses in their search for uniqueness based on local culture, heritage or nature.

Drivers can also be intentional and strategic, fronting the uniqueness of place, qualities related to, for instance, history, landscape, cultural attributes or industrial life, often in combinations and overlapping each other (Granås 2009). Identity politics implies strategic handling of place meanings that are communicable to inhabitants as well as towards external audiences. Such politics also involve the promotion of meaning categories that are embodied with power and have homogenizing effects on place discourses, where some voices are silenced and others are emphasized. It is within this terrain of intentional and strategic place reinvention that place marketing operates. Place marketing is often framed as an intentional practice instigated to enhance the 'attractiveness' and 'market value' of the place. But the task of place marketing is not only that of selling the place, it includes imprinting on identity formations in a wider sense: to sell a place is also to be involved in the formation of the product that is put on the market (Granås 2009). The complex making of tourist places involves for instance different actors drawing into the product different histories, values and lifestyles where some voices are included and others excluded (Førde 2009). New place images may therefore mobilize and legitimate particular sets of actions or policies (Jessop 1997), and there is a potential for new conflicts produced by place brands (Mommaas 2002). Images are social constructions, and as such never neutral. To produce images is to enact power. The construction, and reconstruction, of places is encapsulated within power relations, and is likely to be negotiated and politically contested (Granås 2009,

Holloway and Hubbard 2001). Place promotion policies have become a mandatory part of economic development policies in even the most remote communities in the north, partly as a consequence of tourism but also to attract new inhabitants from more urban regions of Europe.

By 'place reinvention' I refer to continuous and interwoven material, economic, political, social and cultural processes that transform the profile, image and associations connected to a geographical entity, in the way it is experienced and perceived both by people settled within the area and by people from the outside. The re-production of place image is not historically new. However, the strategic promotion of place that I refer to – by use of images – contains elements of newness, both in its spreading as a practice and the forms it takes.

Place reinvention then is a bridging concept, linking planned, intentional and strategic place development and the symbolic imageneering and promotion of place. Place reinvention not only combines the two, but goes beyond and widens the understanding of place transformation as a process that is related to the overall changes going on, also beyond the economic sphere, involving place perceptions, how people identify with place and how place conceptions are manipulated through place narratives.

Places are products of social and industrial activities, but today place development is also a question of choice, and matters for political negotiations, policy making and planning. The strategic work on place-reinvention is characterized by a conscious construction and development of images related to place. Thus, place development tends to be on the political agenda. Places are where structural development patterns intersect to trigger new regimes. And more than before, places are areas for political interests, negotiation and governance. A renewed focus on place development has emerged in a period with major shifts in the political culture, defining new rules of the game for local governments (Clark 2003). New governance regimes appear as consequences of a break with established ways of understanding politics. A number of different processes are intertwined, many beyond formal systems of 'conductors', for instance in the form of public–private partnerships. In order to strengthen attractive narratives of places, and meet the claims to authenticity and uniqueness, images are the objects of negotiation and strategically communicated (Hankinson 2001). Place governance with a planning orientation involves a complex mixture of political activity, technical expertise and moral sensibility (Healey 2010). Healey argues that deliberate place-development and management work involves mobilizing a particular type of imagination, one that 'sees' the places and spatial interconnections and recognizes the complex dynamics through which we experience place qualities as we and they evolve (Healey 2010: 230). (See also, Young, Chapter 23, this volume.) Within the field of local planning and policy performance there is, however, a lack of knowledge about both the effects and the processes of constructing place images and place promotion, as well as the associated negotiations, power relations and diversity of interests. There is a need to broaden the dominant perspective of place identity and representations of places, to analyse representations as part of the discursive structuring of the processes of local and regional development and policies. Place reinvention evolves through complex interactions between external demands and localized practices, ways of thinking and ways of acting which build up over the years, and broader forces which introduce new players, new ideas and new forces to be recognized, interpreted, mediated and struggled over.

References

Anholt, S. 2007. *Competitive Identity: The New Brand Management for Nations, Cities and Regions*. Basingstoke: Palgrave Macmillan.

Appadurai, A. 1990. Disjuncture and difference in the global economy. *Theory, Culture and Society*, 7(2), 295–310.

Augé, M. 1995. *Non-Places: Introduction to an Anthropology of Supermodernity*. London and New York: Verso.

Beck, U. 2000. *What is Globalization?* Cambridge: Polity Press.

Bramwell, B. and Rawding, L. 1996. Tourism marketing images of industrial cities. *Annals of Tourism Research*, 23(1), 201–21.

Brenner, N. and Theodore, N. 2002. Cities and the geographies of actually existing neoliberalism. *Antipode*, 34(3), 349–79.

Castree, N. 2003. Place: connections and boundaries in an interdependent world, in *Key Concepts in Geography*, edited by N.J. Clifford, S.L. Holloway, S.P. Rice and G. Valentine. London: Sage, 153–84.

Clark, T.N. 2003. Globalisation and transformation in political cultures, in *Globalism and Local Democracy: Challenge and Change in Europe and North America*, edited by R. Hambleton, H.V. Savitch and M. Stewart. New York: Palgrave Macmillan, 67–94.

Florida, R. 2002. *The Rise of the Creative Class: And How it's Transforming Work, Leisure, Community and Everyday Life*. New York: Basic Books.

Førde, A. 2009. Creating 'the land of the big fish': a study of rural tourism innovation, in *Place Reinvention: Northern Perspectives*, edited by T. Nyseth and A. Viken. London: Ashgate, 93–110.

Fretter, A.D. 1993. Place marketing: a local authority perspective, in *Selling Places: The City as Cultural Capital, Past and Present*, edited by G. Kearns, and C. Philo. Oxford: Pergamon, 163–74.

Friedman, J. 2002. From roots to routes: tropes for trippers. *Anthropological Theory*, 2(1), 21–36.

Gold, J.R. and Ward, V.W. (eds) 1994. *Place Promotion: The Use of Publicity and Marketing to Sell Towns and Regions*. Chichester: John Wiley.

Granås, B. 2009. Constructing the unique: communicating the extreme dynamics of place marketing, in *Place Reinvention: Northern Perspectives*, edited by T. Nyseth and A. Viken. London: Ashgate, 111–26.

Hall, T. and Hubbard, P. (eds) 1998. *The Entrepreneurial City: Geographies of Politics, Regime and Representation*. Chichester: John Wiley.

Hankinson, G. 2001. Location branding: a study of the branding practices of 12 English cities. *Journal of Brand Management*, 9(2), 127–42.

Harvey, D. 1989. From managerialism to entrepreneurialism: the transformation in urban governance in late capitalism. *Geografiska Annaler B: Human Geography*, 71(1), 50–59.

Healey, P. 2010. *Making Better Places: The Planning Projects in the Twenty-First Century*. London: Palgrave Macmillan.

Holloway, L. and Hubbard, P. 2001. *People and Place: The Extraordinary Geographies of Everyday Life*. Harlow: Prentice Hall.

Hubbard, P. 1996. Urban design and city regeneration: social representations of entrepreneurial landscapes. *Urban Studies*, 33(8), 1441–6.

Jensen, O.B. 2007. Culture stories: understanding cultural urban branding. *Planning Theory*, 6(3), 211–36.

Jessop, B. 1997. The entrepreneurial city: re-imaging localities, redesigning economic governance, or restructuring capital?, in *Transforming Cities: Contested Governance and New Spatial Divisions*, edited by N. Jewson, and S. Macgregor. London: Routledge, 28–41.

Johansson, O. and Cornebise, M. 2010. Place branding goes to the neighbourhood: the case of pseudo-Swedish Andersonville. *Geografiska Annaler: Series B, Human Geography*, 92(3), 187–204.

Kavaratzis, M. 2004. From city marketing to city branding: towards a theoretical framework for developing city brands. *Place Branding*, 1(1), 58–73.

Kavaratzis, M. and Ashworth, G.J. 2005. City branding: an effective assertion of identity or a transitory marketing trick?, *Tijdschrift voor Economische en Sociale Geografie*, 96(5), 506–14.

Kearns, G. and Philo, C. 1993. *Selling Places: The City as Cultural Capital, Past and Present*. Oxford: Pergamon Press.

Knudsen, B.T. and Waade, A.M. (eds) 2010. *Re-Investing Authenticity: Tourism, Place and Emotions*. Toronto: Channel View Publications.

Kotler, P., Armstrong, G., Wong, V. and Saunders, J.A. 1993. *Principles of Marketing*. London: Prentice Hall.

Massey, D. 1994. *Place, Space and Gender*. Cambridge: Polity Press.

—— . 2004. Geographies of responsibility. *Geografiska Annaler: Series B, Human Geography*, 86(1), 5–18.

Minde, H. 2005. The Alta case: from the local to the global and back again, in *Discourses and Silences: Indigenous Peoples, Risks and Resistance*, edited by G. Cant, A. Goodall and J. Inns. Christchurch: University of Canterbury, 13–34.

Mommaas, H. 2002. City branding, in *City Branding: Image Building and Building Images*, edited by V. Patteeuw and Urban Affairs. Rotterdam: Nai Publishers, 32–47.

Moor, L. 2007. *The Rise of Brands*. Oxford: Berg.

Murray, C. 2001. *Making Sense of Place: New Approaches to Place Marketing*. Stroud: Comedia.

Nyseth, T. and Pedersen, P. 2005. Globalisation from below: the revitalization of a coastal Sami community in northern Norway as a part of the global discourse, in *Discourses and Silences: Indigenous Peoples, Risks and Resistance*, edited by G. Cant, A. Goodall and J. Inns. Christchurch: University of Canterbury, 71–85.

Nyseth, T. and Viken, A. 2009. *Place Reinvention: Northern Perspectives*. London: Ashgate.

Pedersen, P. and Viken, A. 2009. Globalized reinvention of indigenuity: The Riddu Riđđu festival as a tool for ethnic negotiation of place, in *Place Reinvention: Northern Perspectives*, edited by T. Nyseth and A. Viken. London: Ashgate, 183–202.

Robinson, J. 2006. Inventions and interventions: transforming cities – an introduction. *Urban Studies*, 43(2), 251–8.

Rose, G. 1995. Place and identity: a sense of place, in *A Place in the World? Places, Cultures and Globalization*, edited by D. Massey and P. Jess. Oxford: Open University, 88–105.

Sandercock, L. 2003. *Cosmopolis II: Mongrel Cities for the 21st Century*. London: Continuum.

Shields, R. 1991. *Places on the Margin: Alternative Geographies of Modernity*. London: Routledge.

Skogheim, R. and Vestby, G.M. 2010. Kulturarv og stedsidentitet: kulturarvens betydning for identitetsbygging, profilering og næringsutvikling. NIBR.rapport 2010:14. Oslo: Norsk Institutt for By og Regionforskning.

van Ham, P. 2002. Branding European territory: inside the wonderful worlds of PR and IR theory. *Millennium: Journal of International Studies*, 31(2), 249–69.

Vik, J. and Villa, M. 2008. Brokete Bygdebilete – Om små bygder med store image, in *Den Nye Bugda*, edited by R. Almås, M.S. Haugen, J.F. Rye and M. Villa. Trondheim: Tapir Akademisk Forlag.

Viken, A. 2008. *Reinvention of Ethnic Identity: A Local Festival as a National Institution on a Global Scene*. Paper presented to Polar Tourism Research Network, Kangersuquujak, 21–25 August.

Ward, S. 1998. *Selling Places: The Marketing and Promotion of Towns and Cities 1850–2000*. London: Routledge.

Zukin, S. 1995. *The Cultures of Cities*. Malden, MA: Blackwell.

Case Study Window – Cultural Quarters and Urban Regeneration

John Montgomery

Cultural Quarters

> *The urban quarter is a city-within-a-city ... it contains the qualities and features of the whole. The urban quarter provides for all the periodic local (daily and weekly) urban functions within a limited piece of land dimensioned on the comfort of a walking citizen, not exceeding 33 hectares in surface and 10,000 inhabitants. Urban functions are zoned block-wise, plot-wise or floor-wise. An urban quarter must have a centre and a well-defined, readable limit. (Krier 1995)*

Cities, their economies and artistic development are inextricably interlinked. Without new work, a diverse division of labour, networks of exporting and producer services and technological innovation, cities become stagnant and may even die. To innovate, however, cities must be creative, in the development and application of new technologies, in bringing new goods and services to the market, and in the wider culture and artistic development (Bontje, Musterd and Pelzer 2011, Montgomery 2007).

Economic prosperity in the future will be dominated by city regions, larger cities but also smaller cities that are 'smart' and which 'punch above their weight'. Many cities and their regions are gearing up to compete as international centres for culture, fashion, finance and innovation. A major weapon in this battle is the quality of life, sense of place and way of living in a city. The cultural life of the city is thus not an add-on but a key point of difference, a specialism. For culture is the means by which cities express identity, character, uniqueness, make positive statements about themselves, what they are, what they do and where they are going. It also, increasingly, is one of the ways they make their living (Florida 2002, Landry 2000). Without a serious commitment to developing its creative economy, a city will be lacking one of the determinants of success (Montgomery 2007). (See also, Miller, and O'Connor, Chapters 3 and 10, this volume.) Cultural quarters are one means of helping to achieve this. (See also, Stevenson, and Sasaki, Chapters 9 and 12, this volume.)

The more recent meaning of the term cultural quarter dates from the early 1980s in the United States, for example in Pittsburgh and Lexington, MA (see Florida 2002: 304–14, and Whitt 1987). Cultural Quarters were proposed in the UK as long ago as 1987 by organizations such as the British American Arts Association (BAAA 1989) and the cultural consultancy Comedia (Bianchini et al. 1988). Culturally-led urban development began to appear as a concept in the urban planning literature from the late 1980s (see Boogarts 1990, Griffiths 1991, Montgomery 1990). Internationally, Sydney is now beginning to get in on the act, with creative industries only recently being taken seriously as economic sectors. The NSW Business Sector Growth Plan (2010) sets out a strategy for developing the creative industries economy in Sydney and across New South Wales. This includes promoting Sydney as a 'digital hub', a *Digital Media Initiative*, financial and business support for small creative enterprises, and the development of creative precincts at Walsh Bay, Ultimo and Sydney's Inner West. In Europe, examples now include the Cable Factory in Helsinki, the Westergasfabriek in Amsterdam, Circus Space in Belfast (Figure 20.1), Wood Green's chocolate factory, Marseilles's cigarette factory, Berlin's Brueri and more. Boston has its MASS MoCA, opened in 1999, and there is now even a similar complex at Factory 798 in Beijing.

Figure 20.1   Circus Space in Belfast
Source: John Montgomery.

Table 20.1 is a summary of the elements one would expect to find in a successful cultural quarter. These are presented under three sub-headings: *Activity, Form and Meaning*. It is important to stress that a good cultural quarter would contain a unique mixture of these elements. Thus a place which has good *Activity* but an inappropriate *Urban Form* will not be a cultural quarter in the sense of being a good place which attracts

Table 20.1   Cultural quarters: necessary conditions and success factors

Activity
• diversity of primary and secondary land uses
• extent and variety of cultural venues
• presence of an evening economy, including café culture
• strength of small-firm economy, including creative businesses
• access to education providers
• presence of festivals and events
• availability of workspaces for artists and low-cost cultural producers
• small-firm economic development in the cultural sectors
• managed workspaces for office and studio users
• location of arts development agencies and companies
• arts and media training and education
• complementary day-time and evening uses

Built Form
• fine grain urban morphology
• variety and adaptability of building stock
• permeability of streetscape
• legibility
• amount and quality of public space
• active street frontages
• people attractors

Meaning
• important meeting and gathering spaces
• sense of history and progress
• area identity and imagery
• knowledgeability
• environmental signifiers

Source: John Montgomery 2003.

everyday users and visitors, but rather a place (most likely) of cultural production removed from the arena of consumption. This means that cultural quarters and indeed the wider notion of city creative economies cannot be considered in isolation from the geography and characteristics of urban places. Similarly, a cultural quarter without *Meaning* will not be much of a place. Nor will it tend to be contemporary, avant-garde or particularly innovative. More than this, a cultural quarter which produces no 'new' *Meaning* – in the form of new work, ideas and concepts – is all the more likely to be a pastiche of other places in other times, or pe;rhaps of itself in an earlier life. A good cultural quarter, then, will be authentic, but also innovative and changing.

To remain successful, a city economy, even an individual enterprise will need to maintain what it is good at but also to be flexible, highly adaptive and embrace change, new ideas, new ways of doing things and new work. Failure to do so will mean that the cultural quarter will disappear entirely, or become simply a collection of publicly-funded venues and facilities, or else an emblem of former culture – 'heritage'. (See Ashworth, Chapter 11, this volume.) Some cultural quarters will, no doubt, deserve to ossify or disappear altogether, to be taken over by other competing uses (offices, apartments) or to become part of the heritage industry. Others might well continue to develop and grow. This brings us to a conundrum, in that at least a proportion of the activity found in cultural quarters might well require governmental support in order to survive *in situ*.

Cultural quarters ought to be places where artists and designers create *new work*, for it is only through the ongoing process of adding new work to old work that culture develops, and that economies grow. Second, cultural quarters will tend to offer a variety of *opportunities for artistic work to be consumed* – in galleries, theatres, music halls and other venues, including public spaces. (See also Bianchini, Chapter 22, this volume.) The work available will tend to cover a range of art forms, through visual and performing arts. The 'products' will be texts, performances, objects, images and sounds. These will be distributed to markets, both locally and globally, via more traditional means as well as IT. They will be delivered within a mixed economy, ranging from fully commercial work, products and services, to varying degrees of government subsidy. Customers can therefore be found in *situ*, across the wider city region, inter-state and internationally. Very often, there will be close links with formal *education institutions* which provide education and training, a flow of new artists and entrepreneurs, research and development.

Table 20.2   Indicators of good cultural activity

- cultural venues at a variety of scales, including small and medium
- festivals and events
- available workspaces for artists and low-cost cultural producers
- small-firm economic development in the cultural sectors
- managed workspaces for office and studio users
- location of arts development agencies and companies
- arts and media training and education
- art in the environment
- community arts development initiatives
- complementary day-time uses
- complementary evening uses
- stable arts funding

Source: John Montgomery 2003.

Cultural quarters also tend to be places of mixed use, providing a range of *complementary activities during the day-time and at night*. Good day-time uses include speciality retailing (book stores, music shops, avant-garde fashion), private galleries and venues, cafes and restaurants. Good evening and night time uses can include all of these, plus music clubs. Cultural quarters are places where art and creative activity is produced and consumed, where people (artists and customers) may be educated and entertained, and where the ambience is such that people come *simply to hang out* and be seen. Cultural quarters will tend to have a lively street life, with varied comings and goings across various times of the day, and will offer a *public realm* or set of spaces which people are attracted to and spend time in. Table 20.2 is a fairly straightforward listing of the types and range of cultural activity (deemed necessary success factors in a cultural quarter). Headings under which the comparisons are made focus mainly on the presence or otherwise of cultural activity, and include those listed in the table.

Temple Bar, Dublin

Temple Bar is sandwiched between O'Connell Bridge to the east, Dame Street to the south and the River Liffey to the north, in central Dublin. The area dates largely from the eighteenth century when cargo was loaded and unloaded on ships docking at the quays on the south side of the Liffey. The future of Temple Bar was under doubt for many years, not least because the state bus company (CIE) proposed to redevelop most of it as a new transportation centre, linking bus and rail. CIE began to buy up property in 1981, paving the way for demolition and redevelopment. Paradoxically, the fall in property and rental values which resulted triggered off a process of revitalization. Activities which could afford only low rents on short licences – or no rent at all – moved into the area. These included artists' studios, galleries, recording and rehearsal studios, pubs and cafes and restaurants, second hand and young designer clothes shops, books and record stores, as well as a number of centres for a range of 'third sector' organizations. They added an exciting mix of ingredients to those remaining existing businesses which had not yet been bought out or evicted by the CIE – the printers, cutlery shops and seedy hotels. During the mid-1980s, networks of small and medium-sized enterprises became established, feeding off each other and larger cultural players such as the Project Arts Centre and the Olympia Theatre. Here we had a rare example of planning blight breathing life into an urban area through low-rent arts activities.

By 1990, Temple Bar had many disused industrial buildings, gap sites, problems of poor east/west permeability, a residue of entrepreneurial activity, and many buildings which were simply falling down. Paradoxically, Temple Bar also had a reputation as a place of discovery, vitality and a wide range of social and economic exchange. It was frequently referred to as 'Dublin's Left Bank', on account of its relatively high density, a mixture of architectural styles, close proximity to the quay, narrow streets and a lively atmosphere deriving mainly from youth culture – recording studios, video companies, artists' studios, theatres and pubs, cafés and restaurants.

It was important that this alternative culture should not be lost by wholesale redevelopment of the area or by adopting a property value-led approach to urban renewal. Much needed to be done in Temple Bar, not least to prevent the building stock falling into greater disrepair. But great care had to be taken not to destroy the sense of place that had already been created by the mix of activities that were based there. Businesses and arts organizations in the area organized themselves to form the Temple Bar Development Council as early as 1989, and began to lobby for the area to be regenerated as a cultural quarter.

The strategy that was adopted to achieve this was, accordingly, a combination of culturally-led urban renewal, physical renewal and local enterprise development, particularly in connection with the cultural industries and the evening economy (Urban Cultures Ltd 1991/2). This was based on:

- adoption of a stewardship ethos and management approach to knitting back together the urban area;
- adoption of 12 cultural projects to act as urban 'chess pieces': localized strategic interventions to create activity and interest – these include a Film House, sculpture gallery, photography gallery, music venues and the old Olympia Theatre;
- provision of business grants and loans to help young cultural and other entrepreneurs set up in business; this was accompanied by a survey of existing businesses in the area;
- a major training initiative in business skills and the various cultural industries, but also in catering and venue operation;
- promotion and stimulation of an evening economy;
- a major initiative to improve permeability and pedestrian flow through the area, involving the creation of two new public spaces, outdoor venues, niche gardens, corners to sit and watch the world go by, culminating in the design of two new public squares;
- a major programme of public art and cultural animation, designed to reclaim and give meaning to the area's public realm
- an overall approach to property management and upgrading based on balancing the need to improve the area's environment with the need to retain existing activity;
- the introduction of vertical zoning linked to the provision of grants and tax relief status;
- design of new buildings by young Irish architects, with the accent on modern design within the context of the historic street pattern;
- a major marketing and information campaign using good modern design.

This strategy, in the form of a flexible framework plan (Temple Bar Properties Ltd 1992) was largely implemented by Temple Bar Properties Limited, a state-owned development company established in 1991. Temple Bar Properties engaged in acquiring properties, renewing them and negotiating rents with occupiers and by undertaking development schemes on its own volition or as joint ventures with private owners and developers. To do this, it was granted an initial £4m from the EC and leave to borrow £25 million privately but with a state guarantee. Monies generated from rental income were ploughed back into the property renewal programme and environmental action, and used to cross-subsidize cultural projects. In the final analysis, a total of public funding for Temple Bar was some IR£40.6 million, the bulk of which (£37 million) was spent on the Cultural Development Programme 1991–2001. A further £60 million has been borrowed and repaid through TBPL's commercial programme. Over 1991–2001, the private sector is estimated to have invested over £100 million in the area.

In 1992 there were 27 restaurants, 100 shops, half a dozen arts buildings (some of them falling down), 16 public houses, two hotels, 200 residents, 70 cultural industry businesses and 80 other businesses in Temple Bar. By 1996, when most of TBPL's own development schemes had been completed, there were five hotels, 200 shops, 40 restaurants, 12 cultural centres and a resident population of 2,000 people. During the construction phase, some 5,000 yearly full-time equivalent jobs had been created in the building industry (most sub-contracted to Dublin companies). By the end of 1996, there were an estimated 2,000 people employed in Temple Bar, an increase of 300 per cent. The final major commercial development to be undertaken by Temple Bar Properties was the Old City, designed to create a significant retail and residential cluster in the area between Parliament Street and Fishamble Street in the western end of Temple Bar. Situated around a new pedestrian street, Cow's Lane, the development consists of 191 apartments, 24 retail units, a crèche and landscaped gardens.

Temple Bar Properties Ltd as a property development company was effectively wound up in 2001. However, an area management body has been left in place to maintain the cultural venues, continue the animation programme and undertake area-focused services such as street cleaning. By that time, Temple Bar was home to some 3,000 residents, some 450 businesses employing close on 2,500, and 12 high quality cultural venues. The whole area had become a showcase for urban design, architecture, design and style (see Figure 20.2).

Figure 20.2   Temple Bar Gallery
Source: John Montgomery.

Success, however, brings its own problems. By the late 1990s, Temple Bar had developed a reputation as the 'Stag Night' capital of Europe. Research commissioned by Temple Bar Properties at that time revealed that these visitors – largely groups of young women and men from England – were beginning to cause other visitors to stay away. This problem was addressed by a coordinated management response by landlords and hoteliers in the area, essentially by refusing accommodation to large same-sex groups. There is also the temptation to assume that the redevelopment of Temple Bar was all plain sailing, more or less straightforward instalments of commercial property development. This view rather conveniently over-looks the fact that the area's other future could have been as a glorified bus depot. Moreover, it is not readily understood that the recession of the early 1990s and the high interest rates of that time, almost led to Temple Bar Properties going into liquidation before it had made much of an impact. The later boom in Ireland's economy certainly played a large part in the success of Temple Bar, but in the early days there was no boom. Just risk.

Cultural writers acknowledge the importance of Temple Bar as an example of using creativity and design to re-establish an area's economy and sense of place. Florida, for example, writing in 2002, refers to Temple Bar as a 'clever and far-reaching strategy of levering authentic cultural assets to attract people and spur economic revitalisation' (2002: 302). That said, it is also true that many artists have left the area to find cheaper rents elsewhere in Dublin, although many remain in the area, notably at Temple Bar Gallery and Studios (Figure 20.2) and other cultural centres.

Sheffield Cultural Industries Quarter

The area of the Sheffield Cultural Industries Quarter is defined in Sheffield City Council's CIQ *Area Action Plan* as extending to some 30 hectares and located just to the southeast of the city's administrative, retail and commercial core. Since 1984 the area has been undergoing a transformation. By the mid-1980s, the CIQ had become a classic *zone of transition*: a marginal area of the city centre which was once a thriving industrial and workshop centre, but had become characterized by vacant and derelict buildings and gap sites. Slowly at first, but with a marked quickening in the pace of development from the mid-1990s, the CIQ is now recognized as a centre for a wide range of cultural production. This includes fine arts, photography, film making, music recording, graphic and product design. Important initiatives within this spectrum include the Yorkshire Arts Space Studios, the Audio Visual Enterprise Centre, the Leadmill night club, Red Tape Studios, the Sheffield Science Park, the Site Photography Gallery, the Workstation managed workspace, and the Showroom Cinema complex. The Quarter is also home to the Yorkshire Art Space Society, the Untitled Photographic (now the Site) Gallery and a cluster of some 300 small businesses related to film, music and TV, design and computers. Important links have been established between the Cultural Industries Quarter and the adjacent Science Park, particularly with regard to the development of new technology in film, photography and recording, while there are now proposals to develop a Culture Campus in the area which would house Sheffield Hallam University's fine art, media studies and design departments.

In the early 1980s the City Council started to develop a cultural industry policy, aimed at supporting these activities and assisting the economic regeneration of a former car showroom. Two resultant building-based projects – The Workstation (Figure 20.3) and The Showroom – are seen as central to the development of the CIQ. The buildings are owned by Sheffield City Council and were developed by a specially formed registered charity, Sheffield Media and Exhibition Centre Limited (SMEC). The Charity set up a development subsidiary, Paternoster Limited, which took a 125-year lease on the building. Paternoster Limited runs The Workstation as a commercial enterprise, charged with operating the building for the benefit of its tenants, and covenants profits to the parent charity (SMEC) for the benefit of the Showroom Cinema operation. The Showroom also receives revenue grant support from Sheffield Arts Department, the British Film Institute and Yorkshire and Humberside Arts.

Figure 20.3   The Workstation, Sheffield

Source: John Montgomery.

Over 70 organizations now occupy units in the building ranging from the Northern Media School, graphic designers The Designers Republic, the Community Media Association, the Yorkshire Screen Commission, and various film production companies such as Picture Palace North and Dream Factory. Typical of the sector, tenant companies are small to medium size, employing from two to six staff members, although certain companies employ 25 and upwards. By 1997, despite the impressive growth of new organizations, facilities and venues, the CIQ lacked a strong sense of place. There were very few shops in the area, and few bars other than some traditional pubs catering for students of Hallam University. An important strand of both the 1998 *CIQ Vision and Development Strategy* and the *Action Plan* for the CIQ (EDAW and Urban Cultures Ltd 1998) was to encourage secondary mixed-use, particularly along ground floor frontages. This was to include small shops, alternative retail, cafes, bars and restaurants.

In early 1999 the CIQ Agency was established to promote and implement an agreed Development Strategy for the CIQ over a five-year period. The Agency's mission was further to develop the CIQ, building on the successes and, importantly, the broad character and nature of the CIQ as a cultural production centre and as an urban place. The aim was to create a thriving cultural production zone with large numbers of small and medium-sized enterprises, a centre for excellence in knowledge creation and creativity, a visitor destination and a largely mixed-use area with various complementary activities throughout the area generating pedestrian flow throughout the day and into the evening, including residences. The Agency was comprised of a non-executive board with members being drawn from local businesses, Sheffield Hallam University, the Science Park and Sheffield City Council. The board is supported by a small, full-time management team.

The Agency's five-year strategy (to 2004) had the following targets:

- 50 active exporting firms in the cultural industries and an overall doubling of the businesses base within the area;
- an additional 350,000 sq. ft of workspace for the cultural industries;
- 4,200 jobs (3,000 direct, 1,200 indirect), of which 2,500 would be net additional jobs;
- 50 new retail, catering and entertainment outlets provided by private investment;
- 500 additional permanent residents;
- completion of a new urban culture campus for Sheffield Hallam University;
- completion of a number of cultural projects, including a Photography Gallery, Fine Art Gallery and Centre for Performing Arts;
- continuation of a two-pronged training and education strategy aimed both at enterprise development and community access to new technologies and the cultural industries;
- considerable upgrading of the urban public realm to provide a more pedestrian-friendly environment, including established access points and routeways and two new urban squares.

Substantial private sector investment has been attracted into the area in the form of bars, night-clubs, restaurants and student apartments. A major visitor attraction, the National Centre for Popular Music, opened in 1999, although this turned out to be a failure, the only one of such a high-profile nature in the area. The SCIQ as a whole continued to grow. A new 1,000-capacity bar and live music venue opened in the summer of 2000, Red Tape Studios launched a new Internet School, Modal's National Music Convention has been attracted to the Quarter, as has the International Documentary Film Festival. Yorkshire Artspace developed a new building, the old Roberts and Belk cutlery factory was developed as a managed workspace with ground floor cafe bar, and the former Leadmill Bus Garage was redeveloped as offices and apartments. Mixed live-work units were also developed, notably The Cube. Estimates by the CIQ Agency revealed that the CIQ was by 2002 home to some 270 businesses and organizations, including film, TV, radio, science and technology, new media, training and education, live performance, music, arts, crafts, metalworking and a range of support producer and consumer services. By 2006 the CIQ was almost completely developed.

Manchester Northern Quarter

Manchester's Northern Quarter lies just to the north and east of the main shopping area around Market Street and the former Arndale Centre. For many years its major streets – notably Oldham Street – were the most fashionable of all shopping streets in the city. Tib Street was once renowned for its choice of pet shops; Smithfield market was once the largest fish and poultry market outside of London. The area began to take shape in the late eighteenth century at which time Stevenson Square was laid out as a mirror to the fashionable St Anne's Square to the west. Oldham Street developed initially as a mix of private dwelling houses and small businesses, but by the late nineteenth century had become a fashionable destination offering shopping and tea rooms which were popular with ladies. By the early twentieth century Oldham Street had two large department stores (Affleck and Brown, and Lomas) as well as a Marks and Spencer penny bazaar and a Woolworths. There was also a range of popular pubs and eating places, including a Yates's Teetotal Tavern.

The area began to lose vitality and meaning from the late 1940s. Initially, two processes were at work: first, the onset of decline in the textiles industry meant that parts of the area and – equally important – Ancoats to the east, began to lose economic activity, businesses and jobs. This had a knock-on effect on the economy of Oldham Street and environs. Second, the wholesale slum clearance of the northern quadrant of the area and much of Ancoats and other adjacent areas effectively removed the resident population that had sustained many businesses in the area. Smithfield Market was relocated to the outskirts of Manchester in the 1970s, at about which time the Arndale Centre was built. The effect was to tear the remaining life out of what had been one of the most vibrant quarters within the city. Many businesses closed or moved away, and few people remained living locally.

By the late 1980s, the area had few businesses left, mainly fashion wholesalers, some specialist shops (prams, pets and pianos) and a few drinking and transvestite clubs. Manchester City Council became concerned that the area was not regenerating as other parts of the city were at that time, notably Castlefield and the Whitworth Street Corridor. This was despite the fact that the area had been granted Commercial Improvement Area status, allowing landlords and shopkeepers to apply for shop-front improvement grants. By 1993, 27 per cent of the floor space in the area was vacant, many gap sites had appeared and a large proportion of the building stock was in serious disrepair. By 1991 only 345 people were living in the area. However, several interesting things *were* happening across the city. Tony Wilson,[1] an executive at Granada Television and music industry entrepreneur, established Factory Records, a recording label for what would become the Manchester Bands: the Happy Mondays, Joy Division, New Order. Wilson also opened the Hacienda nightclub and Dry Bar, a 'new generation' urban bar. There was also the beginnings of regeneration in Castlefield, led by local bookmaker Jim Ramsbottom and a team of young development professionals and architects. As well as the redevelopment itself, this would lead to the emergence of a network of design professionals and architects. The setting up of the Manchester Institute for Popular Culture (MIPC) with a brief to research popular culture, was also important.

---

[1] Tony Wilson was the founder of Factory Records, and also a TV producer at Granada. The film *24 Hour Party People* is about him.

Amongst other things, an early contribution was a major study of 'The Culture Industry' (Manchester Institute for Popular Culture 1990).[2] There was also a timely recognition of the audio/visual industries (film, television, photography, video) as an important sector in a study by Comedia in 1988.[3] All of these events were of crucial importance in what was to follow.

From about the late 1980s, a number of new businesses moved into Oldham Street in particular, including the Afflecks Palace (fashion emporium), the Dry Bar (Figure 20.4), PJ's Jazz Club and others, while some of the other clubs and venues had remained in the area, including Band On The Wall. These businesses formed the Eastside Association in the early 1990s to lobby for improvements in the area.

Figure 20.4   Dry Bar, Manchester
Source: John Montgomery.

---

[2]   It was later re-published by Derek Wynne as *The Culture Industry: The Arts in Urban Regeneration*.

[3]   *Film, Video and Television: The Audio-visual Economy in the North-west*, December 1988, Comedia Consultancy, for the Independent Film Video and Photography Association, North West Arts, Manchester City Council, the British Film Institute, Channel 4 and Lancashire County Council.

In November 1991 Manchester City Council commissioned a major study of the arts and cultural policy of the city (Urban Cultures Ltd 1992). This study proposed a drive towards mixed use and the creation of distinctive quarters, one of which was the Northern Quarter. In late 1993 a development strategy for the area was commissioned jointly by the Eastside Association and Manchester City Council.[4] The main thrust of the report was to retain the existing rag trade but also to grow new businesses in creative activities, alternative shopping and the evening economy. In addition, a major programme of residential conversion of upper floors was advocated. In analysing property ownership in the area it was recommended that the implementation vehicle for the area should be an alliance of landowners, developers and the City Council with a much better resourced Eastside Association (renamed the Northern Quarter Association) having a prominent role in crafting development schemes for the area. A system of sticks and carrots was put in place to help convince developers and property owners to engage with the overall area master plan; these included a series of small and large grants, and sophisticated application of planning controls up to and including detailed design briefs for key sites and properties.

The MNQ development strategy was specifically aimed at encouraging business start-ups and growth in the creative industries, largely as market-based enterprises. Partly to this end the Cultural Industries Development Service (CIDS) was formed in 2000 to help develop sustainable cultural and creative enterprises in Manchester's metropolitan core, including the MNQ and Ancoats (see Blanchard 1999). CIDS' services included: an information and referral service, a student and graduate placement service, business start-up and expansion grants, industry marketing grants, network development, and professional development programmes. CIDS has now been wound up (O'Connor and Gu 2010), but it can point to many successes. The Creative Industries and Digital Media sector in Greater Manchester now employs around 53,000 people with more than 7,000 businesses. This sector is forecast to grow by 19 per cent during the next decade (NESTA 2009). Manchester is now the UK's largest 'creative hub' outside London. The creative industries now account for 6 per cent of all jobs in Manchester.

It took almost ten years for larger projects such as Smithfield to get underway. Even so, many new schemes were successfully completed by 1999, including several mixed-use retail and residential schemes along Oldham Street and individual four- and five-storey buildings throughout the area. By 2002, the Northern Quarter was the location for over 550 businesses and organizations, although a good proportion of these were already trading in the area before 1995. It had become a place for shopping, music, food and drink, entertainment, fashion, living and working. This included 100 clothing and fashion outlets; 70 cafes, pubs, bars, restaurants and clubs; 50 voluntary organizations; 40 arts, crafts and jewellery shops; 20 vinyl, tape and CD shops; ten hairdressers/ barbers; seven newsagents; and another 200-plus unique specialist shops, services and suppliers.

The MNQ is an example where public intervention and monies have tended to be focused on building improvements (via grants), environmental works (including public art) and improving transport, parking and access.

---

[4]   The report was prepared by Urbanistics, a consortium of Johnson UDC, Urban Cultures Ltd, Stevenson Architects and the Manchester Institute for Popular Culture.

Comparatively little additional money has been found for new venues or events. Rather, the tack has been to encourage enterprise development of the creative industries, to help bring properties into active use and to invest in area marketing. This has meant that the MNQ is more a fashionable residential neighbourhood than a cultural industries quarter (Brown, Cohen and O'Connor 2000). There is a sense then that the MNQ did not fully deliver on its early promise as an urban cultural quarter.

Comparisons and Lessons

The development and maintenance of *cultural venues* has been most ambitious and consistent in Temple Bar, with some successes also in Sheffield. The weakest performer is Manchester MNQ, where comparatively few new venues that are open to the public for viewings, screenings or performances have been established. Temple Bar also offers the most varied programme of *festivals and events*. The Northern Quarter Festival in Manchester operates for only three weeks of the year. Sheffield Cultural Industries Quarters has a poor programme of cultural animation. As for *workspace for artists*, the most impressive achievements have been in the Sheffield Cultural Industries Quarter (recording studios, artists' studios, film companies' centre, live/work units) which deliberately focused on cultural production more than on venues. With the exception of Temple Bar Gallery and Studios, the approach in Dublin has favoured cultural venues over workspaces, while Manchester MNQ has been less successful in securing any low-cost spaces available for artists and crafts workers. In Manchester there has also been very little public sector involvement in developing *managed workspaces* for micro and small studio and office-based businesses. This contrasts markedly with the approach in Sheffield, with the development of The Workstation and other developments. Of the quarters, only in Sheffield and Manchester were there active programmes of *arts and cultural industry* business development, and in the case of MNQ this was effectively the cultural programme for the area.

What most if not all the cultural quarters have in common is the presence either within the quarter, or very close to it, of a major *arts education* and/or training institution, whether it's the Northern Media School in Sheffield, or the fact that Temple Bar is adjacent to Trinity College, and the Manchester Institute for Popular Culture has an office in the MNQ. In the case of Sheffield this outcome was quite deliberate, as part of a strategy to strengthen the links between formal education and enterprise development. All the quarters researched had a *public art* policy, and a programme for securing new work, although not always by direct commission. Although to some extent a subjective judgement, the public art programmes in Manchester and Dublin are the most impressive.

The extent to which the various cultural quarters actively promoted *community arts development* was patchy. As in other matters, the approach in Temple Bar is to build a responsibility for community development into the overall remit for several of the cultural venues, particularly in the case of the Ark (a children's art centre). Sheffield CIQ took a more direct approach to this issue, with a cross-programme emphasis on access and participation. Manchester MNQ is weak on this indicator.

Temple Bar from the outset encouraged a fine-grained mix of uses, and especially fashion and other alternative shops, cafes, restaurants and galleries. This has included maintaining low rent levels for certain types of uses. The early aspiration to become a '24-hour city' was toned down, and new residential developments in the Old City are sited away from bars and pubs in particular. One of the problems in Sheffield CIQ was a lack of street life, largely because of a lack of good active street frontages; this problem was addressed by the *CIQ Vision and Development Study* (EDAW/Urban Cultures Ltd 1998), and reasonable progress was made. Manchester's MNQ has more complementary activities – both day and night – than cultural activity *per se*. In Manchester MNQ, the aspiration to develop an evening economy and greater mixed use remains largely just that, an aspiration. With only a few exceptions, most of the new property investment in the area in the past five years has been for mainstream offices and apartments.

All of the cultural quarters discussed here have good *urban place* characteristics. That is to say, they are relatively compact and walkable urban districts, are more intensely developed than suburban neighbourhoods, have a stock of adaptable buildings, of varying ages, are highly permeable, easy to read and range in scale from two-storey to at least five-storey buildings. The Sheffield CIQ had more gap sites awaiting development than the other cultural quarters. The main route remains Paternoster Row, yet good pedestrian traffic is found on Howard Street and Charles Street. The Manchester MNQ offers perhaps the grandest stock of buildings of all the quarters researched. Oldham Street remains the main thoroughfare, although Tib Street has become a more prominent location for design businesses. Temple Bar remains the cultural quarter where the most effort has been invested in improving permeability (to the extent of creating at least one new street and a new pedestrian bridge), while traditional street patterns and build-to lines have been respected and adhered to.

Temple Bar is also the only quarter to have invested positively and quite deliberately in new *public spaces*, Temple Bar Square and Meeting House Square, even where some demolition of existing properties was required. In Sheffield, there are admittedly two new squares, one adjoining the NCPM and the other at the top of Howard Street adjacent to Hallam University. The former was under-funded and is poorly realized; the latter, named Hallam Square, has more to do with the university calling attention to itself than any strategic intervention in the public realm. The Manchester MNQ has seen little in the way of new public space, and even the long-mooted re-design of Stevenson Square has remained on paper. What new public space there is has usually been an adjunct or planning gain from a commercial office or residential development permission. There is no real strategic approach to new public space in the MNQ. What all of the quarters have generated is much improvement in the quality of their main streets and in street life more generally. This is clearly partly to do with the wider acceptance in planning terms of the value of cafe culture and cuisine, and also to the 24-Hour City concept, albeit it (sensibly) watered down in most cases.

Lastly, all of the quarters have succeeded in accommodating *new activity*, some of it cultural, because of the availability of premises at low rent and offering variety in building styles and types, and in floor spaces and layouts. In Temple Bar old warehouses and town houses were converted to new uses; in Sheffield it has been car showrooms and old cutlery works. In Manchester, buildings originally designed as retail stores with apartments above have converted to showrooms, galleries, studios, offices and residential apartments. None of this is perhaps especially surprising, yet there are still towns and cities which make the mistake of developing all-new cultural precincts away from traditionally laid-out urban areas (or indeed demolished traditional street patterns to make way for large architectural envisionings). The lesson here is that people respond better to mixed use areas, with a human scale and a legible townscape; while if there is no cheap property available, there will be no space for artists and start-up businesses to occupy.

Urban Policy, Market Forces and Management Mechanisms

There are clearly some major differences between the case study examples, not just between the places themselves but also in terms of the *types of intervention* (physical, planning, economic development, cultural) pursued by public agencies, and in the level of public funding. Temple Bar is clearly a direct intervention model where a bespoke, state-owned, development company was established – with a strong cultural remit – to drive forward the area regeneration. The result includes 12 new or fully refurbished cultural venues in the middle of the city. If anything, Temple Bar attracts criticism for having succeeded too well, although such an outcome was not guaranteed, certainly in the early years.

Progress in the Sheffield Cultural Industries Quarter has been less spectacular (and over a longer period) but has also been impressive. Here the early emphasis was on economic development of the cultural industries supported by access, training and education programmes: a strategy for increasing cultural production. This has been achieved through a more selective programme of property acquisition and development, notably of The Showroom and The Workstation, but also AVEC and YASS. In later years, attention extended to cultural consumption in the form of encouraging the private sector to invest in retail units, bars and clubs, and also residential accommodation. The commitment to small firm development and training, however, has not been diluted.

The example of Manchester's MNQ reveals that where property market failure can be shown to be temporary (1993 was, after all, the depths of the then property crash and economic slump) then area regeneration can proceed by a process of planned and brief-led development, backed by some modest area branding, public art and a small events programme. Progress in the MNQ was much slower than in Temple Bar and SCIQ, certainly as regards development of cultural facilities and venues. Rather, efforts focused on micro and small-firm development of the creative and media businesses across Manchester. Development in the MNQ has progressed, although some of the larger schemes have taken a long time to get started. MNQ is arguably more successful as a city centre residential and retail neighbourhood than as a cultural quarter *per se*. The most market-oriented example, MNQ, is probably the least successful culturally. The most interventionist initiative was Temple Bar, yet it curiously also seems the most organic and natural. SCIQ took many years to achieve greater diversity, although private sector investment is now at unprecedented levels. The bottom line is that it is possible to plan a cultural quarter, but resources will also need to be committed to a cultural programme, low-cost studio and business space and public realm improvements. Otherwise the cultural life of any such quarter is in danger of being undermined by rising rents.

The overall conclusion is that all of these examples have been strategic in nature. Whether directly intervening in the property market, or opting to self-build particular strategic projects; whether directing the property market through development briefs and planning permissions, or employing a property leasing plan – all have had a vision and a strategic goal to become a cultural quarter. The type of cultural quarter may have varied (mostly consumption led in Temple Bar, production led in Sheffield, design and fashion led in Manchester) but all three explored the relationships between regeneration, creativity and the cultural economy. All these places have succeeded beyond initial expectations. Yet, all too face the problems of success, notably in the guise of land value colonization and gentrification.

Finally, it must be stressed that not every urban area can or should be a cultural quarter. (See also Sasaki, Chapter 12, this volume.) Cultural quarters only work where there are venues, workplaces for cultural producers and working artists. There is little doubt that cultural activity, urban characteristics of place and a little style add greatly to the urban and economic mix. More than this, cities that do not offer lively small business and cultural districts will be left behind. There are complex dynamics at work in such places; they are not all clones of each other.

References

Bianchini, F., Fisher, M., Montgomery, J. and Worpole, K. 1988. *City Centres, City Cultures: The Role of the Arts in Urban Revitalisation.* Manchester: CLES.

Blanchard, S. 1999. *The Northern Quarter Network: A Case Study of Internet-based Cultural Business Support.* Manchester: Manchester Institute for Popular Culture.

Bontje, M., Musterd, S. and Pelzer, P. 2011. *Inventive City Regions: Path Dependence and Creative Knowledge Strategies.* Aldershot: Ashgate.

Boogarts, I. 1990. *A New Urban Planning Tool Kit: Are Investments in the Arts and Culture New Tools for Revitalising the City?* Paper presented at the 6th International Conference on Cultural Economics. Umea, Sweden, June 1990.

British American Arts Association (BAAA). 1989. *Arts and the Changing City.* London: British American Arts Association.

Brown, A., Cohen, S. and O'Connor, J. 2000. Local music policies within a global music industry: cultural quarters in Manchester and Sheffield. *Geoforum*, 31(4), 437–51.

EDAW/Urban Cultures Ltd. 1998. *Sheffield CIQ Strategic Vision and Development Study.* Sheffield: Sheffield City Council.

Florida, R. 2002. *The Rise of the Creative Class.* New York: Basic Books.

Griffiths, R. 1991. *The Role of the Arts in Urban Regeneration.* Paper presented at a seminar on Revitalizing City Centres and Restructuring Industrial Cities, University of Lodz, Poland.

Krier, L. 1995. Charter of the European City. Paper presented to the conference, The European City – Sustaining Urban Quality, Copenhagen, 24–28 April 1995.

Landry, C. 2000. *The Creative City.* London: Earthscan.

Manchester Institute for Popular Culture. 1990. *The Culture Industry.* Manchester: Manchester Institute for Popular Culture.

Montgomery, J.R. 1990. Cities and the art of cultural planning. *Planning Practice and Research*, 5(3), 17–24.

—— . 2003. Cultural quarters as mechanisms for urban regeneration: conceptualising cultural quarters. *Planning Practice and Research*, 18(4), 293–306.

—— . 2007. *The New Wealth of Cities: City Dynamics and the Fifth Wave*. Aldershot: Ashgate.

NESTA (National Endowment for Science, Technology and the Arts). 2009. *Original Modern: Manchester's Journey to Innovation and Growth*. London: NESTA.

New South Wales State Government. 2010. NSW Business Sector Growth Plan. [Online]. Available at: https://www.opengov.nsw.gov.au/publication/11140 [accessed 19 May 2013].

O'Connor, J. and Gu, X. 2010. Developing a creative cluster in a postindustrial city: CIDS and Manchester. *The Information Society*, 26(2), 124–36.

Sheffield City Council. 1998. *SCIQ Area Action Plan*. Sheffield: Sheffield City Council.

Temple Bar Properties Ltd. 1992. *Development Programme for Temple Bar*. Dublin: Temple Bar Properties Ltd.

Urban Cultures Ltd. 1991/2. *Manchester First: How to Become a Cultured City; Arts and Cultural Strategy for Manchester City Council*. Manchester: Urban Cultures Ltd.

—— . 1992. *Temple Bar: Creating Dublin's Cultural Quarter*, prepared for the Irish Government. UK: Urban Cultures Ltd.

Whitt, J.A. 1987. Mozart in the metropolis: the arts coalition and the urban growth machine. *Urban Affairs Quarterly*, 23(1), 15–36.

Wynne, D. 1992. *The Culture Industry: The Arts in Urban Regeneration*. Aldershot: Avebury.

PART 6
Cultural and Planning Dynamics

Preface to Part 6

Greg Young

In this Part the authors explore the dynamics of the intertwining of the fields of culture and planning as the involution of this relationship intensifies through the globalization of networks, economies and social practices. This drawing together is reflected on a number of levels: at the level of the international state, governance policy, principles and instruments have been adopted as intended 'beacons' for the illumination and guidance of positive cultural approaches; in terms of cultural planning policy-makers in all fields using culture instrumentally for broader policy goals occupy a two-way street which could allow their own perspectives to be transformed by contact with local culture; and a cultural paradigm for planning and governance is identified as a major option to address culture in more sensitive, meaningful and locally specific terms regardless of the scale, sector or setting. In spite of these trends formulaic and problematic culture-led strategies exist and prosper as forms of cultural planning recognized by Hillier in her chapter. However, above and beyond this in Bianchini's opinion views about places and cultural resources differ and are the subject of 'a local politics of symbolic contestation' with narratives that can vary according to 'people's age, neighbourhood, occupation, ethnicity and gender'.

In the chapter by Duxbury and Jeannotte the bridging links between global cultural governance and local practices are described as both stimulating and frustrating to analyse as these relationships are often indirect and context-bound in terms of local culture and politics. In case studies of heritage protection and conservation, the creative industries and the creative cities movement and initiatives surrounding culture, diversity and sustainability the authors critically examine the impact of these global frameworks on cultural planning initiatives at the lesser scales. In the case of culture and sustainability, for example, they trace the evolution of international initiatives including the United Cities and Local Government's (UCLG) *Agenda 21 for Culture* that places culture 'at the centre of its mandate' although it lacks specific goals and targets and a mechanism to monitor performance. The authors recognize that new and complex dynamics are increasingly in play in the field whether in terms of ethical dimensions and moral suasion, inter-scalar municipal and global connections, the development of networks and funding mechanisms or information technologies enabling learning and action.

In his chapter Bianchini unpacks the various meanings and practices that are understood as cultural planning, criticizing non-holistic approaches to culture that focus on and privilege specific elements notably the arts. These fragmented conceptions are often the result of conceptual confusion and yet serve to undermine the full mapping potential of culture along the lines of Geddes 'survey' of culture. In addition Bianchini again notes the need to think in terms of a bigger picture to facilitate the implementation of cultural planning strategies and to contextualize 'the mapping of local cultural resources in international terms'. On this latter macro level he believes there is an important need in the resourcing of such strategies 'to build on a city's arts, media, sports, trade and diasporic links worldwide' while on a micro level he identifies a need for specific training for cultural work such as in cultural mapping, in linking cultural planning with place marketing and social policy, and in working in multicultural contexts. As well, he asks difficult questions, for example whether culture and cultural production can be more than simply instruments in the hands of established professional disciplines, or if the cultural sector and cultural planning can 'play the public policy game' while preserving the integrity and originality of what it has to offer'?

In his chapter Young reasserts the strategic importance of culture as an organizing concept or principle for planning and governance and proposes this as a cultural paradigm. To give traction to this vision he utilizes a concept of culturization first proposed by him including through the use of this neologism. He defines culturization as 'the holistic research of holistic culture and its creative, critical, ethical and reflective utilization in governance functions including spatial planning'. Holistic culture and holistic research are further defined and are included in a Culturized Model for Planning and Governance relevant to planning at all spatial scales and across the broad spectrum of planning types. Young argues that core elements of the model are applicable to other governance functions and public policy as the cultural richness and complexity that exists in communities, institutions, cyberspace and markets demands a corresponding richness in the nature of planning and governance.

In the concluding chapter of this Part Hillier follows with a critique of those versions of cultural panning in which the intention is to plan *for* culture rather than to plan *with* it, as is the goal of culturized planning. She proposes a post-structuralist approach illustrated through a discussion of heritage discourse and practices in relation to Dreamland, Margate the historic UK amusement park. In problematizing traditional UK heritage discourse wherein heritage entities are often 'frozen into truth' she seeks to locate Dreamland through a Deleuzean-inspired rhizomic approach in which history is re-thought as historicity and as the critical extension of the past. For Hillier, riding this 'roller-coaster' is the means to enter a different space-time – 'a Deleuzean dream image' – and to see heritage as expressing 'the essence of time as change'.

In their own ways the contributions in this Part reflect the continuous need to interpret and reinterpret culture (including heritage) as recognized in the longstanding philosophical and theoretical tradition bequeathed by, among others, Friedrich Nietzsche and Michel Foucault. This goal can be promoted in many ways: by the mechanisms of global governance devoted to putting culture on a worldwide agenda as represented by UNESCO and the UCLG; by Bianchini's active concept of the cultural planning process engaging the cultural policy-maker, artist or cultural manager on any area of policy-making to enhance the cultural skills of decision-makers; by the cultural paradigm and culturized model for planning and governance as tools to serve all of these needs; and by a fundamental cultural perspective in Hillier's terms based on thinking and re-thinking past-present and future cultural assemblages differently.

Global Cultural Governance Policy

Nancy Duxbury and M. Sharon Jeannotte

Introduction

Tracing the bridges – both theoretical and practical – between the instruments and mechanisms of global cultural governance and cultural planning practices at the local level is both stimulating and frustrating. Stimulating, because there appear to be growing links between what happens on the ground in cities and communities and what happens in forums such as UNESCO, the World Bank and the UN Human Settlements Programme (UN-Habitat). Frustrating, because these links are often indirect, the paths are usually circuitous and the degrees of influence are highly dependent upon place-specific contexts and cultural politics at the local level. Three main dynamics can be observed to be at play:

1. International agencies attempt to reach out to the local level through particular strategies and programmes. UNESCO is the most dominant player here, although public and private foundations that operate internationally are also very active (see below).
2. Individual cities find international 'designations' or relationships attractive – both for their symbolic value and the distinction conveyed, as well as for the financial rewards that frequently ensue. In this type of dynamic, the moral authority of the international agency is seen as a means of enhancing the municipality's reputation and prestige.
3. Locally based networks of municipalities organizing collectively (for example, United Cities and Local Governments) strive to influence global cultural policies, while also using the knowledge-sharing and collective capacities of the network to support local actions and strategies. This dynamic also encompasses *glocalization* actions based on city-to-city relations and cooperation and anchored in the work of local governments, NGOs and international institutions (Savir 2003).

These dynamics, systems and networks are mediated by national and sub-national levels of government, which sometimes reinforce the moral authority or moral suasion of global actors through other types of governance instruments – regulation, grants and subsidies, organizational or capacity-building, and (more rarely) privatization and taxation. They may also be influenced by national or sub-national networks and initiatives.

In addition, many public and private foundations that operate internationally provide direct funding to enable cultural planning initiatives, and indirectly influence the 'global cultural policy' realm through their investments. Many also contribute to configuring and enabling the terrain on which cultural planning initiatives occur. For example, the European Cultural Foundation has recently funded local cultural policy and planning projects in L'viv, Ukraine, and in three Turkish cities – Kars, Antakya and Çanakkale – which have led to further cultural planning initiatives in those countries (see Ince 2011, Knudsen McAusland 2011). The Aga Khan Trust for Culture operates a number of programmes related to culture, architecture, music, planning and building, and historic cities (http://www.akdn.org). These foundations frequently co-invest in projects in cooperation with a network of other similar agencies.

In this chapter, we provide an overview of key international cultural governance/policy initiatives and assess the main linkages that have developed with local cultural planning. Our examination of the policy threads encompasses three thematic areas: heritage; creative industries and creative cities; and culture, sustainability and diversity. We look at how various global governance instruments are incorporated within these three thematic areas, and consider how the precepts outlined in these instruments have been adapted or localized and linked to the practices of local cultural development and planning. The final section examines Agenda 21 for Culture, an initiative of United Cities and Local Governments, through which cities are collectively acting at the international level to link multicultural diversity and cultural rights with cultural planning, urban planning processes and sustainable development. In closing, we reflect on how global cultural governance and local planning seem to be connected, and consider prospects for the future.

Strategies, Tools and Instruments of Governance

While global entities such as the World Bank or the World Trade Organization have played roles in global cultural governance, the primary actor in this policy area over the past 40 or 50 years has been the United Nations Educational, Scientific and Cultural Organization (UNESCO). UNESCO's work is conceptualized on a global level, with international regions and individual nation states as the primary parties in its initiatives. In the face of rapid urbanization globally, however, UNESCO is increasingly cognizant of the role of cities and is gradually working out means to connect with the local level and to encourage inter-city exchanges (in cultural and other areas). It has tended to adopt four main strategies to influence cultural governance at the local or city level: 1) principles and moral authority/suasion, 2) recognition/designations, 3) programmes and funds to incentivize actions and 4) networks of cities.

Principles and Moral Authority/Suasion

UNESCO's global conventions and recommendations are general references for policy development and direction, and tend to filter down to the local level by influencing or being incorporated within national and sometimes sub-national policies and programme frameworks. As well, some cities will independently develop charters or policies that directly reference international documents (for example the 2004 *Montreal Declaration for Cultural Diversity and Inclusion*). UNESCO also publishes guides promulgating good practices in programme development and operational management, for example on management of heritage sites (the World Heritage Manuals, for example Pederson 2002) and cultural/creative industry development (for example Cano et al. 2011). These types of documents are intended to advise, inform and facilitate local, sub-national or national actors to develop initiatives in these areas. While they do not directly address local cultural planning, they may influence it by recommending and popularizing particular types of policies, programmes and support mechanisms. These publications also reflect UNESCO's influential role as information broker and *imprimatur*, retaining experts to observe, select and synthesize 'good practices' from different localities and promoting these grassroots actions globally.

Recognition/Designations

UNESCO is perhaps best known for its World Heritage Site designations, and while the selection process is politically charged and influenced by clashes of different intellectual approaches (Schmitt 2009), the designations continue to influence tourism flows and serve as a basis for local actions regarding site conservation and management. A UNESCO World Heritage Site designation is an impetus to local development activities (for example 'Palestine ...' 2011). While UNESCO 'encourages community-based policies and conservation practices that foster local development while preserving sites' (UNESCO 2010: 4), and provides guidance to local communities to manage these sites for tourism purposes, the connection to local residents in heritage locations may be neglected when city authorities are primarily interested in attracting visitors. In some historical neighbourhoods, local social inclusion-focused conservation/revitalization policies and programmes are emerging to address this issue, such as in the al-Darb al-Ahmar neighbourhood of Cairo, Egypt (Morbidoni 2011).

Programmes and Funds to Incentivize Actions

UNESCO also makes use of more interventionist governance instruments that go beyond the 'moral currency' of persuasion and designation. For example, in 2002 it launched a network of cultural producers, the Global Alliance for Cultural Diversity, 'to strengthen local cultural industries in developing countries through fostering partnerships between private, public, and civil society in "project partnerships"' (UNESCO 2010: 12). These project partnerships generally emerge from local cultural development strategies. For example, 'Nzassa, House of Music and Dance' in Treichville, Côte d'Ivoire, an innovative support structure for cultural businesses, was launched by the local authorities to harness the potential of the local music and dance sector (UNESCO Global Alliance for Cultural Diversity 2011). By 2010, some 50 projects in 30 countries aimed at fostering North–South cultural producers/market connections had been supported.

UNESCO also provides some financial assistance under the World Heritage Fund (created in 1972) to assist states to identify, preserve and promote World Heritage sites. States contribute 1 per cent of their annual UNESCO dues to the Fund, but can also make voluntary contributions. As well, countries can donate funds-in-trust for specific purposes, and additional income is derived from partnerships, sales of publications and private donations (UNESCO 2008b). However, total annual contributions amount to only about US$4 million, which is clearly insufficient to meet more than a fraction of global needs. Therefore, UNESCO must rely heavily on partners for assistance in meeting its objectives

Networks of Cities

While the previous strategies may engage local governments, their influence on local cultural policies, plans and programmes are largely indirect or focused on the implementation of defined projects. More direct linkages with local governments, mobilized through establishing UNESCO Networks, have emerged since the early-to-mid 2000s. The website of the International Coalition of Cities against Racism articulates the rationale for this focus: 'UNESCO chose *cities* as the privileged space to link upstream and downstream actions. The role of city authorities as policy-makers at the local level, is considered here as the key to create dynamic synergies' (UNESCO 2011d). Within these culture-related networks of cities three types of strategies tend to be employed – knowledge-sharing, recognition and technical assistance – mirroring the instruments that are prominent in other types of global governance:

Knowledge-sharing about 'best practices' and strategies
An example of a UNESCO knowledge-sharing network strategy is the International Coalition of Cities against Racism initiative launched in 2004. The network includes six regional coalitions against racism, discrimination, xenophobia and intolerance which have been established in Africa, the Arab Region, Asia and the Pacific, Europe, Latin America and the Caribbean, and Canada with almost 5,000 cities involved (UNESCO 2010). It encourages municipalities (and their NGO partners) to share their practices and strategies to overcome racism, discrimination and intolerance. Each regional coalition has developed its own 'Ten-Point Plan of Action' and signatory cities are encouraged to integrate this Plan of Action within municipal strategies and policies and involve civil society actors in its implementation. The UNESCO affiliation motivates cities to become involved in this network while, from UNESCO's perspective, the network brings cities into the ambit of its broader initiatives on this topic. UNESCO has organized meetings to facilitate exchange between network members, such as a Networking Event at UN-Habitat's third World Urban Forum in Vancouver, 2006. However, on an ongoing basis the network's 'driving force' resides in the affiliated municipalities' actions, with UNESCO playing a minor role in the ongoing dynamics of the regional coalitions.

Recognition to enhance city branding/promotion
Influenced by the effectiveness of its World Heritage Site designations, a growing interest in networks and the 'creative city' movement, UNESCO launched its Creative Cities Network in 2004. To become a member of the Network, a city must apply to be a UNESCO Creative City in one of seven fields (literature, film, design, music, crafts, media or gastronomy), complete an application of 50–80 pages, and be evaluated by UNESCO and NGO experts (UNESCO 2011c). Only 28 cities are members of this network at mid-2011. The notion of creative tourism is highlighted, reflecting the tourism-based returns cities expect from the designation (UNESCO Creative Cities Network 2006). Member cities are encouraged to share ideas and best practices with one another, and selected member cities have organized international conferences to facilitate this exchange (for example Santa Fe in 2008; Shenzhen in 2010). However, the limited number of cities in the network restricts such knowledge-sharing and makes this secondary intent seem to be only a 'sidecar' to the Network's promotional value.

Providing technical assistance and advice (on local governance issues)
A new UNESCO initiative is the Cities for Sustainable Development and Dialogue programme, intended to address 'the challenge of accommodating modernization and transformation in historic cities without compromising their identity and that of local communities, or their role as drivers of cultural creativity and urban regeneration' (UNESCO 2010: 12). The programme will provide 'technical assistance and advice on innovative urban governance approaches' to local and national governments (2010: 12). The initiative will bring together some of the ongoing work of different areas within UNESCO, and intends to incorporate and reorient the Creative Cities Network to 'maximize the contribution of member cities as active partners' in this new programme (UNESCO 2011b: 135). This strategy seems to echo UN-Habitat's Urban Development and Management stream of initiatives focused on training and capacity building in local authority management and urban governance.

In summary, UNESCO has tended to rely primarily on its moral influence to move member states toward desirable goals. In the absence of financial or regulatory power to achieve its objectives, it has made heavy use of partners, especially private philanthropy and civil society organizations, to support cultural activities in developing countries. More recently, it has begun to reach out to sub-national levels of governments, such as cities, by creating global and regional networks to help achieve its aims. Cultural planning-related assistance has, to date, formed a minor portion of these initiatives but as cities gain socio-economic and political power, UNESCO is paying increased attention to an emerging *glocalization* anchored on city-to-city relations and cooperation among various local governments, NGOs and international institutions.

Thematic Case Studies

In this section, we critically examine three areas where UNESCO has influenced the direction of global cultural governance to varying degrees – heritage protection and conservation, the creative industries/creative cities movement and the cluster of initiatives surrounding culture, diversity and sustainability – and assess the degree to which these global frameworks are reflected in sub-global cultural planning initiatives.

Heritage Protection and Conservation

UNESCO's *Convention concerning the Protection of the World Cultural and Natural Heritage* was adopted in 1972 and is arguably the governance mechanism that has had the most widespread global influence on culture policies at the national, sub-national and local levels. Under the *Convention*, signatory countries are encouraged to protect their natural and cultural heritage and to nominate sites for inclusion on the World Heritage List (UNESCO 1972). As of 2010, 186 countries had ratified the *Convention* and over 900 sites had been listed (http://whc.unesco.org/). Sites are chosen for listing by the World Heritage Committee, which consists of representatives from 21 of the state parties to the *Convention*, chosen for terms of up to six years. The Committee meets yearly and is responsible not only for listings, but also for allocation of financial assistance under the World Heritage Fund, examination of reports on the state of conservation of listed sites, and inscription or deletion of sites on the List of World Heritage in Danger (UNESCO 2008b: 9).

The main governance instrument used to promote UNESCO's heritage protection and conservation goals is the World Heritage List, which operates primarily through the force of moral suasion exercised by UNESCO and the World Heritage Committee. UNESCO aspires to what Schmitt (2009: 111) calls 'metacultural production' or what UNESCO itself refers to as 'the overarching benefit … of belonging to an international community of appreciation and concern for universally significant properties that embody a world of outstanding examples of cultural diversity and natural wealth' (UNESCO 2008b: 9). Describing the dynamics of the yearly Committee sessions, Schmitt observes that 'international institutions do not *a priori* reflect the fixed preferences of national states, but on the contrary have a socializing effect and are thus able to change positions, preferences and identities of state actors' (2009: 110).

Socialization is often complicated by the differing world views of the Committee's two principal advisors – the International Union for Conservation of Nature and Natural Resources (IUCN) and the International Council on Monuments and Sites (ICOMOS). As Schmitt observes, the IUCN 'regards outstanding universal value as an inherent quality of properties', while ICOMOS takes a more social constructivist view (2009: 111). In addition, countries at diverse stages of development sometimes have differing views of what is possible in terms of heritage protection and conservation, with delegates from countries in the global South, as Schmitt puts it, 'generally more appreciative of the difficulties facing local and national authorities', while delegates from the global North see governance as 'basically organized in accordance with the homogenizing principles of "Western" (in a historical sense) modernity' (2009: 114). These differing viewpoints are typical of multi-level governance situations, but are exacerbated in UNESCO's case by its reliance on other actors to flesh out its rather limited range of governance instruments.

As mentioned previously, UNESCO does provide some financial assistance under the World Heritage Fund, but must rely heavily on partners for assistance in meeting its objectives. For example, the 26 international safeguarding campaigns that have been launched since 1972 to save sites such as Abu Simbel in Egypt and the Temple of Borobudur in Indonesia have cost in excess of US$1 billion, much of it derived from private sources, such as the Japan Trust Fund for the Preservation of World Cultural Heritage (UNESCO 2008b). Other prominent partners include the Aga Khan Historic Cities Programme, the Fonds Français pour l'Environnement Mondial, the Nordic World Heritage Foundation and the Organization of World Heritage Cities.

While UNESCO assists states to develop comprehensive management plans for listed sites through the provision of advice and technical training, by necessity it depends upon partners at the national and sub-national levels to implement such plans. In developed countries, where systems of cooperation and support are in place, this can be an effective governance strategy. For example, the Rideau Canal, a 202 km waterway in the Province of Ontario, Canada, was inscribed on the World Heritage List in 2007 as 'the only canal in North America dating from the great canal-building era of the early nineteenth century which still remains operational along its original line and with most of its original structures intact' (Canadian Commission for UNESCO 2007). Parks Canada, a federal government agency, maintains and operates the canal and takes the lead in its preservation, assisted by a complex web of partners.

Parks Canada has developed a Rideau Corridor Landscape Strategy that 'works with First Nations, federal and provincial agencies, municipalities, NGOs, property owners and others to build a new vision for the Rideau', and both the National Historic Sites and Monuments Board of Canada and the Ontario Heritage Trust have recognized the canal as an important heritage site (http://www.pc.gc.ca; http://www. heritagetrust.on.ca). Tourism is one of the primary motivations for involvement, with the Province of Ontario featuring the canal in its online travel guide (http://www. ontariotravelguides.com) and many of the communities along the canal providing information on local heritage attractions near the canal (http://www.ottawatourism.ca; http://www.twprideaulakes.on.ca). The National Capital Commission (NCC), another federal agency, maintains eight kilometres of the canal during the winter months as 'the World's Largest Skating Rink' (http://www.canadascapital.gc.ca). Private sector partners are also active in tourism promotion as part of a consortium called the Rideau Heritage Route Tourism Association (RHRTA), which publishes an online travel guide in several languages (http://www.rideauheritageroute.ca). A group of volunteers, the Friends of the Rideau, manages a Rideau Legacy Fund that is used to publish reports and books about the canal (http//www.rideaufriends.com), and an annual Rideau Canal Festival celebrates its listing as a World Heritage Site (http://www.rideaucanalfestival. ca).

In contrast, designations in some countries may encounter political barriers to successful protection and promotion. For example, the Old City of Jerusalem was proposed for listing by Jordan in 1981, and placed on the List of World Heritage in Danger in 1982. A subsequent request by Israel to extend this site to include Mount Zion was reviewed by the World Heritage Committee in 2001, which decided 'to postpone further consideration of this nomination proposal until an agreement on the status of the City of Jerusalem in conformity with International Law is reached, or until the parties concerned submit a joint nomination' (http://whc.unesco.org/). In 2007, the Director General of UNESCO was asked to send a technical mission to Jerusalem to investigate archaeological excavations being carried out by the Israeli Antiquities Authority on the Mughrabi pathway leading to the Haram el-Sharif. These excavations were viewed as illegal by the Islamic Waqf of Jerusalem, which is responsible for the Haram el-Sharif compound. In paragraph 46 of its report, the technical mission noted that 'in the present situation no dialogue exists between the Israeli authorities and the Islamic Waqf' but still felt that 'all parties should be invited to contribute in addressing and solving this issue in a cooperative way' (UNESCO 2007: 5). Schmitt (2009) has observed that inscription of a site on the List of World Heritage in Danger often functions as a means of 'shaming and blaming' which induces states to 'exert pressure on other local and regional actors to take active measures to protect the site' (2009: 118). However, there are instances, such as this one in Jerusalem, where geo-political tensions prevail over the limited moral authority of UNESCO to promote positive change.

Creative Industries/Creative Cities

Since about the mid-1980s, recognition of the economic significance of cultural industries has grown rapidly in Europe, Australasia and North America, which has led to a generalized political-economic imperative to boost the contribution of the arts and an array of other 'commodified creative endeavours' through various interventions (Johnson 2009: 21). Numerous national and subnational initiatives have contributed to shaping a growing trade in cultural products, a market largely dominated by developed countries that are striving to strengthen their individual positions in these economic flows. Within this context, UNESCO has largely focused on creating 'an enabling environment for the emergence of cultural and creative industries' and enhancing their economic development impact in developing countries (UNESCO 2011a: 34). UNESCO's *Convention on the Protection and Promotion of the Diversity of Cultural Expressions* (UNESCO 2005b) forms an overarching framework for concerns about creative/cultural industries development and expression globally. Complementing these efforts, supranational policies, emerging from specific regions, such as Africa, or agencies such as CARICOM in the Caribbean community, also aim to develop cultural products as a platform for economic development. At an operational level, a wide array of training programmes, 'best practice' guidelines and producer partnerships and projects have been devised and implemented for developing countries, although these are usually designed at a national or supranational regional level.

In the early 2000s, cultural/creative industry development and the 'creative cities'/'cultural capitals' movement dovetailed as cities became increasingly recognized as the dominant engines for the development of the world's creative economy. The movement has been fuelled by economic crises, the rise of the post-industrial economy, and the perceived imperative to attract a mobile 'creative class' (Johnson 2009).

International agencies began linking cultural industries development with city development (OECD 2005, Yusuf and Nabeshima 2003). Some efforts to place cultural industries on the Millennium Development Goals agenda also focused on cities and communities. A Senior Expert Symposium was held in Jodhpur, Rajasthan entitled *Asia-Pacific Creative Communities: Promoting the Cultural Industries for Local Economic Development – A Strategy for the 21st Century* (UNESCO 2005a). Subsequently, the Inter-agency Technical Assistance Programme established an Asian Cities Creativity Index, 'to track and measure the effectiveness of policy initiatives in support of cultural industries' (UNESCO 2005a: 7).

The confluence of cultural/creative industry development and cities is also evident in programmes promoting creative clusters and urban regeneration of particular neighbourhoods or districts. For example, within the European Union, cultural planning related projects have been funded by European Structural Funds and through programmes such as URBACT and URBAMECO. Individual cities and networks of municipalities, such as the URBACT Network on Creative Clusters in Low Density Urban Areas, are planning and implementing local cultural/urban development efforts within the framework of both international and transnational initiatives, which directly inform their actions and policies.

Although 'creative city' competition dynamics continue to be prominent, in recent years they have been critiqued as an unsustainable basis for long-term city development, and a search for more nuanced approaches can be observed. The rise of sustainability concerns has led to a gradual morphing of the concept of *creative cities* to one of *creative sustainable cities* (for example Staines 2010). This shift has been accompanied by a general plea for recognition of more humanized and diversified approaches to city building and development (Duxbury and Jeannotte 2011).

To some degree, this transition is reflected in UNESCO's new Cities and Sustainable Development programme, which will reorient the existing Creative Cities Network to this emergent framework (UNESCO 2011b). While details of the programme are not available at time of writing, the 'urban management system' that is envisioned may serve to further integrate cultural considerations into broader local planning schemes and to catalyse North–South city collaboration partnerships among Creative Cities Network members.

Culture, Diversity and Sustainability

UNESCO has been at the forefront of global efforts to address issues of cultural diversity and sustainability over the past two decades, but global governance in these areas is still very much a 'work in progress'. Seminal works in this area include the 1996 report *Our Creative Diversity*, as well as the 2005 *Convention on the Protection and Promotion of the Diversity of Cultural Expressions*. More recently, UNESCO has articulated the relationship between biological and cultural diversity (UNESCO 2008a); proposed measures to operationalize Article 13 of the *Convention*, which deals with the integration of culture in sustainable development (UNESCO 2009); advocated for the inclusion of culture in sustainable development policy and programmes (UNESCO 2011a, 2011b); and created a new Cities and Sustainable Development Programme.

UNESCO's leadership and, to a certain extent, its moral authority have inspired other supranational, national and subnational initiatives on both diversity and sustainability, although this guidance is seldom explicitly acknowledged. Linkages between global, national and local cultural policy and planning can be observed in a variety of recent attempts to situate culture within a local 'sustainable development' context. These initiatives were developed through loosely organized cross-national informal learning and policy-development networks in which initiatives in one location inspired and informed other efforts.

The evolution of these initiatives can be traced internationally through three distinct periods (Duxbury and Jeannotte 2010). During Phase 1, *Differentiating 'culture' from 'social'* (approximately 2000–02), concerns about the relative neglect of cultural considerations in sustainability discourses and conventions gave impetus to grassroots thinking that fuelled the development of a four-pillar model of sustainability (for example Hawkes 2001). In Phase 2, *Focusing on local development* (2004–06), national governments began to establish frameworks for local sustainability planning including a cultural dimension (for example Government of Canada 2005, NZMCH 2006a, 2006b). In Phase 3, *Rearticulating culture within sustainability at national and transnational levels* (2008–09), subnational level initiatives began to emerge (for example Quebec Ministry of Culture, Communications and the Status of Women 2009, SALAR 2008, Thames Gateway North Kent 2006).

More recently, the UCLG *Agenda 21 for Culture* initiative has placed local cultural policy and sustainable development at the centre of its mandate and has adopted the 'network' governance tool as the central instrument in its efforts to integrate culture into local planning frameworks.

Local Cultural Planning at a Global Scale

From a 'bottom-up' perspective, the potential of networks to influence supranational cultural policy is important to acknowledge. As cultural networks proliferate and develop over time, they are becoming a more visible dimension of cultural development, although their role in cultural policy is still uncertain (Cvjeticanin 2011). City networks are proliferating on a wide array of topics. Two main types of city networks with cultural interests can be observed: 1) political networks with primarily advocacy roles, which include culture as an explicit interest, such as the Eurocities Culture Committee, and 2) staff-oriented national or subnational networks with primarily professional development and knowledge networking roles, such as the Creative City Network of Canada and the Cultural Development Network (Victoria, Australia). Through such networks, individual cities share experiences and ideas, adapting policies and practices from one another. The networks also serve as contact points for national and international agencies, other networks and municipalities and may be catalysts and agents for knowledge-sharing initiatives and collective actions. They may directly influence supranational policies through, for example, publishing responses to policy discussion documents and taking other advocacy-related actions. They may also have a more indirect influence through projects and publications, participating in international think tanks and meetings, and facilitating the circulation and promotion of local 'good practices' which may then inform and influence the policy and programme development of international policy agencies.

United Cities and Local Governments (UCLG), launched in 2004, is an international NGO that represents and advances the interests of cities, local governments and municipal associations throughout the world. The UCLG Committee on Culture brings together cities, organizations and networks that foster relations between local cultural policies and sustainable development. The committee serves as a platform for mutual learning and exchange of experiences, for advocating about the role that cities play in cultural policy and practice, and for enabling cities to contribute to global cultural governance (Duxbury, Cullen and Pascual, forthcoming). The document *Agenda 21 for*

Culture is an initiative of the UCLG Committee on Culture, and through it, UCLG is an active player on global cultural issues. Building on this, in November 2010 the UCLG Executive Bureau officially approved a Policy Statement on 'Culture as the Fourth Pillar of Sustainable Development' which advocated for explicitly recognizing culture in sustainable development and including a strong cultural dimension in governance at all levels – local, national and international (UCLG 2010b).

When a local government formally adopts *Agenda 21 for Culture*, it expresses its will to ensure culture plays a key role in its policies, and shows its solidarity with other cities and local governments. In late 2010, over 400 cities, local governments and organizations were linked to *Agenda 21 for Culture*, with geographic coverage growing more extensive over time. It has informed or influenced local cultural strategies in several cities and launched a Fund for Local Cultural Governance in 2010 to assist cities and local governments in Africa, Latin America and the Mediterranean to implement *Agenda 21 for Culture*. However, implementation is often not straightforward or easy. Within *Agenda 21 for Culture*, there are internal conceptual tensions: as Teixeira Coelho (2009) has observed, its broad scope is difficult to synthesize, some concepts are not explained in detail and it fails as a real 'agenda' because it does not provide specific goals and qualitative and quantitative targets. The implementation of *Agenda 21 for Culture* is challenging for local authorities, with a need to link culture to key local planning and sustainability programmes, which usually do not have a cultural dimension. Thus, the local impacts of adherence may be limited in the short- and medium-term. In addition to these difficulties, UCLG is a very young organization, with scarce resources, limited lobbying capacity and communication limitations. With no defined action plan that follows adherence to *Agenda 21 for Culture* the UCLG Committee on Culture is not able to monitor what cities actually do after adopting it.

In terms of global governance, the UCLG can attend the meetings of the UNESCO Convention on Cultural Diversity (Intergovernmental Committee and Conference of Parties), but only as an observer with the status of an NGO. It can take the floor, but cannot suggest amendments to documents. Thus, while cities (through UCLG) are present in global cultural policy fora, they are not full participants at these tables and their influence is contingent on decision-makers' attention to and endorsement of the UCLG's messages and proposals.

In light of these multiple challenges and issues, the *Agenda 21 for Culture* initiative continues to evolve. To address the implementation and post-adherence limitations, the UCLG emphasizes intermunicipal sharing of knowledge and experience about local implementation practices. To improve the document itself, in 2012, the UCLG Committee on Culture will initiate action to write a new Agenda that will involve practitioners (cities and local governments), civil society (academics, activists), and national and international institutions (UCLG 2010a).

Like UNESCO, UCLG relies heavily on moral suasion as a governance instrument, but lacks the resources to move up the hierarchy to more interventionist and, perhaps, effective measures. Its strength lies in its networking capacities, which have enabled it to establish a presence fairly quickly in the sphere of global cultural governance. The next steps that it takes will be critical in determining its ultimate impact.

Concluding Thoughts

Today

The United Nations is increasingly recognizing the essential role of cities in achieving global development goals and is gradually looking to establish partnerships with local governments and their associations on the ground. Gradually, the planning and governance of cities is gaining attention as a topic for project-based actions and assistance initiatives, with the integration of cultural considerations into planning emerging as one of the most recent topics to be added to this framework. At the same time, local governments are strengthening their global lobbying on international issues and forming more strategic partnerships with UN and multilateral institutions.

In the face of growing complexity in the international arena, how the various international agencies and foundations interconnect or overlap with other players in various cultural topic areas is in constant dynamic shift. The field is changing, and the 'globe' is shrinking through technology. Structurally, local governments are not formally recognized at global cultural governance tables, but are involved in the system in various ways: as clients, policy and programme informants, and increasingly as agents catalysing international exchanges. As *sustainable development* and *urbanization* grow as dominant frames for cultural development, the focus on cities is increasing within global cultural governance and will continue to do so.

Cities have financial and other resource constraints, but have the capacity to be nimble networkers, sharing ideas and experiences from other cities through networks and online project profiles. Cities also bring unique strengths and perspectives to the global cultural policy arena, based on the diversity of local practices, multi-sectoral collaborations and partnerships, and modes of implementation for policies and projects. Because of these factors, they can be innovative partners in global governance.

The symbolic top-down city networks created by international agencies appear to have been launched with 'a vague hope' that the members themselves would provide the energies and dynamics to drive the network's knowledge sharing and training functions, but in a climate of 'limited and reduced monetary resources and stretched staff resources, it is difficult for municipalities to take on such "extra" roles' and the networks tend to focus primarily on city branding or marketing (UCLG 2007: 1). As a result, these networks have generally had limited impact, and it is becoming clearer that creating a network is not sufficient to make it functional and effective over time. Careful investments and strategies are required to integrate the enabling vision expressed by international or national bodies with the needs of cities in the areas of training, exchange of good practices, peer review and similar actions. With careful attention to functionality, the particular strengths of local government and planning can be catalysed by networks and network-projects, but only if there are effective working partnerships with other knowledge and policy partners and stakeholders.

As the thematic cases discussed in this chapter have shown, intergovernmental partnerships with UNESCO appear to work most smoothly and efficiently in traditional areas, such as heritage protection and conservation, where the instrument of moral suasion has been finely honed. Yet, even in this area, moral suasion can only go so far in an environment where national and local authorities are either unwilling or unable to carry forward initiatives. If all the players are functioning optimally, their 'natural' governance roles can be mutually reinforcing: the supranational entity provides the inspiration and moral authority, the national entity finances and regulates, and the regional and local entities provide the local intelligence and 'on the ground' expertise to deliver results. However, as the Jerusalem example illustrates, if this synergy is weak, the moral force of the supranational authority tends to be ineffective as well.

In newer policy areas, the effectiveness of partnerships is still at the testing stage. UNESCO has served as a catalyst for networks focused on cultural diversity and creative cities. Because of national governments' interest in the trade-related aspects of cultural diversity, international networks in this policy area have enjoyed relative success as policy instruments. For example, the International Federation of Coalitions for Cultural Diversity has provided a forum where civil society and governments can engage in dialogue on action plans and deliverables. With regard to creative cities and anti-racism issues, however, the picture is more nuanced.

One of the main challenges for local government movements, such as the UCLG, is therefore to ensure that its partnerships with the UN are mutually beneficial and have a real and sustained impact on global policy development. A cohesive and consistent local government approach is required if thematic partnerships are to demonstrate success in advancing global local government objectives (UCLG 2007).

Tomorrow

In an era of urbanization, on-the-ground cultural development efforts will increasingly be planned and implemented in urban areas. Everywhere funds are limited, and partnerships and collaborations will be necessary to implement programmes and projects. International agencies are relatively accustomed to working in partnership with each other, with individual nations and with major private foundations, but have only recently begun to consider the potential advantages of enlisting cities and urban networks in these working arrangements.

Despite restricted resources, local authorities are already playing multiple roles in this milieu. While to date, they have largely been seen as the recipients of support/implementers of projects, they are beginning to act as providers of expertise and experience-based knowledge. This latter role is growing through national initiatives to encourage inter-municipal assistance projects, served by the willingness of local authorities to innovate and to share their experience with others.

In the future, one can envision a larger role for municipality-directed international initiatives, supporting local needs for development assistance but also co-developing larger-scale policies, strategies and programmes in collaboration with global cultural governance agencies. Municipal ideas and strategies articulated within these initiatives may be noticed by global cultural governance agencies, or will be implemented in partnership with them, and these 'bottom-up' strategies could eventually evolve into policies and programmes at the supra-national or global levels.

Municipal networks, particularly those with an explicitly defined interest in culture such as UCLG, can play a key role in marshalling locally based resources and diverse knowledge, assessing and synthesizing on-the-ground contributions, facilitating mutual learning and exchanges, and acting as brokers and interpreters among multiple governance systems, local to global. They are conduits to local actors and can catalyse and enable local authorities to play active roles in informing and addressing issues of global cultural governance. However, international and national bodies will need to recognize that the resources of municipal networks are very limited and can restrict their potential to serve as effective partners.

Global agencies enjoy the strategic advantages of geographic coverage, moral authority and the capacity to mobilize funds to operationalize policy ideas within programmes and project implementation. Information technologies are enabling new coalitions and new channels for mutual learning and action. With these new tools, agencies can augment these advantages by tapping into a flow of expertise that is multidirectional. Networks of cities, with global agencies as catalysts and enablers, can better realize their strategic advantages as facilitators of exchanges between experts 'on the ground'.

This possible scenario will require openness to dialogue, and the development of trusted channels of interpretation between 'levels'. It will call for an expansion of relations beyond traditional means of advocacy, such as publishing written responses to draft policies and green and white papers, to the creation of ongoing institutional frameworks for exchange and dialogue with formal, appropriate spaces for this to occur. Moral suasion can still play a role in such an interconnected policy and planning environment, but strategic investments and clear links between information exchange and policy development will be needed to keep all the players engaged and to maximize the effectiveness of these new and complex dynamics.

References

Canadian Commission for UNESCO. 2007. The Rideau Canal is inscribed on UNESCO's World Heritage List. *Communiqué*. Ottawa.

Cano, G.A., Bonet, L., Garzón, A. and Schargorodsky, H. 2011. *Políticas para la creatividad: guía para el desarrollo de las industrias culturales y creativas*. Paris: UNESCO.

City of Montreal. 2004. *Montreal Declaration for Cultural Diversity and Inclusion*. Montreal: Montreal City Council.

Cvjeticanin, B. (ed.) 2011. *Networks: The Evolving Aspects of Culture in the 21st Century*. Zagreb: Culturelink Network/IMO.

Duxbury, N., Cullen, C. and Pascual, J. Forthcoming. Cities, culture and sustainable development, in *Cultural Policy and Governance in a New Metropolitan Age*, edited by H.K. Anheier, Y.R. Isar and M. Hoelscher. Cultures and Globalization Series, Vol. 5. London: Sage.

Duxbury, N. and Jeannotte, M.S. 2010. Culture, sustainability and communities: exploring the myths. *Oficina do CES*, 353 (September).

——. 2011. Introduction: culture and sustainable communities. *Culture and Local Governance/Culture et Gouvernance Locale*, 3(1–2), special issue on 'Culture and Sustainable Communities', 1–10.

Government of Canada. 2005. *Integrated Community Sustainability Planning: A Background Paper*. Discussion paper for 'Planning for Sustainable Canadian Communities

Roundtable', of the Prime Minister's External Advisory Committee on Cities and Communities, September 21–23.

Hawkes, J. 2001. *The Fourth Pillar of Sustainability: Culture's Essential Role in Public Planning.* Melbourne: Common Ground.

Ince, A. 2011. Çanakkale, *Turkey: One City's Efforts to Build Capacities for Local Cultural Policy Transformation.* [Online: CES/UCLG Committee on Social Inclusion and Participative Democracy, Inclusive Cities Observatory]. Available at: http://www.uclg-cisdp.org/sites/default/files/Canakkale_2010_en_final_0.pdf [accessed 5 April 2013].

Johnson, L.C. 2009. *Cultural Capitals: Revaluing the Arts, Remaking Urban Spaces.* Aldershot: Ashgate.

Knudsen McAusland, L. 2011. *L'viv, Ukraine: Development of a Participative Cultural Planning Framework for the City of L'viv.* [Online: CES/UCLG Committee on Social Inclusion and Participative Democracy, Inclusive Cities Observatory]. Available at: http://www.uclg-cisdp.org/sites/default/files/Lviv_2010_en_final_0.pdf [accessed 5 April 2013].

Morbidoni, M. 2011. *Cairo, Egypt: The al-Darb al-Ahmar Housing Rehabilitation Programme.* [Online: CES/UCLG Committee on Social Inclusion and Participative Democracy, Inclusive Cities Observatory]. Available at: http://www.uclg-cisdp.org/sites/default/files/Cairo_2010_en_FINAL.pdf [accessed 5 April 2013].

NZMCH (New Zealand Ministry for Culture and Heritage). 2006a. *Cultural Well-being and Local Government. Report 1: Definition and Context of Cultural Well-being.* Wellington: NZMCH.

—— . 2006b. *Cultural Well-being and Local Government: Report 2: Resources for Developing Cultural Strategies and Measuring Cultural Well-being.* Wellington: NZMCH.

OECD (Organisation for Economic Co-operation and Development). 2005. *Culture and Local Development.* Paris: OECD Publishing.

Palestine: a land of World Heritage Sites. 2011. *This Week in Palestine,* 155. [Online]. Available at: http://www.thisweekinpalestine.com/i155/pdfs/March%20155-2011.pdf [accessed 23 May 2011].

Pederson, A. 2002. *Managing Tourism at World Heritage Sites: A Practical Manual for World Heritage Site Managers.* Paris: UNESCO.

Quebec Ministry of Culture, Communications and the Status of Women. 2009. *Notre culture, au coeur du développement durable: plan d'action de développement durable 2009–2013.* Quebec: QMCCSW.

Savir, U. 2003. Glocalization: a new balance of power. *Development Outreach* (November). World Bank.

Schmitt, T.M. 2009. Global cultural governance: decision-making concerning world heritage between politics and science. *Erdkunde,* 63(2), 103–21.

Staines, J. 2010. *Sustainable Creative Cities: Role of Arts in Globalised Urban Context.* [Online: Culture360.org]. Available at: http://culture360.org/magazine/sustainable-creative-cities-role-of-arts-in-globalised-urban-context/ [accessed 23 June 2011].

Swedish Association of Local Authorities and Regions (SALAR). 2008. *Culture in the Sustainable Society.* Stockholm: SALAR.

Teixeira Coelho, J. 2009. For an effective and contemporary Agenda 21 for Culture, in *Cities, Cultures and Developments,* edited by J. Pascual. Report no. 5. Barcelona: Committee on Culture, UCLG.

Thames Gateway North Kent. 2006. *Sustainable Culture, Sustainable Communities: The Cultural Framework and Toolkit for Thames Gateway North Kent.*

UCLG (United Cities and Local Governments). 2007. *Background Paper for the Working Group on Partnerships with United Nations Agencies.* Approved at meeting of the Working Group, Paris.

——. 2008. *Agenda 21 for Culture.* Barcelona: UCLG.

——. 2010a. *Committee on Culture: Programme for 2011–2013.* Barcelona: UCLG.

——. 2010b. *Culture: Fourth Pillar of Sustainable Development.* Policy Statement approved by Executive Bureau of UCLG.

UNESCO. 1972. *Convention Concerning the Protection of the World Cultural and Natural Heritage.* Paris: UNESCO.

——. 2005a. *Asia-Pacific Creative Communities: Promoting the Cultural Industries for Local Economic Development – A Strategy for the 21st Century.* Bangkok: UNESCO.

——. 2005b. *Convention on the Protection and Promotion of the Diversity of Cultural Expressions.* Paris: UNESCO.

——. 2007. *Report of the Technical Mission to the Old City of Jerusalem.* Paris: UNESCO World Heritage Centre.

——. 2008a. *Links between Biological and Cultural Diversity Concepts, Methods and Experiences.* Paris: UNESCO.

——. 2008b. *World Heritage Information Kit.* Paris: UNESCO World Heritage Centre.

——. 2009. *Article 13 of the Convention on the Protection and Promotion of the Diversity of Cultural Expressions. Operational Guidelines – Integration of Culture in Sustainable Development.*

——. 2010. *65 Ways UNESCO Benefits Countries All Over the World.* Paris: UNESCO. [Online]. Available at: http://unesdoc.unesco.org/images/0019/001903/190306e.pdf [accessed 16 January 2011].

——. 2011a. *36 C/5 Draft Resolutions 2012–2013. Vol. 1.* UNESCO General Conference. [Online]. Available at: http://unesdoc.unesco.org/images/0019/001919/191978e.pdf [accessed 15 June 2011].

——. 2011b. *36 C/5 Draft Programme and Budget 2012–2013. Vol. 2.* UNESCO General Conference. [Online]. Available at: http://unesdoc.unesco.org/images/0019/001919/191978e.pdf [accessed 15 June 2011].

——. 2011c. *Creative Cities Network: How to Apply.* [Online]. Available at: http://www.unesco.org/new/en/culture/themes/creativity/creative-industries/creative-cities-network/how-to-apply/ [accessed 5 April 2013].

——. 2011d. *International Coalition of Cities against Racism.* [Online]. Available at: http://www.unesco.org/new/en/social-and-human-sciences/themes/human-rights/fight-against-discrimination/coalition-of-cities/ [accessed 5 April 2013].

UNESCO Creative Cities Network. 2006. *Towards Sustainable Strategies for Creative Tourism: Discussion Report of the Planning Meeting for 2008 International Conference on Creative Tourism.* Santa Fe, NM, October 25–27, 2006.

UNESCO Global Alliance for Cultural Diversity. 2011. *Nzassa, House of Music and Dance (Treichville, Côte d'Ivoire).* [Online]. Available at: http://portal.unesco.org/culture/en/files/40911/127358389011Nzassa_factsheet__eng.pdf/Nzassa%2Bfactsheet_%2Beng.pdf [accessed 6 December 2011].

Yusuf, S. and Nabeshima, K. 2003. Urban development needs creativity: how creative industries affect urban areas. *Development Outreach* (November). World Bank.

'Cultural Planning' and Its Interpretations

Franco Bianchini

Introduction

The 'cultural planning' approach originated in the US in the late 1970s–early 1980s. It became better known after it was adopted by Robert McNulty in his work for the Washington-based NGO Partners for Livable Places (later Partners for Livable Communities) in the course of the 1980s, and later spread to the UK and Australia. Versions of cultural planning are used in public policy-making also in Canada, Sweden, Italy and several other countries.

According to my own definition (Bianchini 1999), unlike traditional cultural policies – which are still mainly based on aesthetic definitions of 'culture' as 'art' – cultural planning adopts as its basis a broad definition of 'cultural resources', which encompasses:

- arts and media activities and institutions;
- the cultural practices of youth, ethnic minorities and other local communities of interest;
- the heritage, both tangible and intangible, including archaeology, gastronomy, local dialects and rituals;
- local and external perceptions of a place, as expressed in the cultural representations of a place, myths, tourism literature, media coverage and surveys of residents and visitors;
- the natural and built environment, including public and open spaces;
- the diversity and quality of leisure, cultural, eating, drinking and entertainment facilities and activities;
- the repertoire of local products and skills in the crafts, manufacturing and services. (See also Young, Chapter 23, this volume.)

Secondly, while traditional cultural policies tend to have a sectoral focus – for example policies for theatre, dance, literature, the crafts, film – cultural planning adopts a territorial remit. Its purpose is to harness the contribution of cultural resources to the development of a locality. Two-way relationships can be established between cultural resources and any type of public policy – in fields ranging from housing to health, economic development, education, social services, tourism, urban planning, architecture and cultural policy itself. Cultural planning cuts across the divides between the public, private and voluntary sectors, different institutional concerns, types of knowledge and professional disciplines. In addition, cultural planning strategies encourage innovation in cultural production, for example through intercultural exchanges, co-operation between artists and scientists and crossovers between different cultural forms. It is also important to clarify that cultural planning is not intended as 'the planning of culture' – an impossible, undesirable and dangerous undertaking – but rather as a culturally sensitive approach to urban planning and policy.

Cultural planning is an important method for revealing and valuing hidden and neglected cultural assets, including aspects of popular memory and intangible heritage, alternative social imaginaries and the creativity of marginalized social groups (in some cases, children, young people, immigrants and the elderly).

This chapter examines aspects of the origins, and some of the interpretations of the cultural planning approach. It also discusses the continuing relevance of the approach to some of the challenges cities face in the second decade of the twenty-first century, and some issues concerning professional specializations within the broader cultural planning field. The main focus of the chapter is on the experience of West European countries, but some of its arguments and conclusions may be of relevance also to other regions of the world.

A Brief History of Cultural Planning

Although the term 'cultural planning' has been in use only since the second half of the 1970s, some of its central ideas find their origins in ancient civilizations, although clearly in the ancient world access to public space was severely limited, due to the existence of slavery and the exclusion of women from public life. These ideas included the creation of public squares as foci for social interaction and civic activities, and the integration of public spaces and buildings as venues for theatre and other cultural events into physical planning, for example in the Greek city states (as in Athens under Pericles, from 461–429 BC) and in the Roman Empire. In medieval Tuscany, some city states incorporated aesthetic standards into physical planning through their building regulations, a prime example of this being Siena's municipal statute of 1309 (Tragbar 2006: 3,125). In the Italian Renaissance, the integration of art into urban planning and design was in part due to the work of visual artists who were also engineers, architects and planning theorists. These included Filarete (c.1400–c.1469), Leon Battista Alberti (1404–72), Francesco di Giorgio Martini (1439–1502) and Leonardo Da Vinci (1452–1519). However, Lauro Martines rightly points out that the vision of the 'ideal city' developed in fifteenth-century Italy by these and other artists/polymaths 'was politically ... deeply conservative ... Social preconceptions made it impossible to plan the restructuring of cities except by allotting more space to the powerful and less to the powerless' (2002: 273–4).

Austrian architect and planning theorist Camillo Sitte (1843–1903), in his 1889 book *Der Städtebau nach seinen künstlerischen Grundsätzen* (*City Planning According to Artistic Principles*), argued that there was much to learn from the aesthetic qualities of the agora and other elements of the ancient Greek *polis*, and from the irregularity and creativity of the layout of European medieval cities. Sitte argued that 'the building of cities should not be a merely technical question, but also an artistic issue, in the most precise and noble meaning of this term' (Sitte 2007: 20). He lamented the fact that in the course of the 'mathematical' nineteenth century 'the building and growth of cities … (became) purely technical questions' (Sitte 2007: 200).

The German town planning journal *Der Städtebau* (which had been founded by Sitte, and whose first issue was published in 1904, after Sitte's death) influenced Scottish biologist, botanist, sociologist and planner Patrick Geddes (1854–1932), whose work made an important contribution to shaping the conception of cultural planning which emerged in the late twentieth century (see also Greed, Chapter 5, this volume). Murdo Macdonald describes Geddes as 'a generalist thinker' (2009: 12), who recognized 'the interdependence of arts and sciences' (2009: 3). Macdonald points out that Geddes was part of a Scottish intellectual tradition 'in which one area of knowledge is honoured with respect to the way it relates to others and informs the whole' (2009: 12). Among the important influences on Geddes from different disciplines were those of French intellectuals, including Auguste Comte (1798–1857), one of the founders of sociology, and anarchist geographer Élisée Reclus (1830–1905). Geddes' key role as a cultural planning pioneer is recognized by Colin Mercer. He argues that cultural planners of the early twenty-first century should return to Geddes' three central planning principles:

> planning is not a physical science but a human science. Geddes insisted that all planning must take account of the three fundamental coordinates of Folk-Work-Place … Planners need to be, that is, anthropologists, economists and geographers and not just draftsmen. They need to know how people live, work, play and relate to their environment … Survey before plan … We need to be able to fold and integrate the complex histories, textures and memories of our urban environments and their populations into the planning process … Cities produce citizens … We need to relearn some of the civic arts of citizen-formation if we are to aim not just for 'urban' but for civic renewal. (Mercer 2006: 5–6)

The second of Geddes' three principles as identified by Mercer – 'survey before plan' – is an important antecedent for today's 'cultural mapping'. Mercer himself implies that one of the lessons we have to learn from Geddes is that 'we need to do some cultural mapping – tracing people's memories and visions and values – before we start the planning' (2006: 5). Geddes' idea of the survey is certainly much broader than that normally used in urban planning and policy today. A survey of 'work' would have to encompass the characteristics of the formal and informal economy, what people do to make a living now, and what they could do in the future. A survey of 'folk' would analyse how people express their creativity and spirituality, as well as traditions, patterns and trends in social life, cultural activities, and the tangible and intangible heritage of a place. A survey of 'place' would focus on the physical environment, but also on the flora and fauna, the ecosystems, the geology and the weather patterns of a locality.

According to Lewis Mumford, who could be described as one of Geddes' disciples, 'Geddes and his colleagues in Edinburgh, were perhaps the first to undertake a thoroughgoing civic survey as a preliminary to town planning and municipal action' (1940: 376). Geddes' background as a biologist predisposed him to seeing places holistically, as living organisms. Mumford describes Geddes as 'an ecological sociologist'. He adds that Geddes included in his civic survey 'as matters of first importance, the geographic setting, the climatic and meteorological facts, the economic processes, the historic heritage' (1940: 376). Such a broad, holistic perspective is not yet the norm today in the cultural mapping research on which cultural planning strategies are based.

Geddes, with his colleagues in Edinburgh, was involved in starting modern town planning in Britain, and was a key influence in the development of the 'garden city' concept. However, Geddes' ideas in the sphere of what today we would call cultural mapping and planning were for decades largely forgotten or interpreted fairly narrowly. The logic of professional specialization and division between disciplines prevailed. Macdonald quotes urban geographer Brian Robson, who concludes that 'too often it was "the bare bones, not the spirit" of Geddes' work that was taken up' (2009: 12).

The concept of 'cultural planning' as it is used today in Canada and Australia, for example, originated in the 1970s in the United States. Its invention is normally attributed to Harvey Perloff, who, in his 1979 book *The Arts in the Economic Life of the City* (Perloff 1979) suggested the establishment of 'a cultural element in the general plan for the city, county and Region' (quoted in Kunzmann 2004: 397). The concept was later developed, with slightly different emphases, first by Robert McNulty (McNulty 1991) and Partners for Livable Places in the United States (now Partners for Livable Communities, with whom Perloff had collaborated), in the late 1970s and 1980s, then by my own work in the UK since the late 1980s (Bianchini 1996a, 1996b, 1991a, 1991b, 1990), and by Colin Mercer in Australia from the early 1990s (Mercer 1996, 1991a, 1991b). Despite some differences, all these authors shared a commitment to valuing the contribution of local cultural resources not only to cultural policy, but also to many other areas of public policy, ranging from tourism and economic development to education, health and housing. Colin Mercer, for instance, defines cultural planning as

> the strategic and integral use of cultural resources in urban and community development ... cultural planning has to be part of a larger strategy for urban and community development. It has to make connections with physical and town planning, with economic and industry development objectives, with social justice initiatives, with recreational planning, with housing and public works. (2006: 6, emphasis in original)

This notion of cultural planning was adopted in the 1990s and 2000s by some academics, policy makers, consultants and community artists, mainly in the UK, Scandinavian countries, Australia and Canada. One example is Lia Ghilardi, who uses the interesting concept of 'urban and cultural DNA' in her mapping work, which follows an approach inspired by Geddes (Ghilardi 2009). However, the cultural planning approach remains relatively marginal in public policy-making. This is an illustration of a more general problem recognized by Greg Young:

examples of the systematic and foundational inclusion of culture in urban and regional planning are in short supply all around the globe. What familiarity with culture there is, tends to lie with its limited and often opportunistic inclusion in strategic planning and specific planning sectors, such as tourism, heritage and marketing. (2006: 43)

Different Interpretations

Although it did not use the term 'cultural planning', the 'global cultural development policy' introduced by French Culture Minister Jacques Duhamel (1971–73), was interdisciplinary, and based on co-operation between different ministries. It therefore had some of the characteristics of cultural planning (Girard 1996). Related terms are in some cases used to refer to the cultural planning approach and to urban development in particular. For example, Graeme Evans defines as 'cultural regeneration' a policy model in which 'cultural activity is ... integrated into an area strategy alongside other activities in the environmental, social and economic sphere' (2005: 968), while Hans Venhuizen (2010) uses the concept of 'cultural spatial planning', and both Richard Brecknock (2006: 81–5) and Charles Landry (2010: 33) prefer the notion of 'planning culturally' to 'cultural planning'. Brecknock in particular emphasizes the importance of 'cultural literacy' (2006: especially 47–59 and 86–8) for urban planners, in a context of growing multi-ethnicity and multiculturalism in cities in the West.

Graeme Evans interprets cultural planning as the development of planning norms for the cultural sector, to ensure the planned distribution of cultural infrastructure across a certain geographical area, and the setting of standards for the provision of cultural amenities (Evans 2001, see also Evans, Chapter 13, this volume).

A key related concept is that of the 'creative city' (Landry 2004, 2000, Landry and Bianchini 1995), which is itself understood in a variety of different ways, ranging from narrower policies for the cultural and creative industries at city and regional level, to fully fledged urban and regional strategies, aimed at harnessing people's creativity as a resource, in policy areas going well beyond the cultural sector.

One of the problems with cultural planning is that in some cases the term 'cultural plan' is used to refer to a policy document focusing on the arts and not on the wider notion of 'culture' (Stevenson 2005), leading to an arts or cultural policy, rather than to the more wide ranging and challenging enterprise of culturally sensitive and culture-based policy development in fields which go beyond culture. Some cultural plans as they exist today don't embrace the totality of the cultural resources of a place in an anthropological sense. However, Jason Kovacs' survey of municipal cultural plans in the Ontario province of Canada concludes that 'at least half of all cultural planning initiatives in the province's mid-size cities do not exhibit an arts-focused planning agenda' (2011: 338). Kovacs observes that several municipal plans addressed broader issues of 'intercultural involvement and business practices' and one plan dealt with even wider questions, concerning 'natural heritage, transportation and urban design' (2011: 338).

Reductive interpretations of 'cultural planning' and of the 'creative city' massively curtail the two concepts' potential for public policy innovation. The short book *The*

Creative City (Landry and Bianchini 1995) is inspired largely by psychological research on creativity in individuals and organizations. The book tries to understand if some of the lessons from this research could be applied to urban strategies. The book then makes a series of recommendations on how different kinds of public policy at city level could be made more creative. Evidently this cannot be reduced to the much more focused task of developing a strategy for the cultural and/or creative industries. Creative city strategies have relevance for education, transport, ecology, housing, health and many other policy areas.

As suggested earlier, my favoured interpretation of 'cultural planning' is as a way of thinking culturally and even artistically about public policy. Cultural planning is a culturally sensitive approach to urban and regional planning and to environmental, social and economic public policy making; it is about creating a two-way relationship between cultural resources and public policy. It is a relationship in which people who are in charge of mapping cultural resources should be also able to influence public policy – it should be a dialogue between equals.

Artistic Thinking and Cultural Planning

There are many initiatives and projects which, while not using the term 'cultural planning', highlight the special role of artists as interdisciplinary, innovative and critical brokers of ideas and strategies for 'place making', and more generally for the development of territorially defined communities.

Jonathan Metzger writes that *konst*, the word for 'art' in Swedish, 'derives from the same linguistic root as the adjective "strange"' (2011: 213), which in Swedish is *konstig*. Metzger quotes Swedish artist Lars Nilsson who plays on this double meaning when he says that 'art ... must never be useful but it must be allowed to be artful/strange' (2011: 213). Metzger argues that

> *Nilsson enacts the commonly posited binary opposite between utility and artistry, staging these terms as mutually excluding opposites ... On the contrary art and artistic activities can be very useful, and actually function as a powerful vehicle of communication in the planning process, precisely because of the artistic license that grants the artist a mandate to set the stage for an estrangement of that which is familiar and taken-for-granted, thus shifting the frames of reference and creating a radical potential for planning in a way that can be very difficult for planners to achieve on their own. (2011: 213)*

Metzger aims 'to shift focus away from ... "art in the public sphere" to instead try to generate a discussion on art *as* a public sphere' (2011: 2130). This is a very important point. The idea that because art causes estrangement it is not useful is wrong – it is useful precisely because it asks different and difficult questions, questions which are often not asked by specialists.

Like Patrick Geddes, many artists are generalists, not specialists. In 1966 artists John Latham and Barbara Steveni founded the Artist Placement Group (APG) in London (Eleey 2007). One of Latham's conceptual innovations was the notion of the artist as an 'incidental person', able to challenge what he called the 'Mental Furniture Industry' (MFI also the name of a major British furniture retailer): 'the apparatus of learnt knowledge and received opinion' (see http://www.flattimeho.org.uk/project/45/). John A. Walker writes that, according to Latham, artists possess several advantages:

> they ... operate on a longer time-base than other groups in society; they are skilled in handling conceptually unfamiliar material; and they are noted for their independence and their creativity. Since their activity is 'undefined in its own terms' artists can communicate to many different levels within a hierarchy ... The Artist Placement Group seeks to develop a new professional being whose knowledge and skills extend across disciplines so that he (sic) can make connections between the various specialisms found within organizations. (2010: 4–5)

This kind of thinking about the lateral and potentially innovative nature of artist-led interventions is embodied in a range of different projects. One of them is Cittadellarte ('City of art'), founded in 1998 in the northern Italian city of Biella. This is a former textiles factory renovated and put back into use by Michelangelo Pistoletto, one of the leading figures in the *Arte povera* ('Poor art') visual arts movement of the late 1960s in Italy. The foundation Cittadellarte-Fondazione Pistoletto uses this revamped factory as a laboratory for artists from all over Europe and beyond. Cittadellarte defines itself as

> a new form of artistic and cultural institution that places art in direct interaction with the various sectors of society. A place for the convergence of creative ideas and projects that combine creativity and enterprise, education and production, ecology and architecture, politics and spirituality. An organism aimed at producing civilization, activating a responsible social transformation locally and at global level (see http://www.cittadellarte.it/info.php?inf=2, emphases in original).

The foundation works through 'offices' operating in different fields, including Ecology, Economics, Politics, Spirituality, Food and Architecture. For example, the Politics office of Cittadellarte founded in 2008 the Mediterranean Cultural Parliament, to stimulate intercultural dialogue in a region characterized by political, religious, economic and ethnic tensions.

Artist-led cultural planning has led to interesting experiments, for instance in the field of urban lighting. Lighting installations by artists as part of projects like the Fête des Lumières (Festival of Lights) which started in 1999 in Lyon, France, and Luci d'artista (Artist's lights) which was also launched in 1999 in Turin, Italy, challenged assumptions and mainstream ideas in a field which has traditionally been dominated by engineers. Zenobia Razis (2002) discusses how an artist-led approach to urban lighting has produced in many cases aesthetically stimulating, innovative and sustainable solutions.

Cultural Planning and the Challenges of Urban Policy

Advocates of cultural planning argue that policy-makers in all fields should not simply be making an instrumental use of cultural resources as tools for achieving non-cultural goals, but should let their own mindsets and assumptions be transformed by contact with the soft infrastructures which make up local culture. This can happen if policy-makers learn from the six key sets of attributes of the types of thinking characterizing the processes of artistic thinking and cultural production. This thinking tends to be:

- holistic, flexible, lateral, networking, and interdisciplinary
- innovation-oriented, original and experimental
- critical, inquiring, challenging and questioning
- people-centred, humanistic, and non-deterministic
- 'cultured', and informed by critical knowledge of traditions of cultural expression
- open-ended and non-instrumental

In a cultural planning process, the cultural policy-maker, the artist and/or the cultural manager can become the gatekeepers between the sphere of cultural production – the world of ideas and of production of meaning – and any area of policy-making, also in order to enhance the cultural skills of politicians and decision-makers more generally.

Cultural planning, like the creative city idea, is very much about opening up policy systems to young talent, setting up pilot projects, preparing research and development budgets and not being paralysed by the fear of failure. Unfortunately today, as a result of austerity policies, budgets for experiments, training, study visits and research and development and innovation have been severely cut. Precisely at the time when we need to reinvent an economic base for Europe, and when we have growing unemployment affecting especially young people, policy makers are shutting down possible sources of innovation and policy solutions.

Another characteristic of cultural planning is that it is critical, questioning and challenging. The mapping of the cultural resources of a locality has to include the difficult areas, the taboos, the skeletons in the cupboard, the contradictions and problems. It is often in the areas of conflict that the most interesting resources lie. The artistic team responsible for Linz European Capital of Culture 2009 discovered this, when it had the courage to explore the Hitler legacy for the city, through projects including the exhibition *The Culture Capital of the Führer: Art and National Socialism in Linz and Upper Austria* (September 2008–March 2009, see http://www.spiegel.de/international/europe/hitler-s-culture-capital-linz-tackles-its-past-as-a-fuehrer-city-a-578785.html). The exhibition was well attended, and stimulated public debate.

Cultural planning is also a 'cultured' approach which is required to study cultural resources such as the urban 'image bank' and other key cultural resources. The urban image bank consists of local and external images of a city, including the following:

- media coverage
- stereotypes, jokes and 'conventional wisdom'
- representations of a city in music, literature, film, the visual arts and other types of cultural production
- myths and legends

- city marketing and tourism literature
- views of residents, city users and outsiders, expressed, for example, through surveys and focus groups

It may be appropriate to map this cultural resource by using content and/or discourse analysis, as well as historical reconstructions of how cultural representations of a city have evolved. This would require collaborative work, involving not only architects, planners, engineers, economic development specialists, place marketers and tourism and cultural development officers, but also specialists in urban mindscapes and imaginaries, including urban historians, sociologists, anthropologists, semiologists, psychologists and artists. Knowledge of a city's history, and of people's memories, can be an important source of inspiration for future creative ideas. Learning about local traditions of innovation can inspire new traditions.

This cultured and collaborative approach has stimulated some innovative thinking about place marketing. Chris Murray (2001) maintains that it is not appropriate directly to transfer to city strategies used for marketing products like cars, chocolate or shoes, as often happens in mainstream professional practice. He argues that city marketing is often interpreted very reductively, as the designing and selling of place identities. According to Murray, place marketing from a cultural planning perspective should rather be aimed at revealing, discovering and celebrating the complex identities, 'cultural biographies' (Rooijakkers 1999), mindscapes and imaginaries (Weiss-Sussex with Bianchini 2006) of each locality. It is important to resist the temptation to produce 'official' place marketing strategies based on dominant views. Policy makers should recognize and build upon the local politics of symbolic contestation, and on the fact that views about what is the most important narrative about a place to be promoted can vary, for example in terms of people's age, neighbourhood, occupation, ethnicity and gender.

Cultural mapping exercises should also consider the creative milieus and emerging cultural talent of a city, possibly by associating in mapping processes young people, whose input is often crucial to understand the moving current of local creativity.

The adoption of austerity policies as a response to the deepening economic crisis since the second half of the 2000s in Europe has led to cuts in funding for cultural activities, and in many cases to lower political status for culture. There are also clear limitations to the political effectiveness of advocacy for the cultural sector based on economic and social impact research. Cultural planners and policy makers need to build new alliances, with green movements, and with emerging forms of thinking about possible alternatives to finance-led capitalism as it is presently constituted.

For the success of cultural planning strategies, it is important not only to create effective horizontal, cross-departmental and interdisciplinary institutional implementation arrangements but also to contextualize the mapping of local cultural resources in international terms. It is going to be very difficult for policy makers to resource cultural planning strategies in the future if they do not adopt an international perspective, which could build on a city's arts, media, sports, trade and diasporic links worldwide.

The cultural assets, themes and narratives revealed by mapping can be used in a variety of contexts, ranging from artistic programming, organization of festivals and other cultural events, business development strategies in the creative industries and other sectors of the local economy, as well as policy development in fields ranging from transport, housing, physical planning and tourism strategies. Small, and often under-resourced, cultural planning teams, need to form partnerships with colleagues in the public, private and third sector. This is essential to have the skills, breadth of knowledge, energy and time which are needed to use the results of cultural mapping imaginatively and effectively in urban policy.

There is also a need for specific training, for example in cultural mapping, in linking cultural planning with place marketing and social policy, and in working in multicultural contexts. I will discuss here briefly the latter two professional specialisms, which have not been considered in earlier sections of this chapter.

Two Examples of Professional Specializations in Cultural Planning

The Culture and Social Action/Social Policy Specialist

This professional specialization developed within alternative cultural networks in Western Europe and North America in the late 1960s and 1970s. It is the product of the institutionalization of movements like community arts and community media in Britain, socio-cultural animation in France, and *Sozio-Kultur* in Germany. The main aim of this professional specialization in the 1970s was to encourage the expression of people's creativity in order to raise awareness of conditions of subordination and oppression, and trigger potentially revolutionary processes of social and political change. There was, especially since the second half of the 1990s, a strong drive to institutionalize, depoliticize and de-radicalize this professional specialization. In the British case, for example, the Labour governments led by Tony Blair, particularly from 1997 to 2001, put a strong emphasis on the importance of the link between cultural policy and social inclusion (Smith 1998). The Blair governments highlighted the usefulness of cultural policies and actions to improve people's behaviour and strengthen social stability and order.

The aims of this professional specialization in the 1990s and in the 2000s were influenced by the findings of studies about participatory cultural activities. In 1992 Phyllida Shaw wrote *Changing Places*, a report commissioned by the Scottish Arts Council in association with the Industry Department of the Scottish Office, which described the impacts of participatory arts projects in urban areas in Scotland that qualify for special funding because of their levels of deprivation. She argued that the benefits of these projects included the following: enhancing community identity; improving the image of a place both within and outside an area; improving communication within a community; creating opportunities for teamwork; increasing self-confidence; improving the physical appearance of a place; creating a safer environment; improving health by reducing stress, and providing training for employment (Shaw 1992).

In 1995 the Community Arts Network of South Australia published a study (Williams 1995) which surveyed the social, educational, artistic and economic benefits of a national sample of 'organizers' and 'observers' of community-based arts projects funded by the Australia Council. According to the respondents to Deidre Williams' questionnaires, arts activities had significant long-term value in relation to the following objectives, among others: developing 'community identity' (52 per cent and 52 per cent); establishing 'networks of ongoing value' (54 per cent and 45 per cent); lessening 'social isolation' (45 per cent and 39 per cent); improving 'understanding of different cultures or lifestyles' (42 per cent and 37 per cent), and 'solving problems' (31 per cent and 28 per cent).

There are various contradictions and dilemmas associated with this professional specialization. One of these concerns the definition of 'social inclusion' itself, which presupposes the existence of a 'normal' majority and of a 'deviant' minority whose behaviour has to be made to conform to that of the majority. It may be more fruitful to promote social interaction and communication between different social groups, which values the richness of different types of knowledge, experience, intelligence and creativity.

Another controversial issue is that of the validity of the findings of studies aimed at evaluating the socials impacts of community-based arts projects. For instance, Paola Merli (2002) offers an interesting methodological critique of Matarasso's evaluation of the social impact of arts activities (1997).

The Specialist in Multiculturalism, Interculturalism and Cultural Diversity Issues

Multiculturalism and interculturalism are relatively common experiences at the level of individual consumption (for example, of food, music and crafts objects), but it is much more difficult to encourage groups of people to participate in multicultural and intercultural projects, which are essential to achieve social and cultural sustainability. It is important here to understand the difference between the terms 'multiculturalism' and 'interculturalism'. Urban cultural policies in Britain, for example, since the mid-1970s have aimed at multiculturalism, which generally means the strengthening of the distinctive cultural identities of different ethnic communities, by enabling them to have their own cultural voices. This is a valuable objective, but multiculturalism does not necessarily encourage communication between cultures. On the contrary, multicultural policies can contribute to entrenching particularisms and vested interests. If the aim is to counteract racism, then perhaps more resources could be directed to intercultural projects, aimed at building bridges between different communities and at producing cultural hybrids, which can be important not only in terms of artistic and cultural innovation, but also for their potential applications to social, economic and organizational innovation (for a fuller discussion of the potential of intercultural interventions see Bloomfield and Bianchini 2004, Wood and Landry 2008).

Concluding Observations

It is important to ask whether culture in general, and more particularly the processes of cultural production (despite the debate on cultural planning) continue to be simply instruments in the hands of established professional disciplines like physical planning, local economic development, place marketing, tourism promotion and social policy. Can entirely new professional disciplines be founded through the interaction between ideas and experiences from the cultural sector and the assumptions and canons of established public policy professions? What are the main obstacles to this process? Can the cultural sector play the public policy game, but without compromising its integrity and the originality of its contribution? Can cultural planning help refine and revitalize public policy debates in a way which is comparable to the contribution made by the environmentalist movement since the early 1970s? What should the role of universities and other education and training institutions be with regard to these difficult questions?

One thing is certain: the cultural sector is being invited, in some cases courted even, to play the 'big game' of public policy. It is important that, if people working in the cultural sector accept to play this game they should do so as equal partners and with mutual respect, rather than in their traditionally subordinate position.

Acknowledgements

I would like to thank Greg Young for his patience and support in the preparation of this chapter. Parts of the last three sections of the chapter are based on my article 'Professional specializations in urban and regional cultural planning: an overview', in a special issue of *Journal of the Institute of Theatre, Film, Radio and Television* entitled *Management of Culture and Media in the Knowledge Society: Post Conference Anthology*, Faculty of Dramatic Arts, University of Belgrade, 2012, forthcoming.

References

Bianchini, F. 1990. Urban renaissance? The arts and the urban regeneration process, in *Tackling the Inner Cities*, edited by S. MacGregor and B. Pimlott. Oxford: Clarendon Press, 215–50.

——. 1991a. *Urban Cultural Policy*. National Arts and Media Strategy Discussion Documents, 40. London: Arts Council of Great Britain.

——. 1991b. Models of cultural policies and planning in West European cities. *The Cultural Planning Conference*, Mornington, Victoria, Australia, EIT.

——. 1996a. 'Cultural planning': an innovative approach to urban development, in *Managing Urban Change*, edited by J. Verwijnen and P. Lehtovuori. Helsinki: University of Art and Design Helsinki, 18–25.

—— . 1996b. Rethinking the relationship between culture and urban planning, in *The Art of Regeneration*, edited by F. Matarasso and S. Halls. Nottingham and Bournes Green: Nottingham City Council and Comedia.

—— . 1999. Cultural planning for urban sustainability, in *City and Culture: Cultural Processes and Urban Sustainability*, edited by L. Nyström. Karlskrona: Swedish Urban Environment Council, 34–51.

Bloomfield, J. and Bianchini, F. 2004. *Planning for the Intercultural City*. Bournes Green: Comedia.

Brecknock, R. 2006. *More than Just a Bridge: Planning and Designing Culturally*. Bournes Green: Comedia.

Eleey, P. 2007. Context is half the work. *Frieze*, 111 (November–December). [Online]. Available at: http://www.frieze.com/issue/article/context_is_half_the_work/ [accessed 27 July 2012].

Evans, G. 2001. *Cultural Planning*. London: Routledge.

—— . 2005. Measure for measure: evaluating the evidence of culture's contribution to regeneration. *Urban Studies*, 42(5–6), 959–83.

Ghilardi, L. 2009. *Thinking Culturally about Place and People: The Cultural Planning Approach*. [Online]. Available at: http://www.move2009.org/presentations/LIA_GHILARDI_the_cultural_planning_approach.pdf [accessed 28 July 2012].

Girard, A. 1996. Les politiques culturelles d'André Malraux à Jack Lang: ruptures et continuities, histoire d'une modernisation. *Hermès*, 20, 27–41.

Kovacs, J.F. 2011. Cultural planning in Ontario, Canada: arts policy or more? *The International Journal of Cultural Policy*, 17(3), 321–40.

Kunzmann, K. 2004. Culture, creativity and spatial planning. *Town Planning Review*, 75(4), 383–404. [Online]. Available at: http://www.scholars-on-bilbao.info/fichas/KUNZMANN%20CultureCreativitySpatialPlanningTPR2004.pdf [accessed 27 July 2012].

Landry, C. 2000. *The Creative City*. London: Earthscan.

—— . 2004. *The Art of City Making*. London: Earthscan.

—— . 2010. Think culturally, plan imaginatively and act artistically, in *Game Urbanism*, edited by H. Venhuizen with contributions by C. Landry and F. van Westrenen. Amsterdam: Valiz.

Landry, C. and Bianchini, F. 1995. *The Creative City*. London: Demos.

Macdonald, M. 2009. *Patrick Geddes and the Scottish Generalist Tradition*. [Online]. The Royal Town Planning Institute in Scotland's 2009 Sir Patrick Geddes Commemorative Lecture at the Royal Society of Edinburgh, 20 May 2009. Available at: http://www.scribd.com/doc/73434979/Patrick-Geddes-and-Scottish-Generalist-Tradition [accessed 8 July 2012].

Martines, L. 2002. *Power and Imagination: City-States in Renaissance Italy*. London: Pimlico.

Matarasso, F. 1997. *Use or Ornament? The Social Impact of Participation in the Arts*. Bournes Green: Comedia.

McNulty, R. 1991. Cultural planning: a movement for civic progress. *The Cultural Planning Conference*, Mornington, Victoria, Australia, EIT.

Mercer, C. 1991a. Brisbane's cultural development strategy: the process, the politics and the products. *The Cultural Planning Conference*. Mornington, Victoria, Australia, EIT.

—— . 1991b. *What is cultural planning?* Paper presented at the Community Arts Network National Conference, Sydney, Australia, 10 October.

——. 1996. By accident or design: can culture be planned? in *The Art of Regeneration*, edited by F. Matarasso and S. Halls. Nottingham and Bournes Green: Nottingham City Council and Comedia.

——. 2006. *Cultural Planning for Urban Development and Creative Cities*. [Online]. Available at: http://www.culturalplanning-oresund.net/PDF_activities/maj06/ Shanghai_cultural_planning_paper.pdf [accessed 8 July 2012].

Merli, P. 2002. 'Evaluating the social impact of participation in arts activities: a critical review of Francois Matarasso's 'Use or Ornament?'. *International Journal of Cultural Policy*, 8(1), 107–18. For a response by François Matarasso to this article see Matarasso, F. 2003. Smoke and mirrors: a response to Paola Merli's 'Evaluating the social impact of participation in arts activities'. *International Journal of Cultural Policy*, 9(3), 337–46.

Metzger, J. 2011. Strange spaces: a rationale for bringing art and artists into the planning process. *Planning Theory*, 10(3), 213–38.

Mumford, L. 1940. *The Culture of Cities*. London: Secker and Warburg.

Murray, C. 2001. *Making Sense of Place: New Approaches to Place Marketing*. Bournes Green: Comedia, in association with the International Cultural Planning and Policy Unit, De Montfort University, Leicester.

Perloff, H.S. with Urban Innovations Group. 1979. *The Arts in the Economic Life of the City*. New York: American Council for the Arts.

Razis, Z. 2002. *Reflections on Urban Lighting*. Bournes Green: Comedia, in association with the International Cultural Planning and Policy Unit, De Montfort University, Leicester.

Rooijakkers, G. 1999. Identity Factory Southeast: towards a flexible cultural leisure infrastructure, in *Planning Cultural Tourism in Europe*, edited by D. Dodd and A. van Hemel. Amsterdam: Boekmanstichting, 101–11.

Shaw, P. 1992. *Changing Places: The Arts in Scotland's Urban Areas*. Edinburgh: Scottish Arts Council and Scottish Office Industry Department.

Sitte, C. 2007. *L'arte di costruire le città*. Milan: Jaca Book.

Smith, C. 1998. *Creative Britain*. London: Faber.

Stevenson, D. 2005. Cultural planning in Australia: texts and contexts. *Journal of Arts Management, Law and Society*, 35(1), 36–48.

Tragbar, K. 2006. 'De Hedificiis Communibus Murandis …': notes on the beginning of building regulations in medieval Tuscany, in *Proceedings of the Second International Congress on Construction History*, edited by M. Dunkeld. Queens' College, Cambridge University, 29 March–2 April, 3117–31. [Online]. Available at: http://www.arct.cam. ac.uk/Downloads/ichs/vol-3-3117-3132-tragbar.pdf [accessed 4 April 2013].

Venhuizen, H. with contributions by C. Landry and F. van Westrenen. 2010. *Game Urbanism: Manual for Cultural Spatial Planning*. Amsterdam: Valiz.

Walker, J.A. 2010. *Artist Placement Group (APG): The Individual and the Organization. A Decade of Conceptual Engineering (1976)*. [Online]. Available at: http://www. artdesigncafe.com/Artist-Placement-Group-APG-John-Latham-Barbara-Steveni [accessed 4 April 2013]. Originally published in *Studio International*, 191(980) March–April 1976, 162–4.

Weiss-Sussex, G. and Bianchini, F. 2006. (eds) *Urban Mindscapes of Europe*. New York: Rodopi.

Williams, D. 1995. *Creating Social Capital*. Adelaide: Community Arts Network of South Australia.

Wood, P. and Landry, C. 2008. *The Intercultural City*. London: Earthscan.

Young, G. 2006. Speak, culture! Culture in planning's past, present and future, in *Culture, Urbanism and Planning*, edited by Javier Monclus and Manuel Guardia. Farnham: Ashgate, 42–59.

Stealing the Fire of Life: A Cultural Paradigm for Planning and Governance

Greg Young

In Salman Rushdie's magic fable *Luka and the Fire of Life* (2010) the eponymous hero sets out on a desperate quest to steal the fire of life that can save his father from perpetual sleep. Inspired by the mission of Rushdie's figure, I believe that planning and governance are in a similar need of culture as the metaphorical 'fire of life', if they are to shrug off planning indifference and counter the reigning cultural expedience that characterizes global neo-liberal governmentalities for the most part. Of course, the culture I am speaking of is ontologically holistic and the governance I refer to comprises the full scope of state, market and social institutions and activities that govern planning and administrative outcomes. Under governance viewed in these terms, the sectoral beneficiaries of culture's fire and largesse would be widespread, and include not only spatial and strategic planning but also other key areas such as public administration, health, education, development studies and international relations. In this chapter I seek to highlight some of the conceptual and methodological possibilities for governance in widespread terms, by focusing on planning as a case study for culture, drawing on my book *Reshaping Planning with Culture* (Young 2008b) and other sources (Young 2008a, 2005).

A Cultural Era

I propose to begin, however, by hazarding a brief description of our era as a cultural one, although, as Jameson observed, this remains so 'in some original and yet untheorised sense' (1984: 87). This condition is not, however, without referents as the culture-saturation of our time has a number of identifiable features, as well as recognizable opportunities. Culture has long been recognized as expanding (Williams 1966) and does so at unprecedented rates in an information and knowledge age where lives are lead in a condition in which 'Culture refers to Culture' (Castells 1998: 477). This weight of culture is expressed in many dimensions, from the growing importance of cognitive factors in social life that characterizes a post-industrial world of symbolization and personal and professional representation, through to the fact that cultural consumption and production have become 'the principal activity of Europeans' (Sassoon 2006) and for numerous others around the globe.

In addition, as cultural practices, intangible culture, material culture and indeed our very ideas about culture itself, become objects of their own interest, culture increasingly constitutes its own capital. In this way, culture assumes a role as our preeminent intellectual resource. In this sense, therefore, the 'cultural turn' (Chaney 1994) overlapping from the late twentieth century, and its promise of informed cultural understanding, are poised for greater social incorporation in new and stronger practical terms. Under this scenario, culture and the enabling agencies of governance and planning step forward in shared importance, in spite of neo-liberal forms of governmentality that serve to impede the possibilities for cultural integrity and planning coherence. This is because unlocking planning and governance synergies, and addressing the complexities of key issues such as sustainability in nuanced terms, depend on deeper and richer approaches to culture as defining tools.

Yet although I will argue the possibilities for a cultural paradigm, with culture advanced as the fundamental organizing concept and framework for governance, such a strategy relies on a central distinction I draw in regard to the existence of a dialectic of culture. This dialectic operates between the current mainly commodified inclusion of culture in planning and the ever-present opportunity for culture's integration in planning and governance in more reflexive and humane terms. In this respect Radcliffe argues that in the case of international development thinking 'culture and development are now widely perceived as dialectically related' (2006: 17) and that the field of development is seen as intrinsically one of 'social interaction between multiple conceptions of culture, tradition and modernity' (2006: 24). This is a broader and more 'open' dialectic than the relationship between the commodification of culture in global consumer society, characterized by a number of commentators (du Gay and Pryke 2000, Scott 2000) as a pattern of culturalization, and the contrasted opportunity in social interaction I have previously tagged with the neologism of culturization (Young 2008a, 2008b). The concept of culturization describes a more positive positioning for governance, whereby culture is utilized in specifically creative, critical, ethical and reflective terms. While I utilize the term reflection as a defining quality of culturization and culturized planning, elsewhere (2008a, 2008b) I have also employed the term reflexive. The use of this term serves to highlight key aspects of the nature of values such as the facts that they are fundamentally dynamic, permanently in need of renewal and represent more than an accumulation of empirical knowledge. Giddens takes up this point in noting that 'Only societies reflexively capable of modifying their institutions in the face of accelerated social change will be able to confront the future with any confidence' (Smart 1993: 42).

The Dynamics of Culture

The rise of culturalization is considered by many commentators to be a key aspect of culture since the late twentieth century and is related to post-industrial capitalist production. In this sense, Scott (2000) defines culturalization as a double process in which culture becomes more of a commodity, at the same time as commodities themselves acquire greater cultural and symbolic content. The trend also affects organizational life and technical processes such as planning which can reflect higher levels of culturalization (du Gay and Pryke 2000: 6). As a planning trend, for example, culturalization is manifest across a broad spectrum of commodification whereby cultural themes, values and periods are absorbed in aspects of the development and marketing of commercial and residential sites and complexes, and in the planning and positioning of heritage, tourism and events. As an example, community memories and associations, historic images and traditional customs and values are taken out of community and historical contexts and 'flattened' into stereotypes to create a 'packaged' cultural legitimacy for new developments or refurbished heritage places. In an information and digital age, consumer capitalism easily accesses and appropriates cultural values and materials from their community and intellectual settings, and re-vamps them to create and differentiate new products, meanings and experiences for the marketplace. (See also Ashworth, and Nyseth, Chapters 11 and 19, this volume).

At the same time, it is important to note that culture's assumption of greater cognitive value and a greater need for critical interpretation, is part of the same dialectic that includes its culturalized and commodified expansion. My response to this is to deploy the concept of 'culturization' (Young 2008a) in social discourse in general, and in particular, in governance and planning. Later, I introduce a cultural paradigm and a culturized model for planning and governance (Young 2008a, 2008b, 2005) to promote the convergences between culture, governance and planning along lines that are not exclusively instrumental, commercial or random. At its core, a cultural paradigm has the goal of promoting more humane and robust culturized planning and governance futures.

A Cultural Paradigm for Planning and Governance

The well-known concept of the Kuhnian paradigm (1970) has proved a durable one and has now taken on a more general role in many contexts other than that developed by Kuhn for science itself. For example, the idea of paradigmatic development is accepted by many planning commentators in relation to planning theory (Alexander and Faludi 1996, Allmendinger and Tewdwr-Jones 2002, Taylor 1998) and is described as having influenced new directions in both social and planning theory (Allmendinger and Tewdwr-Jones 2002: 7). Culturization and the culturized planning model share a similar inspiration within planning theory, but go beyond it, to indicate new possibilities for all of governance. The cultural paradigm I envisage applies to governance *simpliciter* and the contours of such a role are becoming more and more apparent on our intellectual horizons. As an example, a culture-based paradigm is recognized and utilized in organizational and management studies (Hofstede and Hofstede 2005, Schein 2004) and for management guidance in cross-cultural and transnational settings (Harris, Moran and Moran 2004).

An over-arching cultural paradigm for governance has the potential to be brought into operation through new cultural models specifically crafted for the diversity of governance functions, sectors and activities. A key benefit to be derived from a development of this kind is the likely generation of new and more creative solutions articulated through the supple culture-based philosophical, theoretical and methodological tools that are capable of addressing shared and overlapping governance issues and problems. For the present, my existing culturized model for planning indicates relevant pathways for developing a range of culturized strategies for broader governance that seeks to accommodate culture in creative, critical, ethical and reflective terms.

A key perspective for governance, in undergoing a developmental journey of this kind, is likely to be the recognition that culture is both the object of governance, and its key procedural means, and that this applies regardless of the governance in question. In the case of the cultural paradigm in respect of planning, for instance, it is possible to view culture as both the object of planning and as its principal 'operative' means. This position accommodates, for example, the deep green view of planning, as necessary to restore full ecological functioning to the Earth. This is a result of the facts that not only are natural processes the subject of cultural understanding, but also their maintenance is increasingly determined by cultural priorities.

I believe my distinction between the object of planning and its operative means parallels that made by Faludi (1987) between procedure and substance in planning theory. The distinction has considerable heuristic value, although it is deemed by Allmendinger (2002: 11) to be an analytical distinction, rather than a real or practical one. In addition, culture can be viewed as the true parameter of planning, rather than as a mere planning variable invoked from time to time in occasional aspects of planning practice. Viewed as the parameter of planning, culture is neither marginalized, as in traditional cultural planning, nor fragmented through the piecemeal insertion and subtraction of cultural elements, even where this is undertaken in the name of cultural sensitization. This is both a liberating insight and one that can be applied across the range of governance sectors and functions I cite. At the same time, culture's decisive role in shaping the quality and value of spatial plans, and in enriching the visioning process in strategic planning, whether in spatial or non-spatial terms, is increasingly the subject of theoretical and practical awareness (Albrechts 2010, Healey 1997, Sandercock 2003, 1998). This includes the admittedly problematical concept of 'The Four Pillars of Sustainability' (Hawkes 2004) adopted by the United Cities and Local Governments (UCLG) organization as policy in 2011 under which culture is added to the list of social, economic and environmental considerations.

Culturization and Planning

In the case of planning, I believe that a culturized approach is the one most likely to assist planners in conceptualizing cities and regions, and their specific identities, wherever they occur around the globe. The specific cultural constellation of each place is the basis and target of planning, and is unlocked by a cultural approach. This includes, as Beauregard reminds us, the fact that planners need to conceptualize the city, in order to act. Without a sense of context, 'plan-making neither enjoys the benefits of limits nor the possibility of implementation' (Beauregard 2008: 33). Although in his work Beauregard is concerned to propose the rhetorical advantages of the 'network city' in focusing attention on the defining qualities of a city, for similar reasons, I would assert the primary importance of culture and culturized planning. The concepts I propose likewise assist in 'managing the complexity of the city' (Beauregard 2008: 254) by 'flagging' and opening up opportunities for intervention. Again, accommodating the nature of a culturally diverse and interdependent world is something that can only be understood and changed, as Castells argues, 'from a plural perspective that brings together cultural identity, global networking and multidimensional politics' (1996: 28).

My approach to planning culturization is therefore clearly, but not uncritically normative in kind. The normative approach I adopt, chimes with a similar tradition in planning history, which has frequently tracked a parallel trajectory with the evolution of key concepts of culture for at least the last 100 years. While the routes of these 'normative journeys' frequently overlapped, in practice, they now visibly converge especially under the impact of contemporary social theory and globalization.

Sketching a Useable Culture

I draw on the concepts of two key twentieth-century thinkers to assist in sketching a useable culture. These concepts are those of the French philosopher Henri Lefèbvre and the British Marxist Raymond Williams. In *The Production of Space* Lefèbvre (1992) outlines a 'trialectics of being' based on a cultural triad of 'spatiality', 'historicality' and 'sociality' or more familiarly, 'space', 'time' and 'society'. This triad provides an abstract ontological categorization that can be correlated to the elements of culture in terms of firstly, the environment, tangible heritage and cultural landscapes; secondly, history and intangible heritage; and thirdly, ways of life. These three elements are essential for the holistic understanding of culture and its practical planning and governance 'uptake'. Each of culture's three moments as Soja argues (1996: 72) 'contain each other (and) cannot successfully be understood in isolation or epistemologically privileged separately, although they are all too frequently studied and conceptualized in this way in compartmentalized disciplines and discourses'.

Allied to Lefèbvre's ontological holism, Williams's much cited definition of culture as 'a whole way of life, material, intellectual and spiritual' (1966: 16) implants for its part a democratic and 'anthropological' perspective to include 'codes of manners, dress, language, rituals, norms of behaviour and systems of belief' (Jary and Jary 1991: 138). Indeed Williams's writings and those of other members of the associated Birmingham School of Cultural Studies furnished many key disciplines such as cultural studies with the modern working tools to approach culture (see also, O'Connor, Chapter 10, this volume). Williams's description of culture became in fact the shorthand definition of choice subsequently woven through research in the humanities and social sciences. The cultural theory, concepts and critical tools of Williams and the Birmingham School gave intellectual purchase to cultural and policy interventions aimed at improving forms and levels of social inclusion and development. (See also O'Connor, Chapter 10, this volume.)

This work built on Edward Tylor's 1871 definition of culture in *Primitive Culture* as 'that complex whole which includes knowledge, belief, art, morals, law, custom, and any other capabilities and habits acquired by man as a member of society' (Tylor 1903: 1) and in focusing on social and environmental improvement also recapitulated Tylor's famous assertion that 'in aiding progress and in removing hindrance, the science of culture is essentially a reformer's science' (Tylor 1924: 453). Culturization is a contemporary expression of the same desire. In spite of this it should be said that perpetuating such a vision is an elusive quest. How for example is it possible to draw on a useable concept of culture for planning and governance as the velocity of culture keeps increasing and cultures themselves continue to vary in seemingly infinite and breathtaking ways as history, the Internet and television daily remind us? As I have intimated, the solution would seem to lie in employing ontological and taxonomic approaches to culture that are valid and relevant, in order to clarify and frame a useable culture. I later spell these out in describing the culturized model for planning.

Sketching a Useable Planning Theory

In sketching a useable planning theory it is first important to note that planning theory encompasses a broad range of types. I am nevertheless, inclined to agree with Allmendinger, (2001) who views these types as fundamentally polarized into neo-modern and postmodern wings. In addition to this, the function and utility of planning theory is hotly debated and such theory is often attacked as lacking relevance, except perhaps to the academic careers of the planning theorists themselves. In spite of this, planning is dependent on theory to refresh and update its thinking. It is also true to say in general terms, that over the last century many planning theories and concepts have been normative in kind, and that this represents a continuing current such that planning may be seen as essentially a normative activity. Culturization broadly relates to the normative tradition and is itself based on a critical, ethical and reflexive vision for planning which the culturized model is intended to facilitate and promote. At the same time, I believe it is important to develop an overall awareness of the strengths and weaknesses of specific planning theories and doctrines in both historical and contemporary terms. For example, each of the four main planning approaches of planning modernism, communicative theory and collaborative planning, postmodern planning and neo-liberal planning has attracted criticism. The agenda of planning modernism is deemed to be naïve with a false confidence in universal reason and one-size-fits-all solutions to problems; communicative theory and collaborative planning are spiked as mere high-minded beneficence in the face of entrenched power and advantage; postmodern planning theory is considered problematic in introducing 'theoretical unsettling' and 'thinking differently' as highly abstract and impractical considerations in planning terms; and neo-liberal planning is considered opportunistic in 'cherry-picking' community cultural assets and meanings and aligning planning more exclusively with the global 'space of flows' (Castells 1991: 350).

Specifically, in regard to communicative planning theory and the new urbanism Fainstein has commented critically that communicative planning theory has 'evaded the issue of universalism by developing a general procedural ethic without substantive content' (2003: 190). In contrast, she believes that the new urbanism has inspired a social movement whose utopianism 'contrasts with communicative planning, which offers only a better process' (Fainstein 2003: 184). In this respect, however, culturization and the culturized model may represent a perspective able to go some of the way towards bridging the gap between universalism, procedure and process. It can achieve this in two ways: first, by enriching the understanding of the substantive object of planning (increasingly urban communities under processes of economic and cultural globalization) and second, by deepening and sensitizing the procedural techniques of planning itself, particularly as they relate to communities and the conservation and development of their inherited environments.

Culturized planning begins with the recognition that culture resides in each and every place, including in its governance regimes, planning traditions and planning cultures. On this understanding, culture is socially, historically and environmentally grounded in places, communities and cultures, so that research, local collaborations, cultural interpretation and theorization all need to reflect these specific constellations. This approach has a special power to reveal the uniqueness of places, communities, institutions and histories, as well as the contemporary impacts of greater cultural flows, multiple cultural derivations and new levels of connectivity. In these respects, Gunder and Hillier (2009: 194) argue that planning needs to engage with local tangible and intangible cultural realities, and be opened to alternative paths to engage with the new, different and unknown.

The inherent creativity, reflexivity and criticality of culturized planning is able to promote this, by pulling in new theorization, hidden and marginal histories and community memories and more powerful thematic approaches to knowledge and interpretation that are themselves seen as part of a condition of dynamic flux. In this way culturized planning, and indeed culturized thinking, can respond to the many ways that there are to be and to become human, by addressing a deeper and richer approach to human needs. It is also capable of promoting more sensitive planning processes across all planning types and scales by linking and 'binding' the cultural integument between geographical scales, not only on the level of the scalar hierarchy of planning instruments, but also in terms of storytelling and the 'production of meaning' (Sandercock 2003: 199) whose patterns overlap the scales of constructed representation. Sandercock recognizes this need when she notes that 'residents and activists may be most familiar with looking at issues from the local or neighbourhood perspective. [However] some stories ask us to take a global perspective' (2003: 199). Harvey presents a related argument in connection with encouraging positive local, regional, national and global action on climate change.

Culturization as a Transformative Practice

The concept of culturization is both a normative tool, as it depends on critical, reflexive and ethical planning reasoning, and a rhetorical tool, designed to give traction to and 'flag' opportunities for the integration of culture in planning, in terms that challenge and unsettle conventional planning solutions and approaches. Unlike the inherent commodification of culturalization, it is also more likely to be community-based and closely related to the values of civic and public cultures. These values may reflect the ethical imperatives and standards of the local, national and international state, as they evolve and are codified, in respect of culture, diversity and human rights.

Culturization is also positioned to embrace the diversity of planning theory and to opt for a theoretical and methodological pluralism that better reflects the uneven modalities of communities and regions around the globe. Planning theory may be 'stretched', and planning practices adapted, to more sensitively address the specific cultural constellation of each and every community and place. It is true to say that most planning theories, planning trends and planning practices present a diversity of individual opportunities and constraints for planning. In each case, however, a culturized emphasis or critique would assist in selecting and 'customizing' the most appropriate theory and practices as well as reflecting the need for changing theory. Culturization as a concept, however, requires the traction of a practical model for its implementation and I now turn to such an outline.

The Culturized Model for Planning and Governance

Having previously developed a culturized model for planning and illustrated it with extensive case studies, my goal in this chapter is to define culturization and to single out the key 'mechanical' features that power the Culturized Model (Young 2008b, 2005) as it may relate not only to planning but also to governance in widespread terms. In general terms culturization may be defined as the holistic research of holistic culture and its creative, critical, ethical and reflective utilization in governance functions including spatial planning.

The Culturized Model is designed to address limitations in terms of current planning theory and practice including those previously described. The Model is tripartite consisting of seven positioning principles for culture; three literacies for planners; and a full-scale Research Methodology. The Research Methodology comprises a concept of Holistic Culture and a concept of Holistic Research. The structure of the Model is illustrated in Table 23.1.

Table 23.1   The Culturized Model for Planning and Governance

Seven Principles for Culture	Planner's Literacy Trio	Culturized Research Methodology
Plenitude	Cultural,	Holistic Culture –
Connectivity	Ethical and	Society
Diversity	Strategic Literacy	History
Reflexivity		Environment
Creativity		Holistic Research –
Critical Thinking		Cultural Data Research
Sustainability		Cultural Collaborations
		Cultural Interpretation

Source: Greg Young 2012.

The goal of the Culturized Model is to make available an accessible template for planning and governance condensed in one place. In addition, clarifying and spelling out the full dimensions and categories of culture and the broad range of research practices may play an educative role in planning, and serve as a corrective to a global pattern that daily privileges access to culture and some forms and dimensions of culture over others. (See also Evans and Bianchini, Chapters 13 and 22, this volume). Research practice also needs to be scoped broadly and holistically, in order to promote the capture of a plurality of approaches and techniques as well as a diversity of theory, to achieve a more reflexive and humane integration of culture than characterizes the lack of accountability associated with the phenomena of culturalization.

Inevitably, the Model is evolutionary in nature, however in the case of planning it is relevant at all geographical scales and across the broad spectrum of planning types. Over time, the development of culturized models for individual sectors and functions of governance beyond planning should see the retention of a number of core elements of the culturized planning model such as its emphasis on a creative, critical, ethical and reflective approach to research. However, other aspects of customized sectoral models may develop with greater diversity.

Seven Principles for Culture

The seven principles for culture are unifying principles designed to support the overall framework of holistic culture and holistic research and to promote the capture of culture in planning. I would see other principles specifically developed for the needs and realities of other aspects of governance developing over time. While all of the seven principles relate across the sectors and functions of governance and planning I single out the Principle of Plenitude and the Principle of Connectivity for the purposes of illustration.

The principle of plenitude

The idea of cultural plenitude directly supports the value of a holistic approach to culture and a plural approach to theory. Simply stated, the concept of plenitude asserts that culture is plentiful and reinvents and expands itself. It is an idea associated with the egalitarian rebellion of cultural studies from the 1960s and 1970s that sought to consider culture as a whole way of life. Culture was no longer seen as a scarce resource, and viewed through new and more democratic lenses, was found to be anywhere and everywhere. Similarly, personal cultural expression did not imply detracting from the culture of another. Foucault's 'plenitude of the possible' (Foucault 1984: 267) was an idea that underwrote the interest in cultural studies in ideas and knowledge, so that 'working-class culture, women's culture, youth culture, gay and lesbian culture, post-colonial culture, third world culture, and the culture of everyday life were all quickly discovered and described' (Hartley 2003: 4). Cultural plenitude is the intellectual, philosophical and creative frame within which optimally a planner operates.

The principle of connectivity

The idea of connectivity is a principle with similar value for cultural and planning considerations and an understanding of ecology and the life world. Raymond Williams argued that the connective operates 'against the frame of the forms' (Bennett 1998: 54). In planning terms connectivity is relevant to planning's scalar hierarchies and its multiple forms. The principle also asserts the connections in culture between thought and feeling, and implies the dangers in isolating instrumental rationality as a solitary tool in decision-making and as a source of knowledge and knowing. It also highlights the value of the humanities, the arts and all of the forms of history, for planning, as variously argued by planning commentators (Forester 1999, Harvey 1990, Hawkes 2004, Healey 1997, Sandercock 1998). Strongly linked to creativity and imagination as espoused for cities and planning by Landry (2000), connectivity parallels both the informational and technological connectedness that characterizes the network society, as well as the multifarious webs of life that constitute ecological knowledge and understanding.

The Culturized Research Methodology

The Culturized Research Methodology includes a concept of holistic culture and a process for holistic research. The typology of holistic culture divides the slippery world of culture into graspable categories, and the holistic research process provides a method to 'scan' the world of potential research for all relevant materials. Together they serve as a heuristic and a 'short hand' tool to 'power' the overall Model. The Model was earlier illustrated in its overall form in Table 23.1 that also shows the three components of which holistic culture and holistic research are each comprised and I now outline these in turn.

Holistic culture

In order to present a concept of holistic culture I utilize Henri Lefèbvre's (1992) earlier ontology of being and expand on its implications for planning culturization. The elements of 'space', 'time' and 'society' may be further reduced in abstraction to constituent elements that may be readily geared to the practical integration of culture in planning. This approach enables an easy recognition of all of culture's elements for planning inclusion to avoid omitting any dimension of culture as a result of a theoretical or practical bias towards one or more of its forms or dimensions.

I would argue that a 'legible' ontology of culture such as this is as important for planning purposes as is Lynch's 'legible city' (1960) in enabling the inhabitants of a city to find their way. Moreover, a concept of culture derived from the abstract is not inimical to the play of competing concepts of culture, power or politics, or for that matter 'the uneven, unequal unfolding of multiple human geographies' (Gregory 1993: 304). Competing definitions of culture, must inevitably work within the frame of the same ontology, regardless of the element of culture they may privilege or favour. For example, the traditional Western middle-class approach to culture tends to privilege the arts and high culture – such as in the activities of National Trusts in many countries – while an international environmental NGO such as Greenpeace may concentrate on environmental values or the adherents of postmodernism as a style favour the playful elements of popular and consumer culture and their semiotic interpretation. In this respect, I accept Soja's description of Lefèbvre's trialectics of being, as 'a statement of what the world must be like for us to have knowledge of it' (1996: 70). The integration of culture in planning depends on useable cultural categories regardless of diverse perspectives on culture and competing views that valorize separate cultural elements in different ways or prioritize one over another according to need or interest as earlier suggested.

The relationship between Lefèbvre's trialectics of being and the three dimensions or categories of culture as reflected in philosophy, key disciplines and in ordinary language is illustrated in Table 23.2.

Table 23.2 Holistic culture in three dimensions illustrated in Lefèbvre's 'trialectics', philosophy, key disciplines and ordinary language

Lefèbvre's 'trialectics of being'	Philosophy	Key Disciplines	Ordinary Language
Spatiality	Space	Geography	Environment
Historicality	Time	History	Heritage
Sociality	Society	Sociology	Society/ways-of-life

Source: Greg Young 2012.

The concept of holistic culture or whole culture is supported by the perspective of Raymond Williams' view of culture as 'a whole way of life, material, intellectual and spiritual' (1966: 16) which provides a valid and accessible approach to all culture for planners, educators and other professional practitioners. (See also Evans and Bianchini, Chapters 13 and 22, this volume).

Within this, each category of culture may be considered in its contemporary manifestation and in its manifestation in history. For example, the geography and environments of previous centuries and millennia were different comprising earlier cultural landscapes and other climates. History also has its own history in terms of past approaches to historiography and previous concepts and practices in history, or influential cosmologies as they existed before historical recording. Societies and ways-of-life in the past have been as diverse as is practically imaginable.

A culturized model for planning utilizing culture as an organizing principle and category has the power to promote and improve the consideration of cultural values and relationships, and cultural diversity and hybridity in all of their manifestations. However, in addition to a more comprehensive and integrated concept of culture, a similarly inclusive approach to the identification, capture and research of culture is also required.

Holistic research

Working in tandem with the power of holistic culture is the more comprehensive access to culture offered through the practice of holistic research. Holistic research consists of three elements, all of which are required to address both contemporary cultural and social diversity and current levels of planning complexity. They include the continuous need to interpret culture as reflected in the longstanding philosophical and theoretical legacy bequeathed by, among others, Nietzsche (Merquior 1985), Foucault (Merquior 1985) and Geertz (1973). A list of examples for each of the three elements is shown in Table 23.3 and is by no means exhaustive.

Table 23.3   Holistic research

Cultural Data Research	Cultural Collaboration	Cultural Interpretation
Online data	Online collaborations	Interpretation/s online
Social media, blogs	Wikis, bulletin boards	Behavioural, psychological and
Census data	Consultations	communication theories
Project Gutenberg	Story-telling in different	Structuralist theories
Historical records	modes	Postmodern social and cultural
Environmental data	Action research	theory/studies
Cultural mapping	Cultural mapping	Cultural mapping
Heritage places	In-depth interviews	Storytelling, discourse analysis
Cultural infrastructure	Oral histories,	and semiotics
Arts and humanities	community histories and	Indigenous understanding
Music, fiction, poetry	memories	Academic history
Religion	Community projects, for	Feminism and queer theory
Emergent culture	example gardens, the arts	
Arts and humanity	and sustainability	

Source: Greg Young 2012

   Each research mode privileges different kinds of data, knowledge, epistemologies, processes and theories in relation to culture but when considered together they provide an overall view of culture and its planning potentialities. While I use ordinary language to describe the techniques of holistic research, specialist terms also apply.

Cultural data research

Cultural data research is best described as quantitative research and is based on information derived from sources such as census statistics, public records, historical records and environmental data. This information is usually in the public domain and is often available online in the form of databases, encyclopaedias, digitized archives and global literatures. These materials are complemented by the dynamic world of specifically online contributions such as social media, blogs, wikis and bulletin boards. A computer search can readily identify the information that is available. Quantitative aspects and reflections of culture are important to planning but need to be better and more imaginatively integrated within the processes of research and development in planning. This research relates to each and every category of holistic culture.

Cultural collaboration

Collaborative cultural research is sourced from more communicative processes that involve community engagement and participation in numerous areas of plan making, using techniques such as action research, cultural mapping, community histories and community projects in fields such as sustainability and tourism. Online projects such as bulletin boards and wikis as well as community techniques can also be used to reveal the power of place through local studies that explore, document and socially utilize public history (Hayden 1995). These collaborations develop stronger community connections to the past and their results can be shared and exchanged online between diverse geographical areas and multicultural groups. Global collaborations online range from the development of open source materials in software and knowledge banks to activist collaborations in the fields of human, animal and environmental rights. They help to produce new knowledge and inspiration that is relevant to more sensitive environmental and social planning, better architectural and landscape design, improved conservation planning measures and new and more culturized community initiatives for tourism and the local marketing of place.

Cultural interpretation

Cultural interpretation values and utilizes the insights and knowledge derived from cultural theory, including postmodern social theory, academic and popular history, political economy, art and the humanities and indigenous understanding and values. For example, it is the basis of authentic access to indigenous culture in the context of bi-cultural planning and management such as in Australia for the World Heritage area of Uluru-Kata Juta National Park (Uluru-Kata Juta Board of Management 2000). Cultural interpretation in heritage planning using popular and academic history, memory and archaeology provides the means to what Balzac called the ability to '"see" time in space' (Sandercock 2003: 199). Behavioural, psychological and communication theories are also being developed and applied to the exponential growth in online social media, blogs, wikis and bulletin boards. Each approach is relevant and indeed essential to the culturization of planning.

A position of theoretical pluralism embraces the insights from each of the main wings of planning theory and is complemented by methodological pluralism – knowing which planning tool to use and in what circumstances. Interpreting cultural knowledge and ideas and national and international ethical protocols supplies the planner with a more creative, critical and reflective role. This is what Habermas means when he reminds us that 'critical pluralism shifts normative weight to the role of the critic in the pluralist practice of democracy' (Smelser and Bates 2001: 2988).

I should point out, however, that not only are the different elements of holistic research related, but that they represent different practical opportunities for varied aspects of the planning spectrum. For example, general cultural research and quantitative cultural information may be widely included in spatial planning at every scale and in non-spatial forms of planning for social, economic and ecological purposes. They provide the base data for the history of places, and for measurable features. This is in essence the basis of the call from the World Commission on Culture (WCC) in its report *Our Creative Diversity* (1996) for the inclusion of foundational culture in all planning and development. Collaborative techniques are the basis for accessing and including the values, perspectives, needs and stories that relate to community diversity, in terms of gender, sexual preference, ethnicity, religion, class and disability. They tap the seemingly inaccessible or what may otherwise remain hidden or inaudible, and give voice to diversity strengthening individual cultural groups and thus the community in its totality. It is the knowledge form most readily identified with neo-modern planning theorists. Cultural interpretation on the other hand comes into play at specific stages in planning and has the capacity to make a major contribution to particular planning topics and to culturally diverse, bi-cultural, multicultural and intercultural planning and management. This aspect frees up different value structures and cosmologies, cultural perspectives and priorities. It enriches strategic planning where theories,

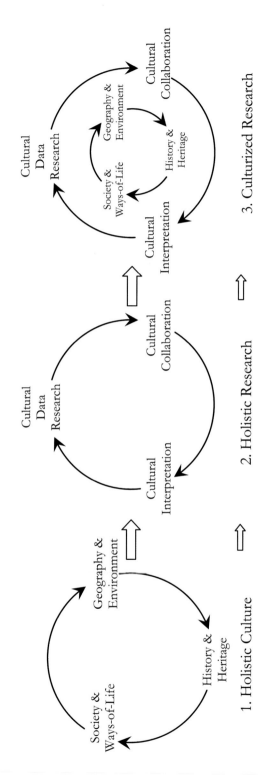

Figure 23.1 The Culturized Research Methodology for Planning and Governance
Source: Greg Young 2012.

artistic considerations and emergent ideas may determine themes and categories in planning documents as is directly apparent in social planning, heritage planning and interpretation, place re-invention and planning for major events. Cultural interpretation is the knowledge strategy most favoured by postmodern social and planning theorists.

What is distinctive about the Research Methodology is that each pathway to culture is considered potentially relevant to planning in some way. Combined with a holistic view of culture, the qualities of each dimension of culture may be animated in order to flow through all aspects of the planning process so that 'no stone is left unturned'. In this way the diverse whole of culture may be woven in a collaborative fashion through the culture of all lives and communities and the governance instruments and practices of the state, market and civil society. While this view is expressed by planning theorists at times it is not accompanied by a developed heuristic for culture or a planning *modus operandi*.

In Figure 23.1 the Culturized Research Methodology and the interrelationship of its components is illustrated in a sequential fashion in order to suggest the dynamics of a culturized research process in one image.

Conclusion

A general cultural paradigm for governance has not emerged until recently embracing all sectors, and within spatial planning itself an incoherent and piecemeal approach to the conceptualization of culture has limited the powerful synergies that lie in the direction of culture. In spite of this, the cultural richness and complexity in communities, governance, institutions, cyberspace and markets demands a corresponding richness in planning and governance responses, if the social and economic relevance of the planning discipline and other governance functions are to be maintained.

I would also argue that a more sensitive and powerful planning and governance depend on culture as a fundamental organizing concept, resource and dialectic. Not only is this so, as a result of the contemporary fact that knowledge intervenes upon itself as the key to higher productivity (Castells 1991: 10) but as Castells also argues, local communities 'must preserve their identities, and build upon their historical roots, regardless of their economic and functional dependence on the space of flows' (1991: 350). In this context, the role of government rises in importance locally, in order to 'reconstruct an alternative space of flows on the basis of the space of places' (Castells 1991: 352–3). In the cultural paradigm and model I propose I view this role as a potential objective for the many forms and levels of governance.

The Culturized Model for Planning and Governance introduces a number of the formal preconditions for good planning and governance in a cultural century so that with both a growing appetite and aptitude for culture, planning and governance may come to fully reflect the opportunities of their times. Under a cultural paradigm, culture is construed as the object of governance and planning, as well as the basis of their procedural means. Taken together, social culture, historical culture and environmental culture constitute the object of governance and planning, while normative approaches to cultural data, cultural collaborations and cultural interpretation are the basis of deepening and sensitizing governance policies, instruments, processes and practices. In the case of planning, for example, this perspective is the antidote to historical swings between a substantive interest in the city, or some other object of engagement, to an almost exclusive focus on the fetish of planning techniques themselves, especially the de-cultured techniques that have tended to dominate planning fashion.

Culturization is a strategy positioned to meet the need for the productive localization and customization of planning and governance theories and methodologies. Culturization also answers calls from the WCC to ensure that 'Any policy for development must be profoundly sensitive to and inspired by culture itself' (Gordon and Mundy 2001: 5) and, from UNESCO, to include 'a cultural perspective in all public policy' (2009: 6) for sustainable development. The key, as I have outlined in the culturized model, lies in the work of dynamic, creative, critical, ethical and reflexive cultural interpretation. Beyond the trend to the commodification of culture a cultural paradigm and culturized models for governance and planning may provide the powerful organizing concepts, principles and enabling methodological tools to return the fires of life and culture to communities, planning, governing and developing themselves wherever they are.

References

Albrechts, L. 2010. Enhancing creativity and action orientation in planning, in *The Ashgate Research Companion to Planning Theory*, edited by J. Hiller and P. Healey. Aldershot: Ashgate, 215–32.

Alexander, E. and Faludi, A. 1996. Planning doctrine: its uses and implications. *Planning Theory Newsletter*, 16, 11–61.

Allmendinger, P. 2001. *Planning in Postmodern Times*. London: Routledge.

——. 2002. The post-positivist landscape of planning theory, in *Planning Futures: New Directions for Planning Theory*, edited by P. Allmendinger and M. Tewdwr-Jones. London: Routledge, 3–17.

Allmendinger, P. and Tewdwr-Jones, M. 2002. *Planning Futures: New Directions for Planning Theory*. London: Routledge.

Beauregard, R. 2008. Planning and the network city: discursive correspondences, in *Contemporary Movements in Planning Theory*, edited by J. Hillier and P. Healey. Aldershot. Ashgate, 245–56.

Bennett, T. 1998. *Culture: A Reformer's Science*. Sydney: George Allen and Unwin.

Castells, M. 1991. *The Informational City: Information Technology, Economic Restructuring, and the Urban-Regional Process*. Oxford: Blackwell.

——. 1996. *The Rise of the Network Society*. Oxford: Blackwell.

——. 1998. *The Rise of the Network Society, The Information Age: Economy, Society and Culture*, Vol. 1. Oxford: Blackwell.

Chaney, D. 1994. *The Cultural Turn*. London: Routledge.

du Gay, P. and Pryke, M. (eds) 2000. *Cultural Economy: Cultural Analysis and Commercial Life*. London: Sage.

Fainstein, S. 2003. New directions in planning theory, in *Readings in Planning Theory*, edited by S. Campbell and S. Fainstein. Malden, MA: Blackwell, 173–95.

Faludi, A. 1987. *A Decision-Centred View of Environmental Planning*. Oxford: Pergamon Press.

Forester, J. 1999. *The Deliberative Practitioner: Encouraging Participatory Planning Processes*. Cambridge, MA: MIT Press.

Foucault, M. 1984. *The Foucault Reader* (edited by P. Rainbow). London: Penguin.

Geertz, C. 1973. *The Interpretation of Cultures*. New York: Basic Books.

Gordon, C. and Mundy, S. 2001. *European Perspectives on Cultural Policy*. Paris: UNESCO.

Gregory, D. 1993. The historical geography of modernity, in *Place/Culture/Representation*, edited by J. Duncan and D. Ley. London: Routledge, 272–313.

Gunder, M. and Hillier, J. 2009. *Planning in Ten Words or Less: A Lacanian Entanglement with Spatial Planning*. Farnham: Ashgate.

Harris, P., Moran, R. and Moran, S. 2004. *Managing Cultural Differences: Global Leadership Strategies for the 21st Century*. 6th Edition. Oxford: Elsevier.

Hartley, J. 2003. *A Short History of Cultural Studies*. London: Sage.

Harvey, D. 1990. *The Condition of Postmodernity*. Oxford: Blackwell.

Hawkes, J. 2004. *The Fourth Pillar of Sustainability: Culture's Essential Role in Public Planning*. Victoria: Cultural Development Network.

Hayden, D. 1995. *The Power of Place: Urban Landscapes as Public History*. Cambridge, MA: MIT Press.

Healey, P. 1997. *Collaborative Planning: Shaping Places in Fragmented Societies*. Houndmills: Palgrave.

Hofstede, G. and Hofstede, G.J. 2005. *Cultures and Organisations: Software of the Mind*. 2nd Edition. New York: McGraw Hill.

Jameson, F. 1984. Postmodernism, or the cultural logic of late capitalism. *New Left Review*, 146, 53–92.

Jary, D. and Jary, J. 1991. *Collins Dictionary of Sociology*. Glasgow: HarperCollins.

Kuhn, T. 1970. *The Structure of Scientific Revolutions*. Chicago, IL: University of Chicago Press.

Landry, C. 2000. *The Creative City: A Toolkit for Urban Innovators*. Stroud: Comedia/ Earthscan.

Lefèbvre, H. 1992. *The Production of Space*. Oxford: Blackwell.

Lynch, K. 1960. *The Image of the City*. Cambridge, MA: MIT Press.

Merquior, J. 1985. *Foucault*. London: Fontana.

Radcliffe, S. (ed.) 2006. *Culture and Development in a Globalising World: Geographies, Actors and Paradigms*. Abingdon: Routledge.

Rushdie, S. 2010. *Luka and the Fire of Life*. New York: Random House.

Sandercock, L. 1998. *Towards Cosmopolis: Planning for Multicultural Cities*. Chichester: Wiley.

—— . 2003. *Cosmopolis II: Mongrel Cities in the 21st Century*. London: Continuum.

Sassoon, D. 2006. *The Culture of the Europeans: From 1800 to the Present*. London: HarperCollins.

Schein, E. 2004. *Organisational Culture and Leadership*. 3rd edition. San Francisco, CA: Jossey-Bass.

Scott, A. 2000. *The Cultural Economy of Cities: Essays on the Geography of Image-Producing Industries*. London: Sage.

Smart, B. 1993. *Postmodernity*. London: Routledge.

Smelser, N. and Bates, P. 2001. *International Encyclopaedia of the Social and Behavioural Sciences*, Vol. 5. Oxford: Elsevier Science.

Soja, E. 1996. *Thirdspace: Journeys to Los Angeles and Other Real-and-Imagined Places*. Cambridge: Blackwell.

Taylor, N. 1998. *Urban Planning Theory since 1945*. London: Sage.

Tylor, E. 1903. *Primitive Culture*, Vol. 1. 4th Edition. London: John Murray.

——. 1924. *Primitive Culture*, Vol. 2. 7th Edition. New York: Brentano's.

UCLG. 2010. *Culture: Fourth Pillar of Sustainability*. UCLG. [Online]. Available at: www.cities–localgovernments.org [accessed 4 April 2013].

Uluru-Kata Juta Board of Management and Parks Australia. 2000. *Uluru-Kata Juta National Park Plan of Management*. Canberra: Commonwealth of Australia.

UNESCO. 2009. *Culture and Sustainable Development: Examples of Institutional Innovation and Proposal of a New Cultural Policy Profile*. Culture 21. UNESCO. [Online]. Available at: www.agenda21culture.net [accessed 4 April 2013].

Williams, R. 1966. *Culture and Society, 1780–1950*. Harmondsworth: Penguin.

World Commission on Culture. 1996. *Our Creative Diversity*. Paris: UNESCO.

Young, G. 2005. The cultural reinvention of planning. PhD thesis, University of New South Wales, Australia. [Online]. Available at: www.unsworks.unsw.edu.au [accessed 4 April 2013].

——. 2008a. The culturization of planning. *Planning Theory*, 7(1), 71–91.

——. 2008b. *Reshaping Planning with Culture*. Aldershot: Ashgate.

Case Study Window – Global Futures: New Opportunities for Creative Cultural Transformation

Jean Hillier

Introduction

> *Culture is a blank space, a highly respected, empty pigeonhole. Economists call it*
> *'tastes' and leave it severely alone. Most philosophers ignore it – to their own loss*
> *... Psychologists avoid it ... Historians bend it any way they like. Most believe it*
> *matters, especially travel agents. (Douglas 1982: 183)*

... and cultural planners; how do they regard culture? Until 2002, anyone throwing their arms in the air and screaming their head off in a listed building in the UK would probably have been summarily ejected for improper cultural behaviour. But in March 2002 the wooden scenic railway at Margate's Dreamland amusement park was Grade II listed by English Heritage[1] – the first time that a fairground ride had ever been listed as an item of 'heritage' value. Dreamland, now on track to become the world's first Heritage Amusement Park, and arguably the home of working-class play, boasts several world-firsts during its 100-odd years of 'vulgar' activity (Figure 24.1).

In this chapter, I argue against what I perceive to be a cultural planning reliance on objects and material representations of pre-identified 'culture' in favour of a poststructuralist approach. I offer a brief introduction to cultural planning, highlighting tensions between what was traditionally regarded as 'high culture' – the liberal arts and sciences – and 'low culture' – 'popular' art, literature and leisure activities. In contrast to the culturalized commodification of culture, I take what Young (2008a, 2008b) terms a 'culturized' view, to refer to a more 'reflexive, critical and ethical use of culture in planning' (Young 2008a: 74).

It is widely acknowledged that a place's culture is inherently linked to its history. Indeed, as Koerner (2010: 138) explains, cultural resource management is often held to

---

[1]   Upgraded to II* in 2011.

Figure 24.1   The Scenic Railway and Heatwave Chair-o-plane (formerly the 'Apollo 14' Orbiter) in the mid-1980s

Source: Reproduced with permission from Nick Laister.

be synonymous with heritage management. I discuss the nexus between heritage and spatial planning as epitomized by the use of heritage 'assets' as a key component of place making. Taking the example of Dreamland amusement park I problematize the UK's traditional heritage discourse as valuing the present's image of an archival past in which heritage planning creates forms as mythical cultural entities 'frozen into truth'. I aim, instead, to re-situate Dreamland in a new, rhizomic approach to cultural heritage which involves rethinking history as historicity.

My aim is to bring the work of Gilles Deleuze to cultural heritage planning in order to liberate it 'from those images which imprison it' (Deleuze 1994: xvii); to stimulate new questions and the invention of new ways of thinking and planning cultural heritage. I invoke Deleuze's concept of style to regard time as a series of virtual lines, each with a past and a projected future. I ask whether heritage can act as a methodology of cultural transformation through which planners can seize the creative power of materiality to generate a new form of historical sense: a generative, affective rethinking of history for visitors by means of sensational encounters which create an evental space for thinking otherwise.

On Cultural Planning

Graeme Evans describes cultural planning as 'the strategic use of cultural resources for the integrated development of cities, regions and countries' (2001: 7). He attributes the rise of cultural planning to a combination of the expansion of cultural production and consumption (especially commercial entertainment as a leisure activity) together with a recognition of the economic potential of the commodification of culture. Cultural planning has embraced not only traditional 'cultural quarters' of art galleries, museums and concert halls (for example Essen, Germany; Perth, Western Australia), but also more recently the '24 hour city' or 'night time economy' (for example Newcastle, England; Madrid, Spain), spectacles (for example Sydney's Gay Mardi Gras) and even 'raves' and 'Pirate Parties' (for example Copenhagen, Denmark) as demand for 'the experience economy' increases exponentially and planners paradoxically attempt to plan what is inherently unplannable (Lovatt and O'Connor 1995, Morgan 2006, Pløger 2010, Sundbo and Darmer 2008).

In 2004, Kunzmann lamented a lack of consideration of cultural dimensions of spatial planning in planning practice, theory and education. Six years later, Markusen and Gadwa (2010: 379) have identified a 'crescendo' of interest, with a multitude of local authorities producing cultural plans and offering incentives to 'cultural districts' in attempts to build what they regard as 'cultural capacity'. As illustrated by authors in this volume, claimed impacts of cultural plans emphasize economic benefits and competitive advantage, together with those of urban regeneration, job creation, enhanced or rebranded city and community identity, enhanced social cohesion and community self-confidence, health promotion and environmental renewal (see also Binns 2005, Evans and Shaw 2004, Smith 2007).

Whilst the validity of such claims may be debatable (for example Evans 2005), cities across the globe continue to push for copy/pastes of perceived 'best practice' culture-led strategies. In this manner, cultural planning risks adherence to a conceptualization of culture as a process 'that is not dynamic, flexible and situational, but linear and linked to a set of clearly defined political and governmental objectives' (Stevenson 2004: 124). It is this formulaic element of cultural planning as culturalization – planning *for* culture – rather than culturization – planning *with* culture, which I seek to problematize below.

In the act of 'selling places' (Kearns and Philo 1993), architectural 'heritage' has often been mobilized to attract tourists and inward investment with an emphasis on protection and conservation of buildings for the cultural benefit of future generations. The importance of conserving historic buildings, or the right of a select elite to determine which buildings should be conserved and how they would be presented to 'the public', was rarely challenged. Prime Minister, David Cameron, for instance, argued the economic importance of British and international tourist visits to 'heritage sites such as castles and country houses, museums, galleries, theatres and festivals' (BBC News 2010). Cameron's reversion to a traditionalist codification of heritage in the sites listed demonstrates both his upper-class background (see Bourdieu 1982) and the inherently political nature of questions of priority between aspects of cultural heritage. His list of places as culturalized, marketable products differs significantly from the culturized, humanistic approach advocated by Young (2008a).

As Evans (2001: 268) indicates, a more sustainable, more equitable cultural policy paradigm should extend to contemporary and 'common' culture as well as to traditional arts and heritage. There is a need for cultural planners to move away from the imperative of marketing and image-branding and a sterile heritage of sanitized museumification which manipulates both history and its visitors (Evans and Smith 2000). I argue below that the politics of absence of the invisible histories within the visible history should be made manifest and should become part of the sign economy (Lash and Urry 1994) of cultural heritage.

On Cultural Heritage

French philosopher, Gilles Deleuze, condemned history 'as an enterprise that stakes out origins and anticipates conclusions' (Flaxman 2000: 24). I suggest that traditional views of heritage could be similarly condemned, particularly when history and heritage are conflated, as they often appear to be. History as heritage tends to define places (such as Dreamland) through an apparatus of capture, referred to as an Authorized Heritage Discourse (AHD) (Smith 2006). Such a limited conceptualization of heritage reflects nineteenth-century upper- and middle-class beliefs, epitomized by David Cameron above, that 'aesthetically pleasing material objects, sites, places and/or landscapes' should be cared for, protected and revered (Smith 2009: 2). Such attitudes suppose that material objects possess innate value. As such, the heritage object is often conflated with the cultural and social values which give it meaning. The AHD thus 'validates and defines what is or is not heritage and constrains heritage practices' (Waterton and Smith 2010: 12). Heritage – in the AHD and for David Cameron – is conceptualized as 'grand', 'tangible', 'old' and 'aesthetically pleasing' sites, monuments and buildings; feel good and comfortable; value inherent; to be conserved as found for future generations; where professional experts have a 'duty' to act for and steward 'a universal past' (from Smith 2009: 3). The Authorized Heritage mentality is a closed system which acts to legitimize and/or delegitimize various cultural and social values. Working-class 'vulgarity', as at Dreamland, would inevitably be non-legitimized as possessing 'absolutely no intellectual or moral content whatsoever' (Jacobsen 2009: 379) by a heritage industry dominated by white, middle-class cultural values.

The UK heritage industry has, however, begun to conserve built forms associated with working-class lives. Initially interested in industrial heritage, such as Ironbridge Gorge (designated a World Heritage Site in 1986) and 'pleasant' workers' housing, as at Saltaire (World Heritage listed in 2001), inclusion of a category of 'culture and entertainment' structures for working-class enjoyment has been a more recent phenomenon. The Heritage Protection Department (HPD 2007: 2) writes that the range of objects listed in this category is wide, 'from buildings of solemn intensity to others of fantasy and delight. What unites them are the factors of pleasure, escapism and self-improvement'; though whether Dreamland could be associated with 'self-improvement' in the HPD sense is rather doubtful.

Beyond solely conserving the exceptional in this category, English Heritage is increasingly recognizing the multivocality of cultures (such as working-class, women's, minority ethnic and so on voices), from erection of blue plaques for Victorian working-class 'celebrities' ventriloquist Fred Russell and muscleman/bodybuilder Eugen Sandow, to listing the Ryhope pigeon cree in Sunderland, the Dreamland rollercoaster in Margate and the entrance to the Kurzaal amusement park in Southend.

Even so, heritage is often reduced to history: an 'it was'. Nietzsche (1965: 4), however, was irritated by 'it was' and aimed to read the past in a generative manner: 'a concept of history that engenders action, not reflection, a history that produces the active rather than justifies the reactive' (Grosz 2004: 114). This is a past with dynamic potential; of being untimely, outside the limitations of path-dependence and objective reality blinkers of the present (Grosz 2004) – an experience of existing in and with the conditions of possibility of a history and its implications: historicity.

Nietzsche (2010) distinguishes three types of history. Monumental history seeks to conserve 'greatness' as a lesson for the present and focuses on great individuals and deeds. Antiquarian history seeks to catalogue and conserve what is traditionally valued from everyday life, respecting and revering the past. Both resonate with the AHD. However, Nietzsche's third type of history – critical history – celebrates an organic, dynamic form of knowledge in which heritage would act as a critical extension of the past, rather than its nemesis. Critical history, or historicity, privileges the conditions of possibility of being, saying and doing, rather than a selective material representation of what was lived, said or done.

Cultural heritage policy, and its orientation towards antiquaria and monumentality through the AHD, have tended to overcome time by freezing it or memorializing it – 'bourgeois memory frozen into truth' (Philo and Kearns 1993: 26) – such that certain people, structures and events become immortalized and others silenced. As Cooke suggests, however, 'one of the ironies of heritage is that its advocates fail to see the historicity of the thing itself … Heritage deals in absolutes' (2007: n.p.). Such heritage policies refuse critical history/historicity and the potentiality of a non-exhausted virtual past, where the virtual is concerned not only with time before, but also time after – the future. I argue that the future of cultural heritage planning requires that it should become open to consideration of the virtual and its promise of 'newness, otherness, divergence from what currently prevails' (Grosz 2010: n.p.). In this way, heritage, linked to present perceptions of visitors, may be released from the clutches of historical categorizations and silences and be mobilized in the cause of another perception oriented to the future. The past, actualized by the visitor as a physical and emotional experience, acts as a stimulant to affect, to thinking and perhaps to action.

So, how might cultural heritage planners transform heritage beyond a historical 'landmark' or 'theme park' to become generative? Can heritage as visitor encounter (an encounter with time; with people, objects, ideas, sights, smells and sounds) stimulate an event in the space for thinking? This is the subject of following sections, assisted by the work of Gilles Deleuze.

Dreamland as Cultural Heritage

Margate, on the south-east coast of England, has several claims to cultural heritage. It is home of the deckchair (invented c1750) and beach donkey rides. It gave working-class tourists one of the earliest switchback railways (1860s) and pleasure gardens with exotic wild animals. The Dreamland amusement park, opened in 1920, boasted several UK-(if not world-)firsts; including the wooden scenic railway (1920), the hydraulic-armed Orbiter (1976), Looping Star, the first ride to loop-the-loop (1984) and the world's largest chair-o-plane (1991).

In the 1800s, visitors arrived by coal/grain hoy from London, for a single fare of 5 shillings, which allowed 'a lower class of visitor to travel to Margate' (HPD 2007: 25). By

1900, 'Merry' Margate had a reputation for the 'vulgarity of its working class patrons' (HPD 2007: 43). From 1920, Dreamland specialized in group outings (known as beanos) from London. Charabancs had their own coach park on the 20-acre site, while Southern Railway offered return tickets from London at the price of a single for groups of eight and over. Dreamland's restaurants seated 3,500 diners at a time.

The art deco cinema complex opened in 1935 including the sea-view café and ballroom. The cinema was air-conditioned and housed a Compton organ (costing £4,850 at 1935 prices) with illuminated waterfall effect and colour-kaleidoscope panels. The opening-night programme described the aim of Dreamland as 'The bringing of happiness to the many at a price they can afford' (Evans 2009: 37). The bars sold only beer (ale or stout), sherry or gin.

The 1930s and post-war 1940s and '50s were the peak times for Dreamland (Figure 24.2), especially on August Bank Holiday Mondays and 'Mother's Day', an annual Monday outing for coach-loads of East End London women. Dreamland offered an urban flight from everyday life; a 'land' of 'innocent fun' for the working class: a place of social inversion, of psychological release from daily tedium and authority, not a place of education or morality. Amusement parks, such as Dreamland, were associated with food and drink in excess, especially 'treats' or festive foods made of sugar (candy floss, lettered seaside rock) or a meal in a fish and chip restaurant.

Figure 24.2   Dreamland in 1950

Source: Reproduced from the Bill Evans Collection with permission from Nick Evans.

Dreamland represents an encounter with excitement (Cross 2006); life – but not life as usual. Instead, Dreamland offered absurdity, freak shows (for example The Headless Lady) and rides such as The Crazy House (later renamed the Fun House). It was a spectacle of sensation stimulated by sight, sound and smell, from the cackling of the Laughing Policeman and the whoosh and screams of those on the Big Dipper, to the smell of candy-floss, the colours of summer clothing, painted stalls and tall, seemingly flimsy, structures on which one could whirl around and defy death in a different manner to how many working-class visitors defied death every day at work. In Dreamland peril became pleasure.

Dreamland represented 'unbuttoned behaviour' (Walton 2000). For instance, wearing 'Kiss-Me-Quick' hats, couples being thrown against each other, or finding darkness (the Caterpillar [Figure 24.3], the River Caves), or an excuse for holding each other tight on 'scary' rides, and air jets directed up women's skirts in the Fun House. It was a meeting place for young people, couples and families: a place of transportation in both physical and mental senses. A place of 'legitimate' excess, which gave adults permission to act like children (Cross and Walton 2005) and to stretch the rules of both bodily capacity and legality.

Figure 24.3   The Caterpillar, replica of the Queen Mary and Motor Boats, mid-1950s

Source: Reproduced from the Bill Evans Collection with permission from Nick Evans.

Unfortunately, Margate became famous in the 1960s for Mods v Rockers battles on the sands, while in the 1970s, Dreamland acquired a bad reputation for drunken behaviour. Many amusement parks around the UK have been dismantled, their charms supplanted by cheap overseas holidays promising guaranteed sun, and their beach-front locations redeveloped for housing and commercial uses. Dreamland, nevertheless, struggled on, being sold twice, eventually to entrepreneur Jimmy Godden who soon recognized that Dreamland's prime location implied a far greater return on his investment if the amusement park could be demolished and redeveloped as luxury beach-front accommodation and retail facilities. The park as a whole was asset-stripped, major rides dismantled and shipped to Scotland, leaving only the rollercoaster standing in a sea of barren concrete.

In 2003, Godden announced that the land was to become a retail and commercial site. There would be no Dreamland. Local residents reacted by forming the Save Dreamland campaign, with some success in that the Thanet Local Plan enquiry reinstated the original policy including an amusement park (Inspector's Report 2005). In 2005 Dreamland was sold to the Margate Town Centre Regeneration Company (MTRC) for £20m, with Godden retaining a 40 per cent share. In 2002 the Big Dipper rollercoaster was Grade II listed by English Heritage, followed in 2008 by Dreamland Cinema (listed Grade II*). By this time, Save Dreamland boasted 14,000 members and the group developed a plan for a Heritage Amusement Park (Figure 24.4) comprising

Figure 24.4   Artist's impression by Jean-Marc Toussaint of the Scenic Railway, Helter Skelter and Cinema complex in the Dreamland Heritage Amusement Park, 2009

Source: Reproduced with permission from Nick Laister. http://www.photographic-library. net/dreamland/DLHAP_scenic_helter[1].pdf.

vintage rides from amusement parks across the UK. The Cinema would house a national centre of working-class culture. Soon after these proposals were unveiled, however, an arson attack destroyed about 20 per cent of the rollercoaster.

Thanet District Council (TDC) did not initially support a Heritage Amusement Park – thinking that luxury flats would sell despite a depressed economic situation. In February 2009 an influential TDC Councillor was reported as telling people to 'forget Dreamland' (Councillor Ezekiel, in Anonymous, *Thanet Times*, 10 February 2009) preferring to support construction of the (private sector) Turner Contemporary Gallery as '[c]ultural tourists and people in that socioeconomic bracket will spend more' (Ezekiel in Hamilton, *Times Online*, 9 May 2009). However, TDC was outmanoeuvred by the heritage network's establishment of the Dreamland Trust with support from the Prince's Regeneration Trust and a Heritage Lottery Fund award towards their £12.4m Park project. English Heritage referred to Margate as a 'breakthrough in heritage-based regeneration' and a Sea Change grant for £3.7m was awarded to the project by the Department of Culture, Media and Sport. In October 2009, TDC performed a U-turn when it realized which way the 'tide was flowing'. When MTRC reneged on its £4m commitment to the Heritage Park in April 2010, TDC agreed to fill the financial gap, its commitment now strong to the project (www.dreamlandmargate.com).

Perturbing the Plane of Heritage Composition at Dreamland

I raise some questions inspired by Gilles Deleuze about cultural heritage planning as a 'balancing act between convention and precedent, between the concerns of a memorialized past and the interests of a not-yet-existent future' (Grosz 2005: 82). I problematize traditional heritage planning as a heritage machine in which the AHD plane of composition leads to presentation of an archival past in which heritage creates stratified, immutable forms. Instead, I aim to perturb the plane, to challenge the absolutes created by the AHD, to re-style heritage plans and planning and to generate insights that are able to open Dreamland and its visitors up to something new.

Invoking Deleuze's concept of style, I challenge the practice of heritage conservation as psychological ontological security: a closed story of the past as 'it was' which projects a set of meanings that allow neither for subtext nor counter-readings. Instead, I regard heritage as embodying potential for re-creation in all its senses and find in Dreamland's dream-image the space for thinking the new. The social life of a 'thing', such as a rollercoaster and other rides, lives through encounters with people and the sensory intuitions and apprehensions of which humans are capable. Jane Bennett introduces the idea of 'thing-power' as 'the curious ability of inanimate things to animate, to act, to produce effects dramatic and subtle' (2004: 351). The rides at Dreamland perform as symbols (the 'screen of dreams' [Ballantyne 2007: 66]) onto which people can project their experiences, memories and thoughts, reperceiving places and events. In discussing experience and memory as a function of the future in a world of presents related to not-necessarily true pasts, I emphasize the generative power of a past-present-future that conserves itself.

For Deleuze, the plane of composition creates the ground required to tell a particular type of story. The AHD thus 'pos[es] and solv[es] the problem for representation that a particular historical situation entails' (Buchanan 2001: 33). Inspired by Deleuze, I seek to perturb the AHD plane of composition, to begin to look beyond the ways in which history and heritage have been claimed and act as reflective screens for selected experiences and memories. I wish to trace the limitations imposed by such claims, in terms of our understanding not only of the past, but also of the present and generative futures.

Deleuze (1983, Deleuze and Guattari 1987, 1994) appropriates Nietzsche's critical history by converting it into a 'becoming/history': 'history amounts only to the set of preconditions ... that one leaves behind in order to "become", that is, to create something new' (Deleuze 1995: 171). If we follow Deleuze, we can begin to re-situate heritage – and Dreamland – not as an assemblage of passive objects, but as having a 'life' of its own and 'characteristics of its own, which we must incorporate into our activities in order to be effective, rather than simply understanding, regulating and neutralizing it from outside' (Grosz 2005: 132).

I regard Dreamland as an assemblage of speeds and slownesses, of materialities and expressivities, practices and encounters, including not only material qualities of architectural interest and rarity, as prized by English Heritage, but also those acoustic, olfactory and haptic qualities typically absented from heritage frames. We can begin to overcome the 'sight-nexus' (Watson and Waterton 2010) of heritage, where description on a plane of composition 'obliterates' tensions between a place's 'material' persistence and its experiential fluidity (Cresswell and Hoskins 2008: 395) and stratifies forms as immobile mythical entities – descriptions, photographs and buildings. In heritage representation, myths are 'resurrected, mined and constantly retold' (Taylor, Preston and Charleson 2000: 189), re- or even newly-imagined. Myth performs living presence, as in the Dreamland case, where the presence of selective mythical histories feeds into the proposal for the new Heritage Amusement Park. However, as Deleuze suggests, it is precisely this 'double movement of creation and erasure' (1989: 44) which offers the opportunity for historicity, rather than history or historicism,[2] to create anew, not only the 'heritage' object, but an understanding of the systems and conditions of possibility with which it may be related – technical inventions, such as electro-hydraulic twisting rotation, repressive industrial relations experienced by visitors in their working lives, conditions of manual labour in the fabrication of rides, ill health from using lead-based paint and so on.

Class is rarely directly apparent at cultural heritage sites. Whether in the 'vulgarity' and 'waywardness' of Dreamland or the camaraderie of the ride operators, the heritage site still supposes a stratified, arborescent[3] structure in which the identification of the working class is fixed and remains lower in the hierarchy, even though valorized as a sort of class-based sociability. Rather than establishing heritage arborescently in a form of pure state, I suggest a Deleuzean-inspired rhizomic approach (see Russell 2010), establishing 'horizontal' connections and challenging the idea that stable meanings and identities can be achieved through historicization (Deleuze and Guattari 1987). I regard heritage not as a 'fixed, singular narrative, but a series of socially constructed

---

2   I here refer to what is known as 'old historicism' after von Ranke and others (see Iggers and Powell 1990).

3   Arborescence for Deleuze implies hierarchical relations of power, thinking and acting (Deleuze and Guattari 1987).

interpretations of the past' (Atkinson 2007: 522). As such, I attempt to re-think a plane of heritage composition which escapes the clichés and doxa that prevent creative thought and which re-presents heritage as a discontinuous, folded territory of possibility. I regard heritage as a question; a provocation which generates thinking differently. Cultural heritage planning asks questions both of heritage planners and citizens; questions 'about our needs and desires, questions above all of action' (Grosz 2005: 132). Cultural heritage is both the result of action and a provocation to action.

Towards a Rhizomic Style of Heritage

Cultural heritage planning thinks in only one dimension of history: the past. I argue that we should rethink history as a process of formation and heritage as a taking-form of history – a material medium for creation. This would entail restyling heritage with an openness not reducible to a single Authorized Heritage Discourse. I concentrate on conceptualization of heritage as a relational, rhizomic process, whereby the qualities of tangible and intangible materialities are not intrinsic as such, but relational to the assemblage of which they are an element. I suggest that we need to perturb and/or displace the 'false depth' of heritage representation, which 'mediates everything but mobilizes and moves nothing' (Deleuze 1997: 67).

Traditional views of cultural heritage have attempted to transcend time by expressing it in some conserved historic state which manipulates it and renders absent several of the elements inherent in its assemblage. The 'timely' postcard view of a heritage place or building (for example Dreamland and the rollercoaster) is given depth and life by movement: the 'rides, slides, tunnels, stairways, rockets, or big dippers which "must lead the visitor-viewer from one particular space-time to another similarly autonomous space-time"' (Deleuze 1989: 58). Historical time becomes immanent to Deleuze's (1989) virtual time in which heritage, as the expression of time, expresses the essence of time as change. Riding the rollercoaster thus becomes the means of entering a different space-time – a Deleuzean dream image.

Deleuze's conceptualization of the dream image challenges traditional views of history and heritage in that chronological history is supplanted by a temporality which 'is eternally splitting into the past and the indefinite future' (Herzog 2010: 263). Instead of the heritage manager's chronological, linear time-line, Deleuze offers us 'coexistent, incommensurate sheets of the past' (Herzog 2010: 263). Encountering heritage with Deleuze would become a means of rethinking history on a virtual plane of rhizomic becoming, generating new affects and thoughts; opportunities for encounter, for social interaction in which multiple possibilities for new heritage/s could emerge. As Russell (2010) asks, is it enough to simply conserve things from the past, or is it just as, if not more, important to generate opportunities for affective and emotional responses to the past which stimulate people to think differently?

Heritage sites may be regarded as 'nodes' in social-cultural-political-economic rhizomes, assemblages or networks. Sites are representations of rhizomic assemblages at the time of their original existence, in the present (for example as economic attractors, agents of social cohesion and so on) and into the future. They are also nodes in the rhizomic assemblages of visitors, maintenance workers and so on: meeting points in networks of relations and pathways with different understandings and meanings.

Rhizomes are creative forces whose particular stage, in this instance, is historicity. Centred in the present, the conditions of possibility and stories of the site may be reconstructed: the various planes of reference and their convergence in the site – their events, structures and agencies. I adapt Negri's (1992: 11) dictum, 'There is no other way to consider [cultural heritage] than to be it, to make it.'

Cultural heritage sites have been sites of multiple encounter throughout their lives. Their current being incorporates the historicity of previous encounters, the present and the potentialities of future encounters which might take place. The sites reflect series of encounters between different assemblages of elements and relations of power and their particular arrangements. For instance, Dreamland reflects encounters between assemblages of force, relations of working-class visitors, various materialities (rides, signs, money, food, drink and so on) and those involved in the construction and maintenance of such materialities, middle-class professionals, politicians and so on. A Deleuzean-inspired rhizomic approach to cultural heritage would encourage visitors to encounter the various assemblages, interacting with the elements sensationally (Deleuze and Guattari 1994: 164) so that although the actualized spatial entity remains, it is accompanied by a multiplicity of virtualities, in terms of relational reconfigurations of meaning which generate their own realities.

Deleuze (1994) stresses the importance of encounter for stimulating new forms of thinking in an event. The Deleuzean event can be described as a 'disruption, violence or dislocation of thinking' (Colebrook 2005: 4), occurring when elements encounter relations, potentials and powers not their own, and which creates new connections, new styles of thinking and new ways of seeing. An event is, therefore, 'the expression of the productive potential of the forces from which it arose' (Stagoll 2005: 88. Events stand for historicity against old forms of historicism or historicization (see Makkreel 1990).

How might cultural heritage stimulate thinking the new: thinking otherwise? There is a need to go beyond landmarks and description to creative moments which perturb or disrupt what is generally taken for granted: for visitors to reperceive places, people, things, occurrences and the conditions of their possibility. Visitors riding the rollercoaster at Dreamland might eventalize[4] public holidays for the working classes[5] and might think differently about the conditions of work of, for example, commercial artists working with lead paint on colourful signs. Visitors might begin to think about the relational assemblages of candy-floss and seaside rock – both of which made sugar treats available to the masses in the late nineteenth century, as well as the force relations of sugar production, sugar barons, slaves, merchant shipping magnates, and the hardships and legislative 'connivances' which made sugar available cheaply in the UK[6]

---

[4]   Michel Foucault described eventalization as 'rediscovering the connection, encounters, supports, blockages, play of forces, strategies and so on which at a given moment establish what subsequently counts as being self-evident, universal and necessary' (1991: 76).

[5]   Bank holidays originated in 1871 when workers were given four days holiday a year when banks were closed in addition to six 'common law' holidays. Weekends with two days free of work became the norm only after the Second World War.

[6]   As Mintz (1985) illustrates, the 500 per cent rise in British sugar consumption, from 234,000 tons in 1800 to 6 m tons in 1890, resulted from an assemblage of British-owned

– and of seaside fish and chips and their origins in the battered fried fish of sixteenth-century persecuted Jewish refugees from Portugal, coming together with the nineteenth-century industrialization of North Sea fishing in Samuel Isaacs' Fish Restaurants (with real cutlery, crockery and table cloths!) in London, Margate and other seaside resorts in southern England. Candy floss, seaside rock and fish and chips all made the trappings of upmarket or 'special' foods affordable to the working classes – at the expense of poor working conditions of sugar plantation workers, sailors and fisherpeople. As Michel Foucault perspicaciously wrote, 'we have to refer to much more remote processes if we want to understand how we have been trapped in our own history' (1982: 210).

I suggest that cultural heritage can place visitors 'in the middle of things' rhizomically, where it can organize creative encounters which encourage visitors to an active engagement with their surroundings and challenge the consistency of the present in that the present becomes inseparable from a presence of a past and a presence of a future incorporated within it. Service's (2010: 21) experience of the sensation of 'rollercoastery' as opening up 'a new universe of creative possibility' resonates with how Deleuze might experience Dreamland; pushing body and ideas to the limit, setting up resonances between elements and mobilizing the conditions of thinking something new. Riding the rollercoaster, eating candy floss and fish and chips at Dreamland could be an experiential event where visitors are affected by their encounter with social, cultural, material and potentially political elements.

McHoul (2009) reminds us that culture and by implication, cultural heritage, is the 'coming-into-presence' through decisions made and decisions still to come. Cultural heritage, therefore, should not be regarded as an empirical presence to be 'inspected' (2009: 70), but as evental – a calling towards potentiality. Culture is never complete: it remains open to becoming. Its 'heritage', likewise, is the past-present-future. For Deleuze, the whole point of history is to 'allow the virtual to come to consciousness … in and through the limited humans who tap into it' (Vitale 2010).

Transforming the Style of Cultural Heritage

Deleuze and Guattari (1987: 110) suggest that we should read texts against the grain, challenging accepted interpretations and meanings. Deleuze further argues the importance of 'digging under the stories' such that 'style becomes nonstyle' (1997: 113), undercutting traditional style – such as the AHD – and escaping the 'death-sentence' (Deleuze and Guattari 1987: 110) of convention. Style makes possible new ways of thinking by establishing transversal links and communication between signs impossible in traditional static representation.

Can heritage act as a methodology through which the creative power of recombinations of materiality, programme, space and cultural narrative can generate the new: a new

---

plantations in the Caribbean, commercial and military naval power, slaving and post-slavery British government legislation which prevented freed slaves from acquiring land and voting. Although what had once been an exotic treat had become an everyday consumable by the end of the nineteenth century, there were distinct social differences in what sugar meant to people.

form of 'historical sense' or Nietzschean/Foucauldian 'effective history'?[7] Dreamland, enfolded into a ride on the rollercoaster, becomes not Dreamland as originally lived, but a new, sensational experience whose meaning the rider is compelled to create. The sign of the rollercoaster can lead us into new worlds – expressed in infinite meanings – and new material experiences. Such methodology, however, demands a transformation of heritage style. In the case of Dreamland, style is sensational.

Style is, for Deleuze, the movement of concepts and the individuating logic of connections between elements which forces us to think. It is style which allows cultural heritage planners to reach beyond signs, beyond narration of history, to the generative effects and affects of historicity. The power of encounter between elements of cultural heritage and between those elements and visitors emerges from style.

A rethought cultural heritage might be guided by principles of encounter and interaction between elements and visitors to stimulate affect and evental spaces for rethinking from which something new or learned[8] can emerge. Cultural heritage elements might be 'readable' or construed through an active interrogation of the forces connecting the signs (Bogue 2008). In this way, one might pass through the subjective and interpretive illusions of AHD-framed historic narrative, to regard the past-present-future as an infinite collection of problematizing assemblages. As Bogue writes, learning involves an engagement with such assemblages and a re-orientation of thought, 'such that thought may comprehend something new in its newness, as a structured field of potential metamorphic forces rather than a pre-formed body of knowledge' (2008: 16). Cultural heritage planners cannot plan or predetermine the truly 'new in its newness', but they can attempt to induce visitor encounters with the new by offering signs and objects in manners such that they can be experienced and problematized through conditions which ask the visitor to critically reinvent themself as a subject and in which it becomes possible (and permissible) to think differently/ in new ways. This will necessitate a critique of orthodox codes and frames such as the AHD and a thinking otherwise about what cultural heritage might imply. Whereas heritage has traditionally been understood as an order of time, fixing time in discrete points, I suggest that a new heritage style could regard time as a series of virtual or floating lines. Individual sequences of images become a series, each with a 'before' or past and an 'after' or projected future, not all of which can or cannot be true. This implies, as Deleuze and Guattari (1994: 41) suggest, a sort of 'groping experimentation' which resorts to measures that 'belong to the order of dreams, of pathological processes, esoteric experiences, drunkenness, and excess' – to holidays at Dreamland perhaps!

Conclusions

It can be argued that culture is merely abstract until actualized at a particular time, in particular sites and circumstances (Shields 1991). Cultural heritage, therefore, is a

---

[7]   Foucault (1977: 154).

[8]   To learn in this sense is to immerse oneself within a new or different element and open oneself to an encounter with disorienting signs which disrupt the habitual functioning of one's faculties (see Bogue 2008: 12).

constructed virtuality, folding together the past, the present and the future. Dreamland, its iconic rollercoaster and other rides, occupy 'a junction between stillness and motion, time arrested and time passing' (Prigogine 1985: 17) – and casting aside the AHD, I would add, time to come. Re-styled heritage 'time' would no longer be overdetermined by succession or organizing frames as an apparatus of capture. Heritage would become not objects to be revealed or known, so much as a plane of potentialities. As Grosz makes clear: 'This is not the abolition of history or a refusal to recognize the past and the historical debt the present owes to it, but simply to refuse to grant even the past the status of fixity and givenness' (2001: 104).

Thinking opportunities for creative transformation of cultural heritage with Deleuze implies a style of heritage which goes beyond memory and dreams, to tap into a virtual plane in which the past survives in itself, not merely as a picture postcard of an 'old present'. This would be a style which allows heritage to become a thought-provoking encounter, 'to tremble, or scream' (Deleuze and Guattari 1994: 176) even. Deleuzean thinking inspires an immanent conception of heritage; a conception which is less traditional archive as container and more relational spatial practice of past-present-future which stimulates visitors to think differently or otherwise on social, political, economic issues. Practices of eventalization and historicity can generate what Shanks and Witmore (2010) term 'unforgetting'; the unveiling of hidden or neglected conditions of possibility of what was (rather than what 'it was') and a past-present-future generative element of time as series.

Heritage elements should be monumental,[9] not in the Nietzschean sense because they conserve or preserve their time, but because they generate a creative, affective rethinking of time and the nature of life itself. We need to tap into the virtual force of heritage, to 'tear it from its constituted time and open its constitutive time, its power to create time, not exist within time' (Colebrook 2006: 82–3). Heritage is both an actuality and a potential to become; what it may have been and what it might still be. A heritage plan, then, would both affirm itself and call itself into question. The aim would be no longer to (re)collect or (re)capture the past in some individual or collective consciousness or identity, but to prevent closure.

Rather than a traditionally framed plane of organization, cultural heritage planning could be about creating a plane of composition in which Dreamland would become a sensational site of embodied practice, where human bodies would encounter the hand-painted materialities of the Big Dipper, the Orbiter, the Looping Star and other rides, together with candy floss, seaside rock and fish and chips. Materialities cannot be reduced to their substance, however. They cannot be completely described, documented or archived. There is always space for sensation; for smell, texture and sound as well as sight. Heritage planning for Dreamland must work with the assemblage of materialities rather than on them. The plan would be an encounter with the non-determined which allows the elements to go beyond their social identities (as professionals planners, as tourists, as working-class people, as vintage rides, as candy floss and so on) in an experiential experiment rather than a stratified plan. It would be 'the coming-into-existence of a prior substance or thing, in a new time, creating beneath its processes of production a new space and a coherent entity' (Grosz 2005: 133).

---

9   A Deleuzean monument is an embodiment of affects that are at once past, present and eternally immanent (Deleuze and Guattari 1994: 176).

Deleuzean-inspired rhizomic heritage plans would modulate the continuous variability of time as series rather than the conservational fixity of time as order. They would be adaptive, reflexive, respectful co-creations of being in which social (self-) awareness is stimulated, old traditions perturbed and disrupted and new 'traditions' allowed to emerge. If we are to unsettle the traditional plane of reference of the AHD and modes of ordering of heritage conservation planning, with their emphases on the middle and upper classes and where time is 'transfixed for our inspection' (Brann 1999, in Dewsbury 2002: 149), perhaps we should rethink what cultural heritage planning is and does.

I advocate a style of cultural heritage planning as a process which resists the 'entombing of a fixed History' (Herzog 2010: 263) of 'correct knowledge' and 'security of understanding' (Lather 2009: 18) and which taps into the thing-power of heritage materialities in ways that stimulate visitors' encounters with time as evental, with creation of space for thinking something new. I call on cultural heritage planners to think beyond conservation, commodification and marketability, to permit Dreamland to escape nostalgic history and to recommence its journey of becoming: a journey of encounter and sensational experiences which last well beyond the encounter itself.

Acknowledgements

I thank John Pløger, Andrew Ballantyne and Jonathan Metzger for valuable comments on an earlier version of this chapter. I also thank Nick Laister, Chair of the Dreamland Trust, for his help with sourcing images and Nick Evans, curator of the Bill Evans collection of Dreamland photographs.

References

Anonymous. 2009. Forget Dreamland and stop living in the past. *Thanet Times*, 10 February. [Online]. Available at: http://www.thisiskent.co.uk/8216-Forget-Dreamland-stop-living-past-8217/story-12006525-detail/story.html [accessed 18 March 2009].

Atkinson, D. 2007. Kitsch geographies and the everyday spaces of social memory. *Environment and Planning A*, 39(3), 521–40.

Ballantyne, A. 2007. *Deleuze and Guattari for Architects*. Abingdon: Routledge.

BBC News. 2010. Tourism must move away from Cool Britannia – Cameron. *BBC News*, 12 August. [Online]. Available at: http://www.bbc.co.uk/news/uk-politics-10950167 [accessed 15 August 2010].

Bennett, J. 2004. The force of things: steps toward an ecology of matter. *Political Theory*, 32(3), 347–72.

Binns, L. 2005. Capitalising on culture: an evaluation of culture-led urban regeneration policy. [Online]. Available at: http://www.ecoc-doc-athens.eu/attachments/390_Binns [accessed 8 November 2010].

Bogue, R. 2008. Search, swim and see: Deleuze's apprenticeship in signs and pedagogy of images, in *Nomadic Education: Variations on a Theme by Deleuze and Guattari*, edited by I. Semetsky. Rotterdam: Sense, 2–15.

Bourdieu, P. 1982 [1979]. *Distinction* (translated by R. Nice). Cambridge, MA: Harvard University Press.

Brann, E. 1999. *What then is Time?* Oxford: Rowman and Littlefield.

Buchanan, I. 2001. Deleuze's 'immanent historicism'. *Parallax*, 7(4), 29–39.

Colebrook, C. 2005. Introduction, in *The Deleuze Dictionary*, edited by A. Parr. Edinburgh: Edinburgh University Press, 1–6.

—— . 2006. *Deleuze: A Guide for the Perplexed*. London: Continuum.

Cooke, P. 2007. Plato's landscape: the quarrel over Lismullen and the Tara/Skryne Valley. *History Ireland*, 15(5). [Online]. Available at: www.historyireland.com/volumes/volume15/issue5/news/ [accessed 8 November 2010].

Cresswell, T. and Hoskins, G. 2008. Place, persistence and practice: evaluating historical significance at Angel Island, San Francisco and Maxwell Street, Chicago. *Annals of the Association of American Geographers*, 98(2), 392–413.

Cross, G. 2006. Crowds and leisure: thinking comparatively across the 20th century. *Journal of Social History*, 39(3), 631–50.

Cross, G. and Walton J. (eds) 2005. *The Playful Crowd*. New York: Columbia University Press.

Deleuze, G. 1983 [1962]. *Nietzsche and Philosophy* (translated by H. Tomlinson). London: Continuum.

—— . 1989 [1985]. *Cinema 2: The Time-image* (translated by H. Tomlinson and R. Galeta). Minneapolis, MN: University of Minnesota Press.

—— . 1994 [1968]. *Difference and Repetition* (translated by P. Patton). London: Athlone.

—— . 1995 [1990]. Control and becoming [interview with Tony Negri], in *Negotiations 1972–1990* (translated by M. Joughin). New York: Columbia University Press, 169–76.

—— . 1997. He stuttered, in *Essays Critical and Clinical* (translated by D.W. Smith and M. Greco). Minneapolis, MN: University of Minnesota Press, 107–14.

Deleuze, G. and Guattari, F. 1987 [1980]. *A Thousand Plateaus* (translated by B. Massumi). London: Athlone Press.

—— . 1994 [1991]. *What is Philosophy?* (translated by H. Tomlinson and G. Burchill). London: Verso.

Dewsbury, J.D. 2002. Embodying time, imagined and sensed. *Time and Society*, 11(1), 147–54.

Douglas, M. 1982. *In the Active Voice*. London: RKP.

Evans, G. 2001. *Cultural Planning: An Urban Renaissance?* London, Routledge.

—— . 2005. Measure for measure: evaluating the evidence of culture's contribution to regeneration. *Urban Studies*, 42(5–6), 959–83.

Evans, G. and Shaw, P. 2004. *The Contribution of Culture to Regeneration in the UK: A Review of Evidence*. A Report to the Department for Culture, Media and Sport. London: London Metropolitan University.

Evans, G. and Smith, M. 2000. A tale of two heritage cities: Old Quebec and Maritime Greenwich, in *Tourism and Heritage Relationships: Global, National and Local Perspectives*, edited by M. Robinson et al. Sunderland: Business Education Publishers, 173–96.

Evans, N. 2009. *Dreamland Remembered*. Margate: Bygone Publishing.

Flaxman, G. 2000. Introduction, in *The Brain is the Screen: Deleuze and the Philosophy of Cinema*, edited by G. Flaxman. Minneapolis, MN: University of Minnesota Press, 1–57.

Foucault, M. 1977. *Language, Counter-memory, Practice* (translated by D. Bouchard and S. Simons). Ithaca, NY: Cornell University Press.

——. 1982. The subject and power, in *Michel Foucault: Beyond Structuralism and Hermeneutics*, edited by H. Dreyfus and P. Rabinow. Chicago, IL: University of Chicago Press, 208–26.

——. 1991 [1977]. Questions of method, in *The Foucault Effect*, edited by G. Burchell, C. Gordon and P. Miller. Chicago, IL: University of Chicago Press, 73–86.

Grosz, E. 2001. *Architecture from the Outside*. Cambridge, MA: MIT Press.

——. 2004. *The Nick of Time*. Durham, NC: Duke University Press.

——. 2005. *Time Travels*. Durham, NC: Duke University Press.

——. 2010. *The Future of Space: Towards an Architecture of Invention*. [Online: *dpi*, 04 Features]. Available at: http://dpi.studioxx.org/demo/?q=en/no/04/future-of-space-towards-architecture-invention-by-Elizabeth-Grosz [accessed 29 October 2010].

Hamilton, F. 2009. Margate regenerating with help of Tracey Emin and J.M.W. Turner. *Times Online*, 9 May 2009. [online]. Available at: http://www.thetimes.co.uk/tto/public/sitesearch.do?querystring=Margate+regenerating+with+help+of+Tracey+Emin+and+J.M.W.+Turner&p=tto&pf=all&bl=on [accessed 13 November 2009].

Heritage Protection Department (HPD). 2007. *Culture and Entertainment Buildings Selection Guide*. London: English Heritage.

Herzog, A. 2010. Becoming-fluid: history, corporeality, and the musical spectacle, in *Afterimages of Gilles Deleuze's Film Philosophy*, edited by D.N. Rodowick, Minneapolis, MN: University of Minnesota Press, 259–79.

Iggers, G. and Powell, J. (eds) 1990. *Leopold von Ranke and the Shaping of the Historical Discipline*. Syracuse, NY: Syracuse University Press.

Jacobsen, M.H. 2009. *Encountering the Everyday*. Basingstoke: Palgrave Macmillan.

Kearns, G. and Philo, C. (eds) 1993. *Selling Places: The City as Cultural Capital, Past and Present*. Oxford: Pergamon Press.

Koerner, S. 2010. Contextualising changing approaches to heritage, in *Unquiet Pasts*, edited by S. Koerner and I. Russell. Farnham: Ashgate, 137–49.

Kunzmann, K. 2004. Culture, creativity and spatial planning. *Town Planning Review*, 75(4), 383–404.

Lash, S. and Urry, J. 1994. *The Economies of Signs and Space*. London: Routledge.

Lather, P. 2009. Against empathy, voice and authority, in *Voice in Qualitative Inquiry*, edited by A. Jackson and L. Mazzei. New York: Routledge, 17–26.

Lovatt, A. and O'Connor, J. 1995. Cities and the night-time economy. *Planning Practice and Research*, 10(2), 127–43.

Makkreel, R. 1990. Traditional historicism, contemporary interpretations of historicity, and the history of philosophy. *New Literary History*, 21(4), 977–91.

Markusen, A. and Gadwa, A. 2010. Arts and culture in urban or regional planning: a review and research agenda. *Journal of Planning Education and Research*, 29(3), 379–91.

McHoul, A. 2009. The being of culture: beyond representation, *Philosophical Papers and Reviews*, 1(5), 67–73. [Online]. Available at: http://www.academicjournals.org/ppr [accessed 11 November 2010].

Mintz, S. 1985. *Sweetness and Power: The Place of Sugar in Modern History*. New York: Penguin.

Morgan, M. 2006. Making space for experiences. *Journal of Retail and Leisure Property*, 5(4), 305–13.

Negri, A. 1992. On Gilles Deleuze and Félix Guattari, a thousand plateaus. *Chimères*, 17 (translated by T. Wolfe). [Online]. Available at: http://libcom.org/library/deleuze-guattari-thousand-plateaus-negri [accessed 9 November 2010].

Nietzsche, F. 1965 [1883]. *Thus Spake Zarathustra* (translated by W. Kaufmann). Harmondsworth: Penguin.

—— . 2010 [1873]. On the use and abuse of history for life (translated by I. Johnston). [Online]. Available at: http://records.viu.ca/~johnstoi/nietzsche/history.htm [accessed 21 July 2010].

Philo, C. and Kearns, G. 1993. Culture, history, capital: a cultural introduction to the selling of places, in *Selling Places: The City as Cultural Capital, Past and Present*, edited by G. Kearns and C. Philo. Oxford: Pergamon Press, 1–32.

Pløger, J. 2010. Presence-experiences: the eventalisation of urban space. *Environment and Planning D, Society and Space*, 28(5), 848–66.

Prigogine, I. 1985 Time and human knowledge, *Environment and Planning B, Planning and Design*, 12(1), 5-20.

Russell, I. 2010. Heritage, identities and roots: a critique of arborescent models of heritage and identity, in *Heritage Values in Contemporary Society*, edited by G. Smith, P. Messenger and H. Soderland. Walnut Creek, CA: Left Coast Press, 29–41.

Service, T. 2010. I had an epiphany on a rollercoaster ride, Critic's Notebook. *The Guardian*, G2, 13 May 2010, 21.

Shanks, M. and Witmore, C. 2010. Memory practices and the archaeological imagination in risk society: design and long term community, in *Unquiet Pasts*, edited by S. Koerner and I. Russell. Farnham: Ashgate, 269–89.

Shields, R. 1991. *Places on the Margin*. London: Routledge.

Smith, L. 2006. *Uses of Heritage*. London: Routledge.

—— . 2009. *Class Heritage and the Negotiation of Place*. [Online]. Available at: http://www.english-heritage.org.uk/upload/pdf/Smith_missing_out_conference.pdf?1245950348 [accessed April 2010].

Smith, M. (ed.) 2007. *Tourism, Culture and Regeneration*. Wallingford: CAB International.

Stagoll, C. 2005. Event, in *The Deleuze Dictionary*, edited by A. Parr. Edinburgh: Edinburgh University Press, 87–8.

Stevenson, D. 2004. 'Civic Gold' rush: cultural planning and the politics of the Third Way. *International Journal of Cultural Policy*, 10(1), 119–31.

Sundbo, J. and Darmer, P. (eds) 2008. *Creating Experience in the Experience Economy*. Cheltenham: Edward Elgar.

Taylor, M., Preston, J. and Charleson, A. 2000. The myth of the matter: parallel surfaces of seismic linings, in *Re-Framing Architecture*, edited by M. Ostwald and R. Moore. Brisbane: Archadia Press, 189–99.

Vitale, C. 2010. Deleuze as logician of the virtual: reading the Cinema books off Hegel's 'Logic', *Networkologies*. [Online]. Available at: http://networkologies.wordpress.com/2010/04/28/deleuxe-as-logician-of-the-virtual/ [accessed 18 May 2010].

Walton, J. 2000. *The British Seaside*. Manchester: Manchester University Press.

Waterton, E. and Smith, L. 2010. The recognition and misrecognition of community heritage. *International Journal of Heritage Studies*, 16(1–2), 4–15.
Watson, S. and Waterton, E. 2010. Introduction: a visual heritage, in *Culture, Heritage and Representation*, edited by S. Watson and E. Waterton. Farnham: Ashgate, 1–16.
Young, G. 2008a. The culturization of planning. *Planning Theory*, 7(1), 71–91.
—— . 2008b. *Reshaping Planning with Culture*. Aldershot: Ashgate.

Afterword: Planning with Culture

Deborah Stevenson

Cities are at the heart of the political, economic, social and cultural processes that shape and define the contemporary world. Indeed, most people on the planet now live in cities. For some commentators, the 'urbanization of the world' has the potential to pull people out of poverty, foster innovation and create the conditions for liberation. Utopian ideals have long been influential in attempts to design and build cities that are equitable and democratic as well as aesthetically pleasing. Many observers, however, are less optimistic, fearing that the dominance of the urban poses a considerable threat not only to the social and environmental sustainability of individual cities, but also to the future of the world. They argue that the degraded and indulgent landscapes of the city are the locales of displacement, disadvantage and inequality, irrevocably depleting the world's resources and putting the very viability of the planet at risk. At the same time as such debates are being played out, however, it is important to remember that cities are also the spaces of everyday lived culture and the most intimate of experiences. They are resonant with memory and their places elicit a range of emotional responses and attachments that are collective as well as individual. Cities and urban development are thus replete with opportunities and difficulties, simultaneously sites of hope and fear. The task for urban scholarship is to engage with such complexities and contradictions in order to understand cities at both their micro and macro levels. The challenge for urban planners is to utilize the insights of this scholarship to ensure their practice is nuanced, sensitive and capable of shaping urban landscapes that are just and sustainable and support the dignity of all urban dwellers.

In negotiating what are often extreme and competing views and approaches, it is necessary to trace the contours of a range of tasks, possibilities and responsibilities that must be confronted by those concerned with the future of cities and the role of planning in crafting that future. This assignment is practical as well as intellectual and theoretical, concurrently imaginative, tangible and experiential; it thus requires sophisticated conceptual and methodological tools. Understanding cities and urban life in all their messiness and in terms of macro and micro processes and systems, therefore, calls for engaged, critical and interdisciplinary scholarship. At the same time, urban planning can no longer be an apolitical, technical activity where planners passively apply a predetermined set of 'solutions' to what are profoundly difficult and dynamic problems. The key step in effecting this shift is to consider both the cultural dimensions of planning and the ways in which planning practice shapes and is shaped by urban culture. What are at stake, therefore, are not only new ways of understanding urbanism, but also a way to build successful and sustainable cities and shape an exciting and innovative approach to urban planning. Pivotal to achieving these ends is the provision of space for interdisciplinary dialogue and the exchange of ideas. And this is the role of this collection of essays, a central concern of which has been to bring together leading scholars from different disciplines to explore important issues in the planning and culture of the city. In seeking to consider cities in terms of the intersection of planning and culture, this volume, as a space for interdisciplinary engagement and reflection, is organized around a number of interrelated themes that are examined through the lenses of different disciplines, roles and sectors. Although each section is structured to grapple with particular aspects of the planning and culture dialectic, they are not self-contained and many narratives and concerns recur.

Cities are shaped very directly by a series of relationships and influences that are simultaneously global and local and one of the most significant of these is diversity and its expression in the spaces and places of the city. Urban diversity takes many forms, with race, age, ethnicity, belief and gender being key elements. It is one of the characteristics that is said to set cities and their social relations apart from those of the small town or village being at the heart of what is understood as urban and conceptualized as urbanism. It is thus unsurprising that diversity is an important theme in this collection, which argues that it is vital for planners to appreciate the ways in which cultural diversity is lived, expressed and encountered in order to create urban spaces that can foster and accommodate difference and support intercultural understanding and dialogue. Acknowledging and working with urban diversity also means recognizing the multiplicity of cities and being able to engage with the significant differences that exist between cities. Not only is no one city a single place or socio-cultural formation but no two cities are the same. Indeed, there are many urban researchers who, recognizing the diversity of cities, go so far as to suggest that the very notion of 'the city' with its connotations of singularity and unity has become redundant. Importantly, though it means that the serial application of 'off the shelf' planning solutions and blueprints irrespective of the city or the space is no longer appropriate if it ever was. It is also important to recognize that ideas and theories about cities and urban planning that have been developed in the global North are rarely applicable to the cities of the South and their diverse cultures. And yet the legacy of colonialism is that the frameworks and approaches that continue to dominate urban planning and urban studies more broadly are those that were shaped in the intellectual and spatial contexts of the rich capital-exporting countries of North America and Europe, in particular. As discussed in this volume, the way forward is to develop perspectives that are comparative as well as interdisciplinary and transnational.

Closely associated with the theme of diversity is the vexed issue of culture itself, which was famously described by Raymond Williams as being one of the most complicated words in the English language. Perhaps, as Williams further suggests, this very complexity is directly related to its importance. Culture is variously understood as a way of life, a set of creative and aesthetic practices and products, and a process of development (improvement). Cultural practices, including those associated with popular pastimes, music and entertainment as well as with the social fault lines of age, class, gender, sexuality, race and ethnicity, are important aspects of the lived diversity that is at the core of the contemporary city. Emerging in recent years as increasingly relevant to urban planning has been the positioning of art and culture as 'industries' which supposedly are capable of making valuable contributions to local economies, animating urban space and fostering lived diversity. As argued in this book, engaging with and navigating such claims are important tasks for urban planners. Increasingly, too, attempts to manage and shape urban space and culture occur through the cultural policy frameworks of government from the local to the transnational. Cultural planning is an important initiative in this context. Indeed, some commentators have suggested that urban planning should be incorporated into an expansive and multifaceted cultural planning agenda that also includes tourism, leisure, heritage and social planning. Such a move may be overly ambitious but at the very least it is important that urban planning articulates with cultural policy and planning and is configured to approach concerns such as city reimaging, urban design and the planning and management of public space in a coordinated and mutually reinforcing way.

The assumption that culture and planning are both comprised and understood in terms of a range of intersecting practices was an important thread running through this collection and was addressed specifically by many of the contributors. Several connected themes stand out from these contributions as important, including public space, governance and place. Culturally informed planning (planning culturally) is that which seeks to make places – and particularly public spaces – that are enlivened and inclusive as well as meaningful to local communities. Place, be it public, private or quasi-public is about identity and built space – iconic architecture for instance often serves as a key marker of a city and its standing in the global hierarchies of cities. There are a number of high-profile examples that come readily to mind but planning is also concerned with the small spaces, the accidental and incidental sites of urban culture that can be important gathering points or the locales of memory and moment. Successful places are those that are meaningful being made through use and imagination. It is the responsibility of planners to create places that are relevant and viable and this means working with, and through, local cultures and communities and being cognisant of the way in which culture shapes the use of space.

In any attempt to map a cultural paradigm for planning, issues of urban governance also come to the fore and so it is unsurprising that this was a central concern of many of the contributions to this volume. This is governance in its broadest and narrowest conceptualizations, which operates at all levels from the local to the international and across the public and non-public sectors. Important are the ways in which cities are formed by processes of governance and with reference to specific legitimating discourses, assumptions and technologies. Also important is the range of cultural processes and contexts that significantly shape the machinery of governance. These are global as well as place-specific. They are simultaneously understated and authoritative, and culture again emerges as a potent organizing concept and resource.

Planning should be a process of creativity and supposition. Too often, however, what is assumed is stasis and predictability and, as was demonstrated repeatedly in this volume, such assumptions can no longer be sustained. Cities and urban cultures are dynamic although to observe the rapidity with which cities and circumstances change is perhaps to state the obvious. It is the case though, as Sophie Watson pointed out in the first chapter of this book, that the world is a place of uncertainty, complexity and contingency. Perhaps it always has been, but what have changed in recent years are the pace of change and the capacity of established theoretical frameworks and belief systems to explain this change. New flexible ways of understanding and imaging cities and their futures are needed and the premise underpinning this volume was that at the centre of this endeavour must be culture – in all its dimensions. It is not enough to say that planning must take local cultures and circumstances into account. Rather, what is needed is a radically different approach to the task and this is one that seeks to combine planning and culture – to plan cities with, and through, urban cultures and complexity.

This book does not present a toolkit for practice or set of ready-made recommendations. Nor does it speak with a unified voice or advocate a particular theoretical position. None of these outcomes and approaches is either desirable or achievable. Rather, what it does is highlight possibilities and synergies as well as significant tensions and divergences. It presents exciting ways of seeing and in so doing makes valuable and original contributions to what is one of the most critical and urgent challenges confronting contemporary urban planning. It is only by engaging with the cultural theories, processes and lived practices that comprise and explain contemporary cities and societies that planning theory and practice will develop the capacity to imagine and shape cities and urban futures that are viable and empowering.

Index